New Directions in Philosophy and Cognitive Science

Series editors: John Sutton, Macquarie University and Richard Menary, Macquarie University.

This series brings together work that takes cognitive science in new directions. Hitherto, philosophical reflection on cognitive science – or perhaps better, philosophical contribution to the interdisciplinary field that is cognitive science – has for the most part come from philosophers with a commitment to a representationalist model of the mind.

However, as cognitive science continues to make advances, especially in its neuroscience and robotics aspects, there is growing discontent with the representationalism of traditional philosophical interpretations of cognition. Cognitive scientists and philosophers have turned to a variety of sources – phenomenology and dynamic systems theory foremost among them to date – to rethink cognition as the direction of the action of an embodied and affectively attuned organism embedded in its social world, a stance that sees representation as only one tool of cognition, and a derived one at that.

To foster this growing interest in rethinking traditional philosophical notions of cognition – using phenomenology, dynamic systems theory, and perhaps other approaches yet to be identified – we dedicate this series to 'New Directions in Philosophy and Cognitive Science.'

Titles include:

Robyn Bluhm, Anne Jaap Jacobson and Heidi Maibom (*editors*)
NEUROFEMINISM
Issues at the Intersection of Feminist Theory and Cognitive

Jesse Butler
RETHINKING INTROSPECTION
A Pluralist Approach to the First-Person Perspective

Massimiliano Cappuccio and Tom Froese (*editors*)
ENACTIVE COGNITION AT THE EDGE OF SENSE-MAKING
Making Sense of Non-sense

Anne Jaap Jacobson
KEEPING THE WORLD IN MIND
Mental Representations and the Sciences of the Mind

Julian Kiverstein & Michael Wheeler (*editors*)
HEIDEGGER AND COGNITIVE SCIENCE

Michelle Maiese
EMBODIMENT, EMOTION, AND COGNITION

Richard Menary
COGNITIVE INTEGRATION
Mind and Cognition Unbounded

Zdravko Radman (*editor*)
KNOWING WITHOUT THINKING
Mind, Action, Cognition and the Phenomenon of the Background

Matthew Ratcliffe
RETHINKING COMMONSENSE PSYCHOLOGY
A Critique of Folk Psychology, Theory of Mind and Stimulation

Jay Schulkin (*editor*)
ACTION, PERCEPTION AND THE BRAIN

Tibor Solymosi and John R. Shook (*editors*)
NEUROSCIENCE, NEUROPHILOSOPHY AND PRAGMATISM
Brains at Work with the World

Robert Welshon
NIETZSCHE, PSYCHOLOGY, AND COGNITIVE SCIENCE

Forthcoming titles

Miranda Anderson
THE RENAISSANCE EXTENDED MIND

Maxime Doyon and Thiemo Breyer
NORMATIVITY IN PERCEPTION

Matt Hayler
A PHENOMENOLOGICAL ANALYSIS OF TECHNOLOGY USE

New Directions in Philosophy and Cognitive Science
Series Standing Order ISBN 978-0-230-54935-7 Hardback
978-0-230-54936-4 Paperback
(*outside North America only*)

You can receive future titles in this series as they are published by placing a standing order. Please contact your bookseller or, in case of difficulty, write to us at the address below with your name and address, the title of the series and one of the ISBNs quoted above.

Customer Services Department, Macmillan Distribution Ltd, Houndmills, Basingstoke, Hampshire RG21 6XS, England

---

Also by John R. Shook

DEWEY'S EMPIRICAL THEORY OF KNOWLEDGE AND REALITY (2000)

PRAGMATIC NATURALISM AND REALISM (*editor*, 2003)

A COMPANION TO PRAGMATISM (*co-editor with Joseph Margolis*, 2006)

F. C. S. SCHILLER ON PRAGMATISM AND HUMANISM: SELECTED WRITINGS, 1891–1939 (*co-editor* with Hugh Mcdonald, 2008)

THE FUTURE OF NATURALISM (*co-editor with Paul Kurtz*, 2009)

JOHN DEWEY'S PHILOSOPHY OF SPIRIT, WITH DEWEY'S 1897 LECTURES ON HEGEL (*co-author with James A. Good*, 2010)

DEWEY'S ENDURING IMPACT: ESSAYS ON AMERICA'S PHILOSOPHER (*co-editor with Paul Kurtz*, 2011)

THE ESSENTIAL WILLIAM JAMES (*editor*, 2011)

PRAGMATIST NEUROPHILOSOPHY: AMERICAN PHILOSOPHY AND THE BRAIN (*co-editor with Tibor Solymosi*, 2014)

DEWEY'S SOCIAL PHILOSOPHY: DEMOCRACY AS EDUCATION (2014)

Neuroscience, Neurophilosophy and Pragmatism

Brains at Work with the World

Edited by

Tibor Solymosi
Mercyhurst University, USA

John R. Shook
University at Buffalo, USA

Selection, introduction and editorial matter © Tibor Solymosi and John R. Shook 2014
Chapters © Individual authors 2014

All rights reserved. No reproduction, copy or transmission of this publication may be made without written permission.

No portion of this publication may be reproduced, copied or transmitted save with written permission or in accordance with the provisions of the Copyright, Designs and Patents Act 1988, or under the terms of any licence permitting limited copying issued by the Copyright Licensing Agency, Saffron House, 6–10 Kirby Street, London EC1N 8TS.

Any person who does any unauthorized act in relation to this publication may be liable to criminal prosecution and civil claims for damages.

The authors have asserted their rights to be identified as the authors of this work in accordance with the Copyright, Designs and Patents Act 1988.

First published 2014 by
PALGRAVE MACMILLAN

Palgrave Macmillan in the UK is an imprint of Macmillan Publishers Limited, registered in England, company number 785998, of Houndmills, Basingstoke, Hampshire RG21 6XS.

Palgrave Macmillan in the US is a division of St Martin's Press LLC,
175 Fifth Avenue, New York, NY 10010.

Palgrave Macmillan is the global academic imprint of the above companies and has companies and representatives throughout the world.

Palgrave® and Macmillan® are registered trademarks in the United States, the United Kingdom, Europe and other countries

ISBN: 978–1–137–37606–0

This book is printed on paper suitable for recycling and made from fully managed and sustained forest sources. Logging, pulping and manufacturing processes are expected to conform to the environmental regulations of the country of origin.

A catalogue record for this book is available from the British Library.

A catalog record for this book is available from the Library of Congress.

Transferred to Digital Printing in 2015

Contents

List of Figures	vii
List of Tables	viii
Preface	ix
Notes on Contributors	xi

Part I Pragmatism, Philosophy, and the Brain

1 Neuropragmatism and the Reconstruction of Scientific and Humanistic Worldviews   3
   John R. Shook and Tibor Solymosi

2 Keeping the Pragmatism in Neuropragmatism   37
   Mark Johnson

3 How Computational Neuroscience Revealed that the Pragmatists Were Right   57
   W. Teed Rockwell

4 Pragmatism, Cognitive Capacity and Brain Function   71
   Jay Schulkin

Part II Cognition, Emotion, and the World

5 The End of the Debate over Extended Cognition   105
   Jeffrey B. Wagman and Anthony Chemero

6 Knowing and the Known: Brain Science and an Empirically Responsible Epistemology   125
   David D. Franks

7 Dewey's Rejection of the Emotion/Expression Distinction   140
   Joel Krueger

Part III Creativity, Education, and Application

8 Finding Unapparent Connections: How Our Hominin Ancestors Evolved Creativity by Solving Practical Problems   165
   Robert Arp

9 Neuropragmatism and Apprenticeship: A Model for
   Education   185
   Bill Bywater and Zachary Piso

10 A Neuropragmatist Framework for Childhood Education:
   Integrating Pragmatism and Neuroscience to Actualize
   Article 29 of the UN Child Convention   215
   Alireza Moula, Antony J. Puddephatt, and Simin Mohseni

Part IV   Ethics, Neuroscience, and Possibility

11 Pragmatism and the Contribution of Neuroscience to Ethics   243
   Eric Racine

12 Pragmatist Ethics: A Dynamical Theory Based on Active
   Responsibility   264
   Markate Daly

13 Moral First Aid for a Neuroscientific Age   291
   Tibor Solymosi

Index   319

List of Figures

4.1	Action and purpose in living a life	75
4.2	Comparisons of primate brains	76
4.3	Social contact in primates is consistently linked to neocortical expansion	77
4.4	Toolbox of cognitive functions	79
4.5	Facial recognition and the cortex	86
4.6	A scan revealing that in even in silence the auditory cortex is activated	91
4.7	Kinds of objects and corresponding brain activation	94
8.1	The construction of a harpoon	180
10.1	Neuropragmatism and the UN Child Convention	217
10.2	Moula's daily-life machine and internal and external organization of action	230
11.1	Comparison of the processes implied in applied ethics and pragmatic ethics. Process of applied ethics	253
11.2	Comparison of the processes implied in applied ethics and pragmatic ethics. Process of pragmatic ethics	254

List of Tables

4.1	Peirce on inference and hypothesis	80
4.2	Lakoff and Johnson on perception and action	89
11.1	Common conceptual arguments against the neuroscience of ethics	249
11.2	Key features of Dewey's ethics	251
11.3	Interpretations of neuroscience's contribution to ethics following strong and moderate (pragmatic) naturalisms	252

Preface

The essays collected here are representative – but not exhaustive – of the connections and advancements to be made between neuroscience and pragmatism. These 13 chapters sort into four major themes that together illuminate what we have been calling *neuropragmatism*. Part I examines the historical, theoretical, and empirical connections between neuroscience and pragmatism, as well as its philosophical import. In Chapter 1, we introduce neuropragmatism in its cultural context, providing core theses, and pointing to new vistas. The importance of neuropragmatism is carefully considered by Mark Johnson, who argues, in Chapter 2, that as exciting as advances in neuroscience are, philosophy, particularly pragmatism, still has value and work to do, but as important as this work is, pragmatists must not ignore these new advances either. W. Teed Rockwell, in Chapter 3, argues that the recent advances in computational neuroscience further corroborate many of the key insights of classical pragmatism, particularly with regard to truth. Jay Schulkin elaborates on the connections between pragmatism, the brain, and cognition in further detail in Chapter 4.

Part II focuses on particular contemporary issues having to do with cognition, emotion, and their place in the world. Jeffrey B. Wagman and Anthony Chemero, in Chapter 5, provide persuasive empirical evidence in favor of extended cognition, while recognizing that despite the evidence, the debate will rage on due to theoretical or definitional disagreements. Such disagreements are anathema to the work being done in sociology and its intersection with neuroscience, as discussed by David D. Franks in Chapter 6. Joel Krueger, in Chapter 7, elaborates the further consequences of John Dewey's view of emotion in light of recent empirical research.

In part III, the relationship between problem solving, creativity, evolution, and education is examined. Robert Arp considers the evolution of creative problem solving with regard to vision and tool development in Chapter 8. Chapter 9 focuses in on the consequences of this evolution by connecting neuroscience, anthropology, and mirror neurons to educational practices, as elaborated by Bill Bywater and Zachary Piso. In Chapter 10, Alireza Moula, Antony J. Puddephatt, and Simin Mohseni apply many of the lessons of the previous two chapters to pedagogical policy, focusing on Article 29 of the United Nations Child Convention.

The final part considers the possibilities in neuroscience and ethics. Eric Racine reviews in Chapter 11 much of the recent work going on in neuroethics that beneficially stems from pragmatism, especially Dewey's ethical theory. In Chapter 12, Markate Daly draws on the work of George Herbert Mead and Daniel Kahneman as well as dynamical systems theory to develop a relational ethic that contrasts with a strict rule-based ethic – both of which are in regular conflict in contemporary America. In the final chapter, Tibor Solymosi not only reviews recent claims about the relationship between neuroscience and ethics, he also develops ethics as a social technology that has its origins in the historical social problems of humans, suggesting a new means of conceiving ethical inquiry in the advance of neuroscience.

Our hope with this volume is to present the promise of pragmatism for further neuroscientific inquiry as well as the promise of neuroscience for further philosophical work. The conversation is already underway, yet we believe much more remains to be discussed. We thank our friends and families for their support. We are very appreciative of James Giordano's help and support from the very start. Lastly, we thank our respective institutions, departments, and colleagues for their ongoing support.

Notes on Contributors

Bill Bywater is Professor emeritus of philosophy at Allegheny College in Meadville, Pennsylvania. His research interests range from critical race theory and feminism to philosophy of science and education. Central to his research on these varying themes is his work explicating a theory of inquiry utilizing affinities between the work of John Dewey and the delicate empiricism of Johann Wolfgang von Goethe.

Anthony Chemero is cognitive scientist and philosopher, who has extensively authored and co-authored papers on the intersection of philosophy and cognitive science, particularly from the perspective of dynamic systems theory. He is author of *Radical Embodied Cognitive Science* (2009). After several years as an associate professor of psychology in the scientific and philosophical studies of mind program of the psychology department at Franklin and Marshall College, Chemero is now a professor of philosophy and psychology at the University of Cincinnati.

Markate Daly is currently an independent scholar. She earned her doctorate at the University of Wisconsin-Madison and taught in several California Universities and colleges before opening a practice in philosophical counseling for individuals and consulting for non-profits.

David D. Franks is Professor emeritus in the Department of Sociology and Anthropology at Virginia Commonwealth University who has brought sociology and neuroscience together through the work of George Herbert Mead. This intersection he calls 'neurosociology,' which is the title of his book length treatment on the issue: *Neurosociology: The Nexus Between Neuroscience and Social Psychology* (2010), which won the American Sociological Society's Biology and Society Subsection Outstanding Book Award in 2012.

Mark Johnson is Knight Professor of Liberal Arts and Sciences in the philosophy department at the University of Oregon. He has published several books on metaphor, bodily experience, imagination, and ethics, including *Metaphors We Live By* (with George Lakoff, 1980/2003), *The Body in the Mind* (1987), *Moral Imagination: Implications of Cognitive Science for Ethics* (1993), *Philosophy in the Flesh* (with George Lakoff, 1999),

The Meaning of the Body: Aesthetics of Human Understanding (2009), and *Morality for Humans: Ethical Understanding from the Perspective of Cognitive Science* (2014).

Joel Kruger is Lecturer in philosophy at the University of Exeter. His main research interests include issues in pragmatism, phenomenology, and philosophy of mind, with a particular focus on empathy and social cognition. His articles have appeared in *Consciousness and Cognition, Phenomenology and the Cognitive Sciences, Journal of Consciousness Studies, Inquiry, Philosophy East and West*, and *Contemporary Pragmatism*.

Simin Mohseni works on neurobiology at Linköping University, Sweden, where she is Senior Lecturer in the Department of Clinical and Experimental Medicine, Cell Biology.

Alireza Moula is University Lecturer in the Department of Social and Psychological Studies at Karlstad University, Sweden. He researches social medicine, social work, and education.

Zachary Piso is currently enrolled in the PhD program in philosophy at Michigan State University. His interests focus on the intersection of ecology and philosophy, particularly where ecology is understood as the science of systems. More specifically, he is working on environmentalism, dynamic systems theory, and ethics from a pragmatist perspective.

Antony Puddephatt is Associate Professor of sociology in the Department of Sociology at Lakehead University, Canada. His research interests include social pragmatism, symbolic interactionism, and related schools of interpretive theory.

Eric Racine is Director of the Neuroethics Research Unit, as well as an associate research professor, Institut de recherches cliniques de Montréal, and in the Department of Medicine and Department of Social and Preventive Medicine, Université de Montréal, and in the Departments of Neurology and Neurosurgery, Medicine & Biomedical Ethics Unit, McGill University. He has written numerous articles on neuroethics, especially from a pragmatist standpoint. His recent book is *Pragmatic Neuroethics: Improving Treatment and Understanding of the Mind-Brain* (2010).

Teed Rockwell is an interdisciplinarily-trained philosopher who has brought Dewey's thought to recent discussions in neurophilosophy, particularly with regard to the interest in dynamic systems theory. He is author of *Neither Brain Nor Ghost: A Nondualist Alternative to the Mind-*

Brain Identity Theory (2005). He is an instructor in the department of philosophy at Sonoma State University.

Jay Schulkin is trained in both philosophy and neuroscience. He has published extensively in both, explicitly from a pragmatist perspective. His several books, such as *Roots of Social Sensibility and Neural Function* (2000), *Bodily Sensibility: Intelligent Action* (2004), *Effort: A Behavioral Neuroscience Perspective on the Will* (2006), *Cognitive Adaptations: A Pragmatist Perspective* (2009), *Naturalism and Pragmatism* (2012), and *Reflections on the Musical Mind: An Evolutionary Perspective* (2013), elaborate upon the insights of classical pragmatism through recent neuroscientific research. He is a research professor in the Department of Neuroscience, Center for Brain Basis of Cognition, at Georgetown University, Washington, DC.

John R. Shook received his PhD in philosophy from the University at Buffalo, was a professor of philosophy at Oklahoma State University from 2000 to 2006, and then joined the faculty of the Science and the Public online EdM program at the University at Buffalo, where he also is research associate in philosophy. In recent years, he has also been a visiting fellow at the Institute for Philosophy and Public Policy at George Mason University, and at the Center for Neurotechnology Studies of the Potomac Institute for Public Policy in Virginia. Volumes he has edited include *Pragmatic Naturalism and Realism*, *the Blackwell Companion to Pragmatism*, and *The Essential William James*. His recent articles range across social neuroscience, neurophilosophy, moral psychology, and neuroethics.

Tibor Solymosi is Assistant Professor of Philosophy at Mercyhurst University in Erie, Pennsylvania. He graduated *magna cum laude* from Allegheny College and earned his PhD at Southern Illinois University, where he studied at the Center for John Dewey Studies. His writings focus on consciousness, free will, neurophilosophy, scientific inquiry, and ethics. They have appeared in several academic journals and interdisciplinary volumes. His website is neuropragmatism.com.

Jeffrey Wagman is Professor in the Department of Psychology at Illinois State University. His research focuses on situating traditional psychological phenomena within a contextual cycle of an organism's perceiving opportunities for behavior and an organism's goal-directed control of its behavior. He is author or co-author of several scientific articles in peer-reviewed journals and anthologies.

Part I
Pragmatism, Philosophy, and the Brain

# 1
Neuropragmatism and the Reconstruction of Scientific and Humanistic Worldviews[1]

John R. Shook and Tibor Solymosi

> The question of the integration of mind-body in action is the most practical of all questions we can ask of our civilization. It is not just a speculative question; it is a demand: a demand that the labor of multitudes now too predominantly physical in character be inspirited by purpose and emotion and informed by knowledge and understanding. It is a demand that what now pass for highly intellectual and spiritual functions shall be integrated with the ultimate conditions and means of all achievement, namely the physical, and thereby accomplish something beyond themselves. Until this integration is effected in the only place where it can be carried out, in action itself, we shall continue to live in a society in which a soulless and heartless materialism is compensated for by soulful but futile and unnatural idealism and spiritualism.
>
> John Dewey[2]

Neurophilosophical pragmatism, or *neuropragmatism*, is a scientifically informed treatment of cognition, knowledge, the body-mind relation, agency, socialization, and further issues predicated on sound judgments about these basic matters. Neuropragmatism is capable of grappling with philosophical questions arising at many levels, from synapse to society. There is much at stake, as the epigraph by Dewey states. With its firm grounding in science, neuropragmatism may be the philosophy best equipped to deal productively with the challenges facing our culture, as developments in neuroscience and neurotechnology bring about both better means for dealing with old problems, and new ways of creating and dealing with the problems of today and tomorrow.

The amazing progress of the behavioral and brain sciences has confirmed many of pragmatism's core claims, culminating in a resurgence of neopragmatism, and then its fresh flowering in neuropragmatism. The recovery of the concept of dynamic embodied and embedded cognition, and the renewed appreciation for the brain's systems as evolved functions, have together carried many researchers toward the tenets of neuropragmatism. Scholars bold enough to draw conclusions about the nature of mind, the dynamic nature of human knowledge, and the practical criteria for judging epistemic success unite the cognitive strands of neuropragmatism. Searching for such a comprehensive reunion of science and philosophy should not be disdained. In the words of the editors of a recent book on embodied cognitive science,

> We need to put together conceptual analyses of the notions of representation, computation, emergence, embodiment, and the like, with empirical work that allows us to bring together ecological, dynamic, interactive, situated, and embodied approaches to the scientific study of cognition.[3]

Neuropragmatism offers a philosophical intersection for coordinating this pluralistic effort. The prefix 'neuro' portends no reductionist agenda. Quite the opposite: the anti-reductionist, pluralistic, and interdisciplinary tradition of pragmatism remains securely at the heart of neuropragmatism. All the same, a philosophical position on cognition and mind must cohere with the best neuroscience available.

We begin with a brief history of pragmatism and the sciences of life and mind. From this history, we update pragmatism in this neurophilosophical form by introducing twelve theses of neuropragmatism. These theses emphasize the connections between pragmatism and the sciences of life and mind, and experimentally propose research programs for engaging scientific researchers as well as for navigating the consequences of research for the larger public.

Classical pragmatism and neuropragmatism

Pragmatism has from its origins formulated philosophical theories about culture, intelligence, and knowledge in ways that respect biology, anthropology, and cognitive science. Classical pragmatism was the original American cognitive science and neurophilosophy. Charles Peirce,

William James, John Dewey, and George Mead were all experimental psychologists who tried to reform philosophy in light of evolutionary biology, experimental psychology, and brain science. Indeed, most of the early American psychologists and sociologists had strong pragmatist leanings. Essentially, pragmatism is vitally interested in entirely naturalistic accounts of intelligence and agency, so that all other fields of philosophy – from epistemology to ethics – can be reformed in turn. By integrating science and philosophy, pragmatism attempts to prevent both scientism and speculation from inflating debilitating dualisms.

Pragmatism has always viewed itself as essential to a complete and consistent naturalistic worldview. Any naturalism has to explain how rationality, intelligence, and science are possible within the natural world. Pragmatism has serious opponents not interested in advancing naturalism. At the turn of the 20th century, major philosophical options were few: common sense empiricisms, neo-Kantian rationalisms, phenomenologies, and neo-Hegelian idealisms. Common sense empiricism sought pure sensory impressions or sense data ideas that carry information about nature untainted by any thought, so that cognition simply combines and rearranges that original information into knowledge systems. Neo-Kantian rationalisms, noticing empiricism's deep problems, postulated non-empirical rational principles to account for scientific knowledge. However, such rationalism fed into anti-naturalism and dualism, as did the phenomenologies that prioritized qualitative experience over nature or biology. Reconciling empiricism and rationalism by adding historicism, neo-Hegelian cultural psychologies stumbled onto the way that knowledge gradually grows from the interfusion of evidence and reasoning in social contexts. John Dewey and George Mead further naturalized this cultural historicism by incorporating Darwinian evolution and experimental psychology.[4] They proposed a pragmatic naturalism in opposition to naïve empiricism, static representationalism, reductive materialism, methodological individualism, and animal behaviorism. To accomplish this pragmatic naturalism, pragmatists explored more metaphysical issues such as radical empiricism and direct perception, teleological accounts of living systems, non-reductive emergent naturalisms, and perspectival and process ontologies. Not surprisingly, neurophilosophers, and especially neuropragmatists, have been gradually re-engaging these wider issues.

Pragmatism went into eclipse in philosophy departments by the 1930s, due to the ascendency of analytic and linguistic philosophy, along with imports from European positivism. Yet pragmatic ideas continued to flourish in the social sciences from psychology and linguistics to sociology and anthropology. The neopragmatism of the 1970s and 80s, especially in the hands of Richard Rorty, was well known for its linguistic and epistemic conventionalism, but not for its congruence with the latest brain science. Hilary Putnam's meaning externalism and pragmatic realism[5] also helped to make actual human cognition relevant to philosophical debates. Some philosophers, inspired by W. V. Quine's kind of naturalism (which sustained the Deweyan point that cognitions and knowings must be natural events amenable to scientific study), demanded continuities between science and philosophy and pulled analytic philosophy back from pure rationalism.[6] As the new cognitive and brain sciences emerged in the 1980s and 1990s, they had grown many of the seeds of pragmatism, and when analytic philosophy began to take the brain seriously once again, it encountered these pragmatic ideas. Rationalist analytic philosophers, strong AI proponents, and excessively cognitivist researchers rebelled against such pragmatism; for example, Jerry Fodor has called pragmatism 'the defining catastrophe of analytic philosophy of language and philosophy of mind.'[7] However, some analytic philosophers have been returning to parts of pragmatism in various ways, driven by respect for science and its discoveries.

Scholars such as Mark Johnson and the late Francisco Varela recognized in the 1990s that pragmatism was receiving much reconfirmation in the brain sciences. A younger generation, such as Anthony Chemero, W. Teed Rockwell, and Tibor Solymosi – fluent in both classical pragmatism and the latest neuroscience – was in the best position to take stock of matters. Solymosi recently coined the term 'neuropragmatism.'[8] From its grounding in the current behavioral and brain sciences, neuropragmatism confirms many core views of traditional pragmatism. Neuropragmatism continues to reform philosophical views about such things as the mind-body relation, the function of intelligence, the nature of knowledge and truth, the nature of voluntary agency and responsibility, the function of social morality, and the ethical ways for dealing with new technologies. Along the way, it distinguishes itself from other neuroscience-based philosophical outlooks.

Twelve theses of neuropragmatism

This section offers 12 theses of an ambitious neuropragmatism that deals with core philosophical issues. The first three are grounded in biology and anthropology. Many theoretical views across cognitive science and neuroscience regard these theses as foundational.

1. Animals are goal-oriented organisms, and their nervous systems function to sustain life in various practical ways.

2. Cognition in all its manifestations (intelligence/mind/consciousness) is embodied and not explicable apart from that bodily context.
3. Human cognition in all its modes should primarily be studied and comprehended in terms of its practical service for the ways that humans live.

Neuropragmatism emphasizes four additional theses, supported by behavioral and brain sciences, which enlarge the significance of the first three.

4. Cognitive systems are dynamically adaptive to organism-environment interactions, to deal with shifting conditions of situations as practical goals are pursued.
5. Under pressures from dealing with the environment, the brain modifies its neural connections to improve practical performance. The measure of this neural learning is improved habitual efficiency at specific routine tasks.
6. Complex cognitive processes are the brain's work of effectively coordinating behavior for reliably achieving variable goals in a changing environment.
7. Human intelligence has so many cultural features for facilitating cooperative aims that it should primarily be studied and evaluated largely in terms of its service for social goals.

Five more theses of neuropragmatism remain to be mentioned, but we pause here for some elaboration of the first seven.

Neuropragmatism is tightly allied with theories of neuroplasticity, the vast unconscious, reason-emotion-volition integration, embodied cognition, and the extended mind. All these theories have prototypes in the works of classical pragmatists. Combating any philosophy of mind that depicts it as fundamentally passive, receptive, representational, cognitivist, or mechanistic, the classical pragmatists sought to understand the mind in its biological medium. All of the brain in all of its functioning for life must be taken into account. William James lent scientific respectability to the notion that the fringes and margins of consciousness extend deep down into entirely unconscious emotional and intuitive cognition. The pragmatists affirmed that cognition is basically about applying learned habits to ongoing situations demanding immediate active responses from the organism. Since the environment is never the same, cognition therefore depends on continuous learning, which is the dynamic development of specific habits through the brain's modifications, as the brain's neurons grow or modify their interconnections, as the organism perceptually manages its situated experiences of interacting with its world.[9] Also recognizing how centers of the brain are typically involved in many kinds of coordinated tasks, the classical pragmatists resisted the notion that each part of the brain deals only with narrow tasks or specific sorts of representations. As integrated phases within the continuity of brain processes, the traditional schema of perception, reasoning, emotion, and will cannot be mechanically separate and only temporally related in a series leading to action. Sensation, thought, feeling, and volition are interfused; they are discriminable but not separable aspects of the continuous flow of neural activity.[10]

Neuropragmatism continues pragmatism's emphasis on the way that human cognition is not just geared with the external world but tightly interwoven into the organism's interactions with the environment, forming an organic whole. This fusion makes it impossible to draw a thin clear line where the external world stops and cognition begins. Although the brain is obviously the locus of cognition, it does not necessarily follow that only brain events suffice to account for all the functions and features of cognition. William James's notion of radical empiricism depends on treating mind and world holistically, and John Dewey's empirical naturalism finds mind embodied and embedded in organism-environment transactions. In a chapter of Dewey's 1925 *Experience and Nature*, entitled 'Nature, Life and Body-Mind,' he writes,

> Every 'mind' that we are empirically acquainted with is found in connection with some organized body. Every such body exists in a natural medium to which it sustains some adaptive connection...The natural medium is thus one which contains similar and conjunctive forms. At every point and stage, accordingly, a living organism and its life processes involve a world or nature temporally and spatially 'external' to itself but 'internal' to its functions.[11]

The organism's effective coordination of modifying its environment (natural and social) exemplifies cognition. Pragmatism has always refused to treat neurons (or any other brain cells such as glia that may modulate brain activity) as the exclusive place where cognitive meaning is enacted; neurons are essential to, but not entirely constitutive of, cognition. Neuroscience properly studies the interrelated processes of brain activity, but cognitive neuroscience cannot help explain the processes of learning and knowing by referencing brain activity alone in isolation from any context. Philosophy, for its part, will be unable to show how to integrate body and mind if knowledge is examined quite apart from any bodily context. Pragmatism's resistance to atomistic and reductivist naturalisms is nowhere more evident than in its treatment of experience and mind as dynamic, systemic, contextual, ecological, and social.

Biology cannot study life with utter disregard for the environment; nervous systems qua biological systems must not be studied any differently. The same goes doubly for the functions in which such systems take part, such as cognition. Cognition, therefore, is not to be solely done within the head in the end but is rather understood in terms of life and living within environments. Grounding mind in biology takes life seriously. What are the existential truths of life? As Michael Schwartz and Osborne Wiggins describe life, there cannot be any firm or fixed divisions between organic bodies and their environment. Schwartz and Wiggins offer the following existential truths about life:

1) Being vs. non-being: Always threatened by non-being, the organism must constantly re-assert its being through its own activity.
2) World-relatedness vs. self-enclosure: Living beings are both enclosed with themselves, defined by the boundaries that separate them from their environment, while they are also ceaselessly reaching out to their environment and engaging in transactions with it.
3) Dependence vs. independence: Living beings are both dependent on the material components that constitute them at any given moment and independent of any particular groupings of these components over time.[12]

What is true of life is also true of mind: Mind cannot be comprehended except through what it does, and what mind does is transcend itself by ceaselessly modifying its lived environment. By studying those modes of modification, the mind is studied and nowhere else. At no time do an organism's activities or cognition deal with some 'external world' that can be specified independently from the organism. An organism can neither perceive nor interact with 'the world at large,' but only confront its own 'life-world' that it can experience and modify. There is no point in first specifying what the external world is like and then asking how an organism cognizes that world. Neuropragmatism, like classical pragmatism before it, studies cognition as it actually transforms the lived environment. The organism's environment is not the same as the external world. Jacob von Uexküll used the term *Umwelt* for the 'life-world' that a species tries to grapple with. Dewey's conception of 'experience' as doing-undergoing, Heidegger's use of *Erlebnis*, and Richard Lewontin's environmental constructivism similarly point to this conception of the available life-world within which cognition does its work.[13]

In a basic sense, the sciences all realize how cognition is localizable to organic bodies dealing with their environments, and that cognition cannot be spiritually or Platonically independent from organic matter. Pragmatism and neuropragmatism tend to agree with recent theories about 'embodied cognition' that offer more specific implications of this organic embodiment for humanity. As Wilson expresses embodied cognition's claims, cognition is situated by taking place in the context of a real-world environment and inherently involves perception and action.[14] Wilson recounts the ways that cognition is for action. The function of the mind is to guide action and things such as perception and memory, which must be understood in terms of their contribution to situation-appropriate behavior. Cognition must be understood in terms of how it functions under the pressure of real-time interaction with the environment.

The invention of symbolic representation and written language takes advantage of the way that cognition specializes in dealing with transactions with deliberately modified aspects of the environment. Human cognition can off-load cognitive work onto the symbolic environment so that it holds or even manipulates information for us, and we harvest that information on a need-to-know basis. That makes the environment part of the cognitive system; the information flow between mind/brain and world is so dense and continuous that, for scientists studying the nature of cognitive activity, the mind/brain alone is not a sufficiently meaningful unit of analysis. This statement means that the production of cognitive activity does not come from mind/brain alone but rather is a mixture of the mind/brain and the environmental situation that we are in. These interactions become part of our cognitive systems. Our thinking, decision-making, and future are all impacted by our environmental situations.

These core views of neuropragmatism and (non-representational) embodied cognitive science can be extended to form judgments on classical philosophical problems about the mind-body relation, the natural basis for the highest cognitive functions, and the cultural origin of creative reasoning. For human cognition, the most important part of the lived environment to manage is society: the other humans that one must constantly deal with. Distinctively human cognition is from birth (and perhaps before birth) a matter of brains cognizing in concert. For humans, experience is culture – cognizing the environment is thoroughly shaped by the transmitted modes of cultural activities engaging human brains.

Additional theses of neuropragmatism, together distinguishing it from most other neurophilosophies, suggest ways to handle these issues.

8. Cartesian materialism still pervades too much psychology and philosophy of mind by demanding strict localization of rationality, prioritization of self-consciousness's powers, and the quest for perfect representational knowledge of a fixed external world. The brain exhibits much dedicated modular architecture, but massive parallel and networked processing is dominant. The brain is not hierarchical, but more democratic. Nerve centers across the brains are intricately interconnected with each other, so most any part of the brain has some direct or indirect systemic link to every other part of the brain. There is no inner Cartesian theater where all information is gathered and simultaneously experienced; experience at best displays rough continuities. There is no executive command center giving orders to the rest of the brain; deliberation at best guides habitual motor action. Ordinary cognition does not primarily aim at static representation in general but dynamic adequacy in specific situations.

9. The most sophisticated modes of human cognition are developments and assemblages of lower-level cognitive processes. These complex modes of thought, seemingly far from mere matter or biology, remain embodied and functional for practical success. Higher self-conscious cognitive processes (reflection, inference, hypothesis testing) are socially invented and taught capacities to attentively focus on ways to generalize practical habits for flexible use. These higher social capacities serve to coordinate group cooperative practices where some creativity is needed to maintain efficiency in the face of unstable conditions. Among these social practices are linguistic communication, symbolic representation, and logical inference. As our notion of the 'self' is bound up with these capacities, the self must be another socially constructed artifact of culture.

10. Imagination and deep memory add a contemplative 'space' where techniques can be experimentally attempted on related new problems. Even pure imagination, conceptual play, and aesthetic contemplation are creative capacities existing to refine practice, even though we can also do them in isolation from practical concerns. These creative modes permitted, among other things, the fixation of concepts and select relations among concepts, leading to reasoning. The most

complex modes of rational thinking (i.e., logic, scientific method) are refined developments from integrating component cognitive processes. Such things as logic, science, and all sophisticated modes of creative intelligence are culturally designed and educationally transmitted technologies.

11. Knowledge is the result of experimental problem solving, and the epistemic criteria for knowledge is the technological test of practicality. Scientific knowledge is continuous with technology and ordinary practical skill. Much of human experience, most of morality, and all of knowledge is an emergent feature of social epistemic practices. All *a priori*, conceptual, and linguistic truths are internal to one or another social epistemic practice, and cannot be directly used to criticize some other practice. Because no *a priori* conceptual rigidity can dictate terms of empirical adequacy, only the practical adequacy of a knowledge system is relevant to its validity. For example, no folk belief system rules over any scientific field, and scientific fields should respect pluralism and seek coherence, not unity. By avoiding epistemic dualism and reductivist monism, both epistemology and ethics can be naturalized, by showing how they fit in the natural world of encultured humans.

12. What seem to be '*a priori*' and necessary truths are only habits of cognition so habitually ingrained that our brains either use them unconsciously or our thinking relies on them so thoroughly. Evolution produced the infant human brain capable of speedily acquiring crucial functional habits because all humans need them, and additional functional habits are acquired when culture indoctrinates them into children. Habits are not unyielding reflexes; advanced learning is capable of questioning and amending any *a priori* truth through empirical inquiry and science. Because the *a priori* does not float freely from actual brain development, learning, and language, there is no logic-practice gap. Reason can be naturalized, because its processes and results can be shown to fit in the natural world of embodied and encultured humans. Everything that naturalism requires for justification is entirely natural, including the cultural technologies of intelligence that perform the justifications, so this pragmatic naturalism is an internally coherent, complete, and self-sufficient worldview.

These twelve theses of neuropragmatism permit it to offer an ambitious neurophilosophy. Having stated core theses of neuropragmatism, we may step back and survey wider intersections of neuroscience and philosophy. To establish itself as a fully legitimate neurophilosophy with a claim to some leadership role, neuropragmatism's mode of dealing with the mind must be scrutinized.

Neuropragmatism and mind

Leaving behind reductionism and eliminativism, pragmatism has always sought ways to show how to avoid dualism and representationalism. The Cartesian claim that mind and body have entirely different properties is demonstrably false. Lingering claims that consciousness has unnatural properties similarly rest on philosophical confusions and ignorance of brain science. The vast similarities between the twin functionalities of mind and brain indicate their interdependence and perhaps identity.

Neurophilosophy and neuropragmatism can show how to coordinate the functionalities of thought with the functionalities of nervous systems. For example, thinking and nerve activity both have temporal durations; they are both found in localized living centers rather than diffused through all of nature; they both consist of relational continuities rather than atomic accumulations; they are both dynamic rather than static; they both display growth and decay; they both function in attending to practical dealings with the environment; they both primarily aim at maintaining the organism's well-being. Even the most 'subjective' parts of consciousness, such as the feelings and qualia noticeable in self-consciousness, are aspects of the dynamically functional flow of thought. No pragmatism would seek to 'reduce' felt qualia to nervous activity or anything else to prove that they are natural. The old metaphysical formula demanding identity of all properties for genuine identity was rejected early on by pragmatism and is no longer taken seriously beyond armchair philosophy. For science, functional identity is quite sufficient: where two phenomena are perfectly correlated and display the same functionalities, the two phenomena are rightly regarded as the same natural process observed from different perspectives. Qualitative feelings happen where nervous systems achieve certain degrees of complexity. Subjectivity need not be treated as anything spookily 'unnatural.' The mysteriousness of subjectivity quite vanishes. Subjectivity and perspective is precisely what would be naturally expected when discrete brains generate discrete experience. You have a very different perspective on your brain than anyone else, since you are directly experiencing what it is like to be a brain of a certain complexity.

The lived experience supplied by cognition reflects its neurological basis. Unscientific philosophies point to features of experience or thought allegedly lacking dynamic functionality or integration with action. Worse, anti-naturalistic philosophies further claim that scientific naturalism can never integrate them with energetic matter. However, neurological investigations (much less any sound phenomenology, such as that of pragmatists) have not been able to confirm such static and aloof features of consciousness. Interestingly, such supposedly 'pure' or 'inert' parts of experience (sense data, intense qualia, and the like) are actually detectable by those seeking them only after the most intense cognitive effort to distill them from the ordinary flow of active experience. There simply is no avoiding dynamic and creative cognition. Consciousness is intensely qualitative, to be sure, precisely because the brain puts so much work into that phase of experience. Theories of mind comfortable with taking purity, passivity, receptivity, or representation as basic modes of cognition must be rejected as incompatible with neuroscience. All the same, neuroscience is at liberty to develop specialized theories about micro and macro brain systems, borrowing and modifying terms as it may require. No folk psychology or linguistic conventionalism can dictate terms of scientific inquiry into the brain-mind. The dream of the unity of science having dissipated, teleological and intentional terms can be legitimate features of successful empirical studies at every level from the social to the synaptic, although mechanistic causality dominates at molecular levels. Indeed, the choice between teleological and mechanistic modes of explanation may not be forced. Some varieties of naturalism, like Dewey's, propose that mechanism is visible in teleological systems when analyzed closely enough, but that only means that teleology requires mechanistic parts even while no mechanistic explanation could ever suffice for the whole. After all, wholes typically have genuine powers and properties that no aggregate of parts could have. This is not duplication of causal powers, as reductionists fret, but only the recognition of compatible kinds of causal powers at different scales and systems of nature. The pluralistic stance of pragmatism and neuropragmatism is hospitable to continuities of terminology and causality at multiple levels of brain science.

Higher human cognition can occasionally achieve sustained reflective passivity, open receptivity to experience, and sophisticated representations of the so-called external world. Neuropragmatism cannot deny that humans can do these things. Yet it must undertake explanations for their existence without permitting them to assume any fundamental role in ordinary cognition. Neuropragmatism tends to favor the idea that sophisticated symbolic capacities of human intelligence are scaffolded on the extended mind of linguistic sociality. Basic cognition is not symbolic or representational, but human societies design their environments in ways that offload cognitive work onto the manipulation of external symbols. Rationalism in general makes it difficult to account for cognition and knowledge in any natural terms. Cartesianism was the height of presumptive rationalism by taking our most sophisticated forms of communication (replete with analytic meanings and necessary truths) as essential to all consciousness and cognition. Later representationalisms sustained this obsession with static symbols, rendering it difficult to naturalistically explain even how children acquire linguistic competence.

Neither static nor computational representation characterizes ordinary cognition. Reliance on representation leads to a postulation of foundational perceptions. However, experience is not 'built up' from purer building blocks of direct information from nature. Connectionism comes closer to dynamical and distributed cognition but may still contain aspects or elements of representationalism. Neuropragmatism, like other neurophilosophies, takes close notice of the way that the brain rapidly merges diverse streams of stimuli from all sources in order to guide effective action in the lived moment. All cognitive processes (and hence all conscious experiences too) fuse information about about external sensations, motor control processes, and internal feedback from the body. There is no pure sensation, no pure will, and no pure feeling. There are no dichotomies between sensation, emotion, and reason – these aspects of cognition work together as they guide behavior. Even in the simplest case of behavior, these fusions are evident. Simplistic associations are inadequate because organic circuits create new wholes that are not merely sums or sequences of their parts. In a genuine organic circuit of perception, action, and consequence, the meaning of the perception includes the prior action done to gain that perception (the turning of the gaze toward an object); the meaning of the action includes both a desire (to touch that object) and more perception (to guide the reaching); and the meaning of the consequences of the touching includes the guided action of touching (the felt pain is not just felt pain, but the pain of touching that object). The next time the child sees the flame, he sees a *hot* flame, and when he reaches for that flame, he *reaches for a painful touch*. From now on, for that child, an idea of touching that flame simultaneously contains the idea of pain.[15]

In general, most of the meaning in perceiving things consists of anticipations of potential reactions upon dealing with those things. Organic circuits result in holistic organic wholes of experience. Experience is thoroughly imbued with prospective values of action. That is why we directly experience meanings and values in the world around us.

If meanings or values were only interior mental states, then our experience of an external object would be stereoscopic, a sort of double perception. We would observe the external object as a meaningless material thing, and simultaneously observe it as a useful object to be employed, as if one 'eye' saw the world as it is in itself, while the other 'eye' saw objects as meaningful and valuable. Does lived experience ever seem like this? Hardly – we immediately and directly observe significant, meaningful, and valuable objects without any double 'vision' or contrast between an external world and an internal world. Meanings and values are where they appear to be: embodied in the things that we know how to use. Meanings and values are instances of achieved practical knowledge through learning. Knowledge is built up from our experimental attempts to productively manage our deliberate modifications to the environment. Static representationalism, correspondence theories of knowledge, and Cartesian materialism are not viable theories of the mind and intelligence. Neuropragmatism allies easily with theories of active perception;[16] somaesthetics;[17] naturalizing intention;[18] ecological psychology;[19] ecological cybernetics;[20] social cognition and social epistemology;[21] neurosociology;[22] extended mind;[23] neurophenomenology;[24] and radical embodied cognitive science.[25] Even aspects of connectionism and dynamic systems theory may contribute to the proper synthesis of these positions if excessive representationalism is avoided.[26]

To ask, 'Is mind just in the brain?' is problematic. 'Mind' is ambiguous: it can refer to the localized centers of cognitive processing, or it can refer to the networked channels of meaningful information. Localized mind is where brains are; philosophical options are common substantial cause, or dual aspect monism, or outright ontological identity. Networked mind is wherever brains are coordinating action through communication, and therefore much of intelligence is an emergent feature of human communities modifying environments. Mind is dependent on brains, and cognitive functions are brain functions, either of single or multiple brains. Neurons are all about systemic communication, across synapses and across the room. Many cognitive functions (and all higher cognitive functions) only operate through brains in communication with each other about the common environment. Human psychology must be social and ecological. The 'theory of mind' ways of trying to explain how humans try to understand each other's beliefs and motivations take matters exactly backwards. We do not really start from our own concepts of what constitutes the mental life and tentatively test them against the empirical data of others' behaviors.[27]

Individuality, like mentality, is an emergent social category, not a biological or metaphysical category – no one is born as an individual self. Like every other role, one learns how to be an individual only within a community (and that is why different cultures apply differing notions of individuality). The way that even babies have personalities is not a refutation, but a confirmation of this social theory of the self, since the growing infant learns how to be treated as an individual by being treated in ways particular to her personality (and only later on will she realize that she has a personality). Although there are numerous broad continuities between animal and human cognition, as would be expected given evolution, human cognition displays some notable discontinuities from animal mind because we are now such intensely cultured animals.[28] By taking higher cognition and self-consciousness, like all human communication, as fundamentally social, neuropragmatism is aligned with Peircean semiotics,[29] the social mind,[30] symbolic interactionism,[31] developmental consciousness,[32] and biosemiotics.[33]

Cognition and culture are thoroughly natural. The biological evolution of the human species, and the cultural evolution of complex human associations, suffice to explain all features of cognition. The two modes of evolution are not disjunctive – no form of cognition is independent from either mode, although most complex forms of human cognition are primarily cultural in origin and function. Nothing spiritual or supernatural is needed to account for mind. The highest modes of human cognition aim at social competence, technological expertise, and knowledge of reality. Culture educates members of society into various forms of responsible intelligence, and expects their satisfactory use for group goals. These cognitive modes amount to technological skill and ultimately answer to pragmatic criteria of success set by societies. Essentially, culture is technology; social learning and teaching was the first technology, and all else followed.[34] All epistemology must be social and technological; no philosophical theory of reason, knowledge, or truth can float freely apart from learning's origins in education and experimentation, or avoid answerability to practical social justification within cultural contexts.

Cognition, deliberation, and the role of system three

Reflective deliberation is no illusion or irrelevant luxury. It is a useful imaginative function for specialized human cognition for problem solving. Responsibility in turn is the degree to which one can successfully use reflective deliberation to guide conduct in socially appropriate ways.

Recent work in psychology by Daniel Kahneman on system 1 and system 2, recent interest in revitalizing representationalism in cognitive science, and recent use of the concept of information in the science of consciousness all suffer from a creeping Cartesianism that blocks the road to inquiry. Neuropragmatism offers a way through this hurdle by emphasizing the contextual situation in which inquiry develops. The neuropragmatic sketch of experience, habit, mind, consciousness, and inquiry provided here is used as a framework to reconstruct the important data we consider from psychology, cognitive science, and the science of consciousness. The shortcomings of these empirical studies are overcome by system 3, which is the dual-process of enculturation that situates systems 1 and 2 and provides the means of their further transformation through the work of creative intellectuals, whose task is to imagine and discover new possibilities for lived experience. The introduction of system 3 by neuropragmatism is a philosophical hypothesis intended to effect further philosophical discussion and scientific consideration.

Our hypothesis is that once the Cartesianism underlying the psychological constructs of systems 1 and 2 (as recently popularized by Daniel Kahneman),[35] the debate over what to do about so-called 'mental representations,' and the import of the nebulous concept of information is eradicated, we contend that a reconstruction of the two systems, of representations, and of information yields a third system that resolves the difficulties faced by cognitive scientists suffering from creeping Cartesianism. System 3 is chronologically the most recent of the systems, the most fragile, and the most important for understanding the mental life of the individual yet social human animal. The stereotypical Cartesian and Humean concerns over intentionality and the 'external' world are shown to be more properly conceived as sociocultural events and not strictly (neuro)biological 'things' of individual brains (or minds). So conceived system 3 is cultural insofar as it produces the means by which the first two systems are capable of doing their work in a *specific situation*. System 3 as a cultural mode is also experiential: it draws its power from the long evolutionary history of the experience of *Homo sapiens*.

For pragmatism and science alike, experience is a prominent concept that carries considerable authority. *Prima facie*, the appeal to experience may not seem problematic in itself. However, when it comes to the science of the mind, experience is both that which is to be explained and the means by which an explanation gains some authority. This circularity is even more troubling when we consider that the conception of experience at hand is vague and often an equivocation between what the Germans refer to as *Erfahrung* and *Erlebnis*.[36] The latter refers to the sensationalistic empiricism of David Hume and, subsequently, the logical positivists and empiricists. On this conception of experience, there are mental or experiential *states*, each of which is easily discernible from another. The philosophical and the scientific literature abounds with talk of states of blue or cold, red or hot, etc. This classical view of experience is faced with the problem of *representing* the world external to the experiencing mind. This view relies on an ancient distinction between sensation and perception. Briefly, sensation consists in the bodily sense organs (i.e., eyes, ears, nose, tongue, skin) sensing the external world and transmitting its data about the world to the mind, where it is then perceived.

Dewey called this view the 'spectator theory of knowledge,'[37] and Daniel Dennett has christened it as 'the Cartesian Theater.'[38] The central epistemological and metaphysical issue here is that the mind *is* a thing that passively receives sense data about the world, and that this is how the mind *knows* about that world. Among the several problems with this conception of experience is what is known as 'the veil of ideas or appearances.' This veil divides the world in two, into the mental and inner world and the physical and external world. Somehow these sense data of which the veil is made connect with the external world and thereby represent that world to the inner world. This indirectness of experience presents the problem of knowing *anything* with certainty about the world, which raises further questions about how scientific knowledge is reliable enough for human action to depend upon.[39]

In light of Darwinism, the classical pragmatists found good reasons for rejecting this duality between mind and world. Instead of conceiving of experience as *Erlebnis* (i.e., sensationalistic), they promoted the conception of experience as *Erfahrung*. Experience of this variety is at play when someone asks if you have experience with a skill, like skiing. It is another way of asking if you have *familiarity*. And just as the etymology suggests, there is no real divide – mind and world are of the same source, just as siblings are of the same parents. This intimacy of experience also provides the means of knowing about the world. Instead of experience being a sequence of atomistic states, the pragmatists considered it a continual process of learning. Education occurs through a familiarization – an ongoing transaction between the learner and that which is learned. There are differences but not divides. Of the experiences had, *the differences that prove to make a difference* in future experience are particularly important.

The phase of experience we know as 'inquiry' is able to make the *functional* distinction between organism and environment. This distinction between organism and environment – while often made at the skin – must be functional and not ontological because the two are inseparable: if there is an organism, then there is an environment; if an environment, then an organism. Consider the etymology of 'environment': it is that which 'environs,' *surrounds*, something – in this case, the organism. Recent work in evolutionary biology and developmental systems bears this out. Griffiths and Grey have argued that this coupling of organism and environment is so tight that the proper unit for evolution is the single unit of *organism–environment*, or as Griffiths and Grey suggest, the symbol Œ.[40] From here, Dewey's conception of experience as organism–environment transaction can be restated as Œ-transaction. This conception implies that experience is old: it has a long evolutionary history, most of which is a series of events that are simply had – experiences that are known is a much more recent affair.[41]

Experience as Œ-transaction implies that any attempt to localize experience in any part of the transaction is doomed to failure. Furthermore, experiencing and the products of experience are not exclusively found inside the organism. This transactionalism requires that experience modify both the organism and the environment. Dewey referred to this joint modification as *adjustment*. This is a *dual process* of the organism's *adapting* to the environment and the organism's *alteration* of the environment.[42] Among the consequences of these adjustments are the development of *habits*, the dependable and regular behavioral dispositions to act without foresight or deliberation. Given the time pressures within Œ-transactions, the development of habits comes as no surprise. Some habits become so good at keeping life and limb together they become generic traits in a species. One such trait is plasticity, an individual organism's ability to learn new habits through its interaction with its environment.

Another kind of flexible habit is the active organization of one's environment so that one's firm habits are more effective. For Dewey, this ecological niche construction develops a niche filled not just with transient and fleeting gestures and sounds that communicate the here and now but also with signs and symbols that persist beyond the momentary use. This phase of experience Dewey conceived as mind*ing*. Instead of mind being some sort of individual ephemeral thing that somehow interacts with a physical body, the body *minds* its environment. Minding, on this view, is the dynamic organization of habits of the organism and of its environment that afford meaningful behavior. While dynamic, this scaffolding has far greater stability than does conscious activity, which occurs when the regular flow of habitual activity becomes disturbed and thereby uncertain.

A minding organism goes about its environment with expectations of how this transaction will go. For this reason, pragmatists conceive of belief as a *habit of action* – not a representation or reflection of how the world is independently of human activity, viz., of a reality behind the veil of appearances. When an organism's habits are conducive to activity that maintains life and limb, there is no need for adjustment of the organism nor of its environment. However, when this dynamic equilibrium of Œ-transaction is disrupted, some adjustment is necessary.

In light of recent advances in our understanding of non-human animal life, especially in its continuity with human life, we reserve experience as a larger category than culture. Experience as Œ-transaction is deep, going back millions of years. Culture refers to the idiosyncratic Œ-transactions that define symbolic and sapient – which is to say human – life. Some species are communicative and have just those sorts of experiences, but they do not know it. Others communicate through symbols and signs and not just gestures and sounds, but they are not aware of their semiotics, nor can they inquire into them. Culture grows out of such populations when its instrumentation becomes deliberatively innovative and thus consciously selective.

Recall that adjustment is a dual process that modifies organism and environment alike. This general phase of Œ-transaction develops into a powerful process with the evolution of culture. The introduction of cultural artifacts affords humans the means of deliberate innovation, specifically in using them to discover new strategies for getting about the natural and cultural environment *and* for transmitting the successful strategies to the rest of the culture, thereby reforming it. This process of discovery is undertaken by a small number of inquirers. These researchers have a greater disposition toward fallibilism and stronger attention spans. They have developed a set of habits that are conducive to performing cutting-edge inquiry. The projects taken up by artists and scientists alike – the creative intellectuals – demand an openness and willingness to be self-critical, not only to develop new solutions to problems but to reconsider both the solutions proposed and the articulation of the problems addressed. This degree of critical reflection not only requires above-average conscious attention to the complex situation but also a community that encourages and effects this highly sophisticated sort of inquiry.[43] Culture in a community is a dual-process system that actively promotes discovery via experimentation and deliberately modifies the cultural environment in light of these discoveries.[44]

In his recent book, Kahneman elaborates a dual-process theory of human cognition. He takes up the nomenclature of Stanovich and West that distinguishes between a fast response, system 1, and a slow one, system 2.[45] Even though Kahneman is careful to note that systems 1 and 2 are umbrella terms covering several different subsystems, he nevertheless sets them in nearly perfect dichotomous opposition. Where 1 is fast, automatic, effortless, and always operating, 2 is slow, lazy, and rarely operating. Plus, system 1 is metabolically efficient, whereas system 2 drains energy. The metabolic contrast is well illustrated by the commonsensical descriptions of each system. System 1 is the set of habits or intuitions or instincts that quickly respond to immediate problems a person may face. Usually, the system does a good enough job at reacting, but mistakes are made regularly enough that a fail-safe is beneficial. System 2 evolved to be what Kahneman describes as the conscious self that is, on occasion, capable of pushing back against habit or instinct. This tension is perhaps better assuaged when a person is not in a situation that demands immediate response. That is, system 2 is capable of modifying system 1 through the deliberate intervention into one's lived experience that is intended to change a person's habits. As some recent research suggests, it could be as many as three years of diligence before system 2's efforts to adjust system 1 take hold, rendering moot the need for the conscious self to intervene.[46]

There is a need for system 3 to orient the processes of not just the two systems but of the Œ-transaction as well. System 1 operates often in tension with system 2; the immediate and habitual responses often conflict with the interests of the conscious self. Yet Kahneman's account finds this self to be lazy and often blind to what is really going on, making mistakes of its own. His illustrations of this laziness and blindness are not default traits of system 2 but products of its conflict with another system, the cultural situation. In these experimental illustrations of laziness and blindness, however, the culture is artificially constructed for the purpose of effecting what Kahneman finds to be absurd responses. Yet if the cultural situation changes, the results are likely to be different; the absurdity, in other words, is not a result of system 2 but of system 3 in tension with system 2 (especially when we consider that the particular test subjects' system 2 developed within a different system 3 than the artificially constructed one of the experimental model). The parameters of system 3 shape the nature of the Œ-transaction. This shaping of the trajectory helps elucidate how Œ-dynamic systems anticipate without relying on representations within the brain/mind, as the *Erlebnis* conception of experience requires.

Kahneman admits that he focuses more on system 1 than on system 2. What is even more lacking is how system 2 can anticipate the future. Indeed it is unclear whether such a task is in the purview of system 2. Though it seems reasonable enough, at least to our commonsensical view, the conscious self is capable of anticipation. Many cognitive scientists take it as a matter of commonsense that our mental activity is representational, and that if there is any doubt about this, the clear fact that we anticipate future events requires that we presently represent that future.

How is 'coordinating with the future' handled from a pragmatist perspective? If we are to continue to use the word *representation* with regard to *Erfahrung*, it makes little sense to talk about representational states qua *Erlebnis*. For this conception of experience (as *Erfahrung*) is not one in which states have a role. The best way of using this word then, is as a *re-presentation* of the world, in that the present world is presented anew for the precise purpose of effecting such a world out of the present world – a 'taking aim at' a new world. In order to anticipate, an organism needs information about the world's regularities, patterns of change, etc., so that appropriate action may be taken to bring about the *re-presented* – or, better still, *imagined* – world.

Conscious intervention in the habits of Œ-transaction is integral to this sort of anticipatory adjustment of body and world. However, we still lack an account of how or why representations as ideals (or ends-in-view) could help guide system 2's operations. Kahneman does not recognize (at least in print) the need for this; representationalist cognitive scientists vary on the need as well as the account.[47] We believe this is due to the creeping Cartesianism at play, in at least two ways here. First, modeling cognitive anticipation without drawing on representations presumes that representations are to be found strictly internally to the organism or mind; whereas the pragmatist emphasis on the dynamic transaction of organism and environment considers representations qua ideals (or ends-in-view) as being neither internal nor external but transactional. This transactional conception of experience implies that all inquiry must take place within a situation, within a cultural context. Thus we see the second way in which Cartesianism creeps its way in. The Cartesian ideal of pure inquiry outside of a cultural context is not only impossible to attain, it also blocks the road to inquiry – a cardinal sin, if there ever was one, for pragmatism.[48] This blockage can be overcome by recognizing the need for system 3 as that which provides symbolic affordances, shared aims in action and in inquiry, and, in short, the creative means for anticipating novel ways of living.

The reduction of uncertainty through the cultivation of information is what conscious activity (system 2) strives toward. System 1 lacks the information required for resolving new problems that arise in unorthodox situations. The means by which information is cultivated is not a strictly or exclusively individual act as the Cartesian conceives it. The cultivation of information – what Dewey called *education* – is a social activity that aims at the production of healthy inquirers. System 2, on its own accord, cannot resolve problematic situations. Guidance is required and is provided by the larger system 3, the cultural landscape that provides the values and ideals that orient an individual to the world such that one's interactions with one's environment can be more meaningful than the experiences that have come before – experiences that are not simply unique to the individual but are shared through tradition and education as well.[49]

Kahneman fails to see that his examples situate or frame the inquiries he asks of his subjects in such a way that they are simply not ready for doing awkward financial arithmetic or for anticipating something absurd to happen while focusing on a very specific task. Change the conditions, and the experience will change. How then are we to understand the role of information with regard to systems 1 and 2, and representations and intentionality? We propose that a third system, culture, is the best way to orient ourselves.

System 3 concerns the situational context through which a dynamic system, such as a conscious human, can anticipate by using previously learned skills, previously learned data (from facts to tropes), and previously learned methods of inquiry, to create novel ways of living and doing. These ways, of course, do not appear *ex nihilo*. They grow out of and are thus continuous with the previous ways. Such ways, however, are not so clearly available to a researcher who seeks to strip away culture and context.

Kim Sterelny has done valuable work in the philosophy of nature[50] that argues that humans are unique among primates because we have evolved to be learners, specifically apprentices to each other.[51] Our sociocultural organizations, our scientific and religious institutions – even neonate curiosity – reflect this uniquely human feature. Sterelny has much to say from the perspectives of evolutionary biology and anthropology, specifically about our development of tool use and innovation. But where his account is lacking is in the neural means of apprenticeship (to be clear, this is not the only means, either). Bill Bywater has started this work, from an explicitly neuropragmatist standpoint.[52] He situates recent work on mirror neuron systems with Sterelny's conception of apprenticeship. Take this view with similar work by neurosociologist

David Franks (2010) (who is working from the pragmatist perspective of George Herbert Mead),[53] and we have the basic tools and methods for bridging the work of Kahneman with anthropology.[54]

With the pragmatist sketch of experience introduced here, we hypothesize that system 3 is the means by which human experience qua Œ-transaction becomes oriented to the world and thereby appropriates information in a plurality of ways. From the general traits of systems 1 and 2 as components of the human dynamic system of Œ-transaction, we believe further research along these lines can help elucidate questions about how different cultures learn, how information is selected and passed down through various traditions, and how to resolve tensions between the three systems. Just as system 1 can conflict with system 2, system 3 conflicts with system 2. Consider an example of a dieter. His system 1 wants pie, but his system 2 says spinach is better. In his case, his culture may be one in which delicious but calorie-laden food is everywhere to be had, making the goal of spinach eating quite difficult, if not impossible. But system 3 could also be one in which spinach is easily available, but the cultural ideals – the guiding parameters – emphasize an extremely thin body type that is physiologically and psychologically unhealthy.

Our hope is that by investigating the import of culture in this fashion, we not only resolve or evade theoretical difficulties in the cognitive sciences, but that we also offer a way for utilizing the results of these sciences, along with other inquiries, especially the arts, to address practical concerns for achieving the ever tenuous democratic culture, so well imagined by James, Dewey, and Rorty.

Neuropragmatism and conflict over image

As philosophers from John Locke to John Dewey and Daniel Dennett have argued, our capacities for practical deliberation, normative conduct, and degrees of moral freedom naturally grow together and remain culturally fused together. The intense degree of human sociality accounts for the way our species encourages normative conduct using normative moral responsibility in addition to the older primate emotional motivations of love, kindness, and charity. However, the intense sociality of human life requires the thoughtful management and adjustment of multiple social roles and responsibilities, in turn requiring dynamic moral problem-solving about what to do from situation to situation. Moral concepts such as responsibility, freedom, autonomy, and blame have distinctive functional roles in creatively sustaining the community life of human societies.

Can philosophy reconcile the two opposed 'images' of humanity – the scientific on the one hand and the humanistic on the other? While there is some disagreement on the nature of this reconciliation – generally understood as the conflict between eliminativism and constructivism – the neuropragmatist solution to the conflict is to reconstruct the philosophical notion of science's aims and results that leads to competition between the two images in the first place. This conflict, however, is not merely a theoretical problem for philosophers. It has manifested itself socially in the academy as the two cultures described by C. P. Snow.[55] There is a desperate need for reconciliation of some sort, as there are real life consequences across the life sciences and out beyond the ivory tower into areas like public policy.

Despite great similarities between mainstream neurophilosophy and neuropragmatism, there is a crucial difference between them. This difference resides in contrasting conceptions of experience; it subsequently sets up distinct conceptions of science, and therefore various resolutions to the conflict between the scientific image and the humanistic or manifest image.

The philosophical project of *rapprochement* is taken up in various ways by the many philosophical traditions. The specific differences between mainstream neurophilosophy and neuropragmatism come down to how the problem is articulated, and thus, how it is solved in light of that articulation. Generally speaking, however, the conflict is a genuine one felt by most parties. The concern is that the scientific image ultimately shows the humanistic one to be illusory, thereby bringing into serious doubt genuinely human concerns about dignity, freedom, responsibility, and living a good and meaningful life. Science, it is feared, will rob us of our humanity.

For mainstream neurophilosophers, like Paul and Patricia Churchland, Owen Flanagan, and Daniel Dennett, the conception of science differs in significant respects from the neuropragmatists' view. Moreover, the conception of cultural tradition, what Wilfrid Sellars influentially called 'the manifest image,' similarly differs between the two camps. The main distinction is in how each camp conceives of experience, and subsequently of science. Patricia Churchland articulates the problem in terms of scientific theory versus folk theory, and then, as she often does in the latter work, refers to Quine and his pragmatism.[56] The neuropragmatism we advance here is similar to this branch of neopragmatism but, as will become clearer, stands in stark contrast to the conception of science based on an inadequate conception of experience. The Churchlands continue this discussion in terms of folk psychology versus scientific psychology, and mention the origins of these ideas in Sellars.[57] Paul Churchland further distances himself from pragmatism in his recent book.[58] Flanagan's recent statement of his philosophical project is in these terms but with a greater pluralism, extending the Sellarsian dyad to a sextet.[59] Dennett also makes a clear and accessible statement of the problem, even as he has affirmed most of the neuropragmatist materials for its solution.[60]

While both camps see the manifest or humanist image developing first and providing the framework out of which science and its image develop, mainstream neurophilosophers see the two images as competing with each other for the truth. The truth of science is taken as value-free and objective, whereas the truth of the manifest image is value-laden and subjective. Notice that this conflict is yet another version of mind-body dualism, in which the properties of each – science and culture – are mutually exclusive. Sellars articulates the question that philosophy faces as, 'How, then, are we to evaluate the conflicting claims of the manifest image and the scientific image, thus provisionally interpreted to constitute *the* true and, in principle, *complete* account of man-in-the-world?'[61]

This conflict is generated for mainstream neurophilosophy largely due to residues of logical positivism, which is based on a Humean conception of experience. Like Descartes's rationalistic view of the soul, Hume's empiricism fits the model of the spectator theory of mind that Dewey criticized. Today we recognize such a view as Cartesian materialism. While neurophilosophers like the Churchlands, Dennett, and Flanagan would balk at being called Cartesian materialists, they succumb to the modified account of it (as described by Rockwell).[62] There may not be one specific place in the brain where experience all comes together, but they suppose that there is a specific space delimiting experience: the brain itself. The neuropragmatist denies this limited range of experience or mentality. Mentation goes beyond the cranium, suspended in a cultural medium of communicating humans. Neuropragmatism would not achieve the naturalization of consciousness and mentality by limiting it to a single brain, ignoring how human brains become distinctively human only when wired together. If other neurophilosophers cannot see the 'wires' of sight and sound, a too-narrow scientism has already rendered those into meaningless physical entities. One might as well do that to all the signaling wires of the nervous system and be done with meaning altogether. Avoiding that eliminative dead end, the only alternative is to take seriously the way that both the phenomenology of lived human experience and the physicality of brains interacting with each other and the environment exist in natural spaces much larger than the confines of any cranium taken singly. It seems like we are directly experiencing the external world *because we really are*. The unsurprising fact that complex natural systems of brains and environments can be distorted and deceived into illusions and hallucinations no more proves that consciousness is all in one's head than hacking a computer network proves that the world wide web is all in one's computer.

Even where some mainstream neurophilosophers would not deny that experience and intelligence is partially social, they have not dealt with the full implications of viewing humans and all their cognitive products as encultured. Another problematic residual aspect of Humean experience in logical positivism is the maintenance of the fact/value dichotomy.[63] This issue, too, is complex, as each of the aforementioned neurophilosophers has held varying views throughout his/her career. Regardless, this dichotomy fits the general pattern that neuropragmatism seeks to eliminate. Among the reasons mainstream neurophilosophers have such difficulty in their efforts to reconcile the manifest image with the scientific image is the question of what to do with value (or mentality) in an ontology of value-free facts (or bodies)? Eliminativism is one strategy; constructivism is another. The former fails to keep the sacred aspect of the manifest image, which many find a dissatisfying, if not a terrifying proposal. The latter is left making qualification upon qualification about what is meant by manifest terms like 'consciousness' in ways that end up making their readers wonder whether consciousness is real or illusory. This, too, is unsatisfying. The residues of ordinary language philosophy and the 'linguistic turn,' which is based on a neo-Kantian view of cultural mind, have not helped matters. By encouraging some philosophers to suppose that they have privileged access to analytic truths grounded in enlanguaged culture, a battle arose between linguistic a priorists and neurophilosophers over who had the right to dictate the nature of the self. This battle only sustained the dualistic terms of the debate into the late 20th century, as neurophilosophers felt pushed into viewing culture as a competitor to the scientific image of humanity. Ironically, humanists fearful of scientism have only perpetuated the worry over an inhuman theory of self that an improved cognitive neuroscience would prevent.

Neuropragmatism evades these problems of dualism by integrating science and culture. Neuropragmatism conceives of science (like all modes of intelligence) as an inherently evaluative and thus value-laden method that provides provisional instrumental truths as guides to practical action in the world – not a method of justifying static propositions that objectively mirror or correspondingly represent the non-human external world. In his articulation of the conflict between science and common sense (i.e., the humanist or manifest image), Dewey argues that the subject matter of both science and common sense is one and the same experience, conceived as the dynamic interaction of organism and environment: 'Things interacting in certain ways *are* experience';[64] experience is 'the manifestation of the interaction of organism and environment' or simply 'an interaction of organism and environment.'[65] What distinguishes science from common sense is the mode of inquiry, specifically the experimental method developed into the sophisticated technological and industrial affair that produces the most secure knowledge humanity has about the world to date.

Dewey argues that the humanities or common sense is concerned first and foremost with 'practical uses and enjoyments' of our existential situation, 'with "the ordinary affairs of life", in the broad sense of life.'[66] Another important point Dewey makes about common sense is that it is not static and fixed but always changing in response to the dynamic environment. We see this progression in the history of the humanities, broadly speaking, from myth to mythology to dogma and scripture to Chaucer and Shakespeare through to contemporary poetry, novels, films, and so forth. In one way or another, these affairs are concerned with our everyday lives, not as isolated events but as living experiences, as social interactions with each other in a world, actual and imagined. Through them we see how life could be lived and could be experienced. They not only affect our consciousnesses but bring about qualities in both familiar and novel ways so as to encourage or admonish specific ways of life. They are at the heart of our moral lives. In abstracting beyond the particulars of common sense, Sellars and others end up stopping or freezing a dynamic, living process. Snapshots have their place, surely, but to take the snapshot for the whole is to lose out on the entirety and the richness of life.

Science develops out of the same subject matter as common sense, with a concern for practical affairs of ordinary everyday life. When wholly successful, the results and the methods developed by science feed back into the commonsense world 'in a way that enormously refines, expands and liberates the contents and agencies at the disposal of common sense.'[67] Unfortunately, Dewey notes, this feedback has not been nearly as successful as it needs to be, never amounting to more than providing new tools for upholding tradition, and never fully critiquing tradition. This is due in part to the tendency of the practitioners and outside observers of science to finalize the results and methods of science. Sellars does this in setting up the opposition between the manifest and scientific images as though they both could be *the* complete and *the* final word on matters. Dewey describes the dissolution of the problem of reconciliation when we see that '[s]cientific subject-matter is intermediate, not final and complete in itself.'[68] Science is a provisional and ongoing cultural technology, one of the most humanistic endeavors humans undertake.

Taken and frozen at any intermediate stage, however, the products of scientific inquiry seem to be isolated objects, set apart from the situations in which they were originally encountered. As science progresses, it becomes increasingly removed from practical affairs as its proximate goal is to develop knowledge for its own sake – not to develop within the lived-in environment of ordinary life. This is not its only goal: the products of science are empowering when properly integrated into the humanities and ongoing cultural life. Science, when seen as just a phase within the interaction of organisms with their environments in the process of life, has consequences and applications outside of itself, in the commonsensical world, with which the humanities are primarily concerned. The neuropragmatist conception of experience thus seeks to establish and cultivate the continuities between science and the humanities, between the scientific image and the manifest image, to improve the richness of living experience in a never-ending process of growth.

Pragmatism started off at a time of significant scientific and technological change. The industrial and Darwinian revolutions, as well as the American Civil War, brought about both a sense of crisis and a vision of hope for what humans could do should they work together toward a common goal. Today we are still wrestling with the consequences of Darwinism and industrialization. Yet we have further difficulties with which to wrestle than the classical pragmatists. For among the consequences of Darwinism and industrialization is a globalized information society that has the means to yield life-saving, life-improving medical care *and* the willful creation of biological warfare *as well as* the inadvertent diseases effected by industrial life and life in an information society. The successful scientific models that inspired the classical pragmatists were those of physics, chemistry, and early biology. Neo-Darwinian models of life and the impressive rise of the cognitive and behavioral neurosciences[69] provide new inspiration, new tools, new hopes – and new challenges.

The consequences of these new sciences for our understanding of our world and ourselves are not only undeniable and promising, they are also more threatening. Physics provided a cultural transformation in how we alter our environments and generate energy. But it did not seem to threaten our moral, spiritual, and intellectual lives with any significant conceptual change. Indeed, the changes were seen initially as liberating, until much more recently. With physics, the moral threats came from increased pollution of our environment, and, with the bomb, the very real possibility of mutually assured destruction. Chemistry likewise gave us new materials and fuels as well as chemical warfare and new means of substance abuse. Biology similarly brought benefits and dangers, from longer life spans to biological warfare. But biology brought with it a renewed sense of crisis for the human self-conception. Physics may have displaced the center of the universe from the Earth, but the belief in Cartesian dualism left the human soul seemingly intact. Biology, especially after Darwin, opened 'the gates of the garden of life' to experimental methods.[70] Now opened, the challenge to pragmatism is the threat science, especially the neurosciences, poses to our cherished ideals. For the challenge is not only to bring the products of neuroscientific inquiry to bear on morals and politics, as so many researches are eager to do today, but to use such data in order to bring the experimental method and attitude to morals and politics as well.

Notes

1. The following is a synthesis of parts, drawn from and modified, of two separate articles: Tibor Solymosi and John Shook, 'Neuropragmatism: A Neurophilosophical Manifesto,' *European Journal of Pragmatism and American Philosophy*, 5(1) (2013): 212–234; and Tibor Solymosi and John Shook, 'Neuropragmatism and the Culture of Inquiry: Moving Beyond Creeping Cartesianism,' *Intellectica* 60(2) (2013): 137–159.
2. John Dewey, 'Body and Mind' in *The Later Works of John Dewey*, Vol. 3, ed. Jo Ann Bodyston (Carbondale: Southern Illinois University Press, 1927/1984), pp. 25–40.
3. Paco Calvo, and Antoni Gomila, eds. *Handbook of Cognitive Science: An Embodied Approach* (Amsterdam: Elsevier, 2008), p. 15.
4. Gary A. Cook, *George Herbert Mead: The Making of a Social Pragmatist* (Urbana: University of Illinois Press, 1993) and Jerome A. Popp, *Evolution's First Philosopher: John Dewey and the Continuity of Nature* (Albany: State University of New York Press, 2007).
5. Hilary Putnam, *The Threefold Cord: Mind, Body, and World* (New York: Columbia University Press, 1999).
6. For example, Daniel C. Dennett, *Consciousness Explained* (Boston: Little, Brown & Company, 1991).
7. Jerry Fodor, *Hume Variations* (Oxford: Oxford University Press, 2003).
8. Solymosi, 'Neuropragmatism, Old and New,' *Phenomenology and the Cognitive Sciences*, 10(3) (2011): 347–368.
9. See James's statement of the brain's plasticity in William James, *The Principles of Psychology*, 2 vols. (New York: Henry Holt, 1890), Chapter 4.

10. Michael S. Gazzaniga, *Nature's Mind: The Biological Roots of Thinking, Emotion, Sexuality, Language, and Intelligence* (New York: Basic Books, 1992); Antonio Damasio, *Descartes' Error: Emotion, Reason, and the Human Brain* (New York: Avon Books, 1994); and Damasio, *The Feeling of What Happens: Body and Emotion in the Making of Consciousness* (New York: Harcourt Brace, 1999).
11. Dewey, *Experience and Nature*, in *The Later Works of John Dewey*, Vol. 1, ed. Jo Ann Bodyston (Carbondale: Southern Illinois University Press, 1925/1981), p. 212.
12. Michael A. Schwartz, and Osborne P. Wiggins, 'Psychosomatic Medicine and the Philosophy of Life,' *Philosophy, Ethics, and Humanities in Medicine* 5(2) (21 January 2010) at http://www.peh-med.com/content/5/1/2.
13. Jacob von Uexküll, *Theoretical Biology*, trans. D. L. MacKinnon (London: Kegan Paul, Trench, Trubner, 1926); Richard Lewontin, 'The Organism as Subject and Object of Evolution' in Lewontin and R. Levins, *The Dialectical Biologist* (Cambridge, MA: Harvard University Press, 1985), pp. 85–106; Peter Godfrey-Smith, *Complexity and the Function of Mind in Nature* (Cambridge, UK: Cambridge University Press, 1998); Evan Thompson, *Mind in Life: Biology, Phenomenology, and the Sciences of Mind* (Cambridge, MA: Harvard University Press, 2007); A. Berthoz and Yves Christen, *Neurobiology of 'Umwelt': How Living Beings Perceive the World* (Berlin: Springer, 2009).
14. Margaret Wilson, 'Six Views of Embodied Cognition,' *Psychonomic Bulletin & Review* 9(4) (2002): 625–636.
15. This sort of example is discussed in James, *The Principles of Psychology*, and Dewey, 'The Reflex Arc Concept in Psychology,' in *The Early Works of John Dewey*, Vol. 5, ed. Jo Ann Boydston (Carbondale: Southern Illinois University Press, 1896/1972), 96–109.
16. Susan Hurley, *Consciousness in Action* (Cambridge, MA: Harvard University Press, 1998); Alva Noë,. *Action in Perception* (Cambridge, MA: MIT Press, 2004); Ralph Pred, *Onflow: Dynamics of Consciousness and Experience* (Cambridge, MA: MIT Press, 2005).
17. Richard Shusterman, *Body Consciousness: A Philosophy of Mindfulness and Somaesthetics* (New York: Cambridge University Press, 2008).
18. Franck Grammont, Dorothée Legrand, and Pierre Livet, eds. *Naturalizing Intention in Action* (Cambridge, MA: MIT Press, 2010).
19. J. J. Gibson, *The Ecological Approach to Visual Perception* (New York and London: Taylor and Francis Group, 1986); and Harry Heft, *Ecological Psychology in Context: James Gibson, Roger Barker, and the Legacy of William James's Radical Empiricism* (Mahwah, NJ: Lawrence Erlbaum Associates, 2001).
20. Gregory Bateson, *Steps to an Ecology of Mind: Collected Essays in Anthropology, Psychiatry, Evolution, and Epistemology* (Chicago: University of Chicago Press, 1972); Jesper Hoffmeyer, ed. *A Legacy for Living Systems: Gregory Bateson as Precursor to Biosemiotics* (New York: Springer, 2008).
21. Steve Fuller, *Social Epistemology* (Bloomington: Indiana University Press, 1988); Robert A. Wilson, *Boundaries of the Mind: The Individual in the Fragile Sciences: Cognition* (Cambridge: Cambridge University Press, 2004).
22. David D. Franks, *Neurosociology* (New York: Springer, 2010).
23. Andy Clark, *Being There: Putting Brain, Body, and World Together Again* (Cambridge, MA: MIT Press, 1997); Clark, *Supersizing the Mind: Embodiment, Action, and Cognitive Extension* (New York: Oxford University Press, 2008);

Alva Noë,. *Out of Our Heads: Why You Are Not Your Brain, and Other Lessons from the Biology of Consciousness* (New York: Hill and Wang, 2009); Richard Menary, ed. *The Extended Mind* (Cambridge, MA: MIT Press, 2010).
24. Shaun Gallagher, *How the Body Shapes the Mind* (New York and Oxford: Oxford University Press, 2005); Thompson, *Mind in Life*.
25. Anthony Chemero, *Radical Embodied Cognitive Science* (Cambridge, MA: MIT Press, 2009).
26. William Bechtel, and Adele Abrahamsen, *Connectionism and the Mind: Parallel Processing, Dynamics, and Evolution in Networks*, 2nd ed. (Malden, MA: Blackwell, 2002); Walter J. Freeman, *How Brains Make Up Their Minds* (New York: Columbia University Press, 2001); W. Teed Rockwell, *Neither Brain Nor Ghost: A Nondualist Alternative to the Mind-Brain Identity Theory* (Cambridge, MA: MIT Press, 2005).
27. John R. Shook, 'Social Cognition and the Problem of Other Minds,' in *Handbook of Neurosociology*, ed. David D. Franks and Jonathan H. Turner (New York: Springer, 2012), pp. 33–46.
28. James H. Fetzer, *The Evolution of Intelligence: Are Humans the Only Animals with Minds?* (Chicago: Open Court, 2005); Jesper Hoffmeyer, *Biosemiotics: An Examination into the Signs of Life and the Life of Signs* (Scranton: University of Scranton Press, 2008).
29. Charles S. Peirce, *Peirce on Signs: Writings on Semiotic*, ed. James Hoopes (Chapel Hill: University of North Carolina Press, 1991); Thomas Sebeok, *Signs: An Introduction to Semiotics*, 2nd edition (Toronto: University of Toronto Press, 2001).
30. Jaan Valsiner, and René van der Veer, *The Social Mind: Construction of the Idea* (Cambridge, UK: Cambridge University Press, 2000).
31. Harold Blumer, *Symbolic Interactionism* (Englewood Cliffs, NJ: Prentice-Hall, 1969).
32. Radu Bogdan, *Grounds for Cognition: How Goal-Guided Behavior Shapes the Mind* (Hillsdale, NJ: Lawrence Erlbaum, 1994).
33. Marcello Barbieri, ed. *Introduction to Biosemiotics: The New Biological Synthesis* (Dordrecht: Springer, 2008).
34. See Kim Sterelny, *The Evolved Apprentice: How Evolution Made Humans Unique* (Cambridge, MA: MIT Press, 2012).
35. Kahneman has given a forceful presentation of these so-called systems of cognition in his recent work, *Thinking, Fast and Slow* (New York: Farrar, Straus and Giroux, 2011). System 1 is characterized as the automatic, quick response – habits, instincts, etc. – that humans have that allows them to act without deliberation in time-sensitive situations. System 2 is the slow, deliberative, and conscious attention humans sometimes engage in to override the impulsiveness of system 1.
36. Here we follow the useful distinction made by neopragmatist Robert Brandom, 'The Pragmatist Enlightenment (and its Problematic Semantics),' *European Journal of Philosophy*, 12(1) (2004): 1–16. Despite the utility of the description of classical pragmatism in the first part of Brandom's article, the second part on semantics is extremely problematic, as Larry Hickman has critically addressed (see his 'Some Strange Things They Say About Pragmatism: Robert Brandom on the Pragmatists' "Semantic Mistake",' *Cognitio* 8(1) (2007): 105–113).

37. Dewey, *The Quest for Certainty* in *The Later Works of John Dewey*, Vol. 4, ed. Jo Ann Boydston (Carbondale, IL: Southern Illinois University Press, 1984).
38. Dennett, *Consciousness Explained*.
39. Indeed, by the time of Kant, the problem of knowledge had become how knowledge was even possible in the first place.
40. P. E. Griffiths, and R. D. Gray, 'Darwinism and Developmental Systems,' in *Cycles of Contingency: Developmental Systems and Evolution*, ed. S. Oyama, P. E. Griffiths, and R. D. Gray (Cambridge, MA: MIT Press, 2001), pp. 195–218.
41. To be clear, this recognition does not imply that only cognitive experience is of value; rather, that there are far more varieties of experience than those which are known, that such experiences are often significant, and that without them there can be no cognitive experience in the first place. On this point, specifically within Dewey's thought, see Larry A. Hickman, *Philosophical Tools for Technological Culture: Putting Pragmatism to Work* (Bloomington and Indianapolis: Indiana University Press, 2001), pp. 17–20. A reader may further retort that putting experience in phylogenetic terms is unorthodox because it leaves no obvious space for the subject of experience: it seems odd, if not absurd, to place the subject in the species and not in its members. Yet this way of talking about experience, as though there must first be a subject who is capable of undergoing experience in the first place, prior to any experience whatsoever, is a non-starter. Pragmatists from James and Dewey to Rorty and Dennett reject this conception of experience that presupposes a subject. Rather, the individual subject or self develops out of experience as Œ-transaction. Integral to this achievement of being able to talk to oneself without others' hearing it is an ecological niche in which others talk to each other first and foremost. For more on this point, see Dewey *Experience and Nature*, p. 135, and Dennett, *Consciousness Explained*, p. 195. Lastly, it would behoove the reader to remember that for Cartesians, consciousness, mind, self, and subject are conceived as one and the same thing. We deny such equivocation and consider each of these terms as specific phases within the process of experience.
42. Hickman, *Philosophical Tools for Technological Culture*, p. 21.
43. To appreciate the disproportionality, consider that all humans are familiar with the problem of thirst and its easy resolution. But most humans lack the unique traits required for experimental inquiry performed by a research team in a laboratory.
44. This feedback serves to modify the upbringing of the next generation of creative intellectuals, for the problems and the resources available for inquiry will have been modified and expanded as well.
45. Daniel Kahneman, and Shane Frederick, 'A Model of Heuristic Judgment,' in *The Cambridge Handbook of Thinking and Reasoning*, ed. Keith J. Holyoak and Robert G. Morrison (Cambridge and New York: Cambridge University Press, 2005), pp. 267–293; K. E. Stanovich, and R. West, 'Individual Differences in Reasoning: Implications for the Rationality Debate?' in *Heuristics & Biases: The Psychology of Intuitive Judgment*, ed. T. Gilovich, D. Griffin, and D. Kahneman (New York: Cambridge University Press, 2002), pp. 421–440.
46. Claudia Dreifus, 'A Mathematical Challenge to Obesity: A Conversation with Carson Chow,' *New York Times*, May 15, 2012, D2, available online at:

http://www.nytimes.com/2012/05/15/science/a-mathematical-challenge-to-obesity.html.
47. For an excellent survey of how the term 'representation' is used within cognitive science, see Giovanni Pezzulo, 'Coordinating with the Future: The Anticipatory Nature of Representation,' *Minds & Machines*, 18 (2008): 179–225. See our examination of Pezzulo's conclusions in Solymosi and Shook, 'Neuropragmatism and the Culture of Inquiry.'
48. See Charles Sanders Peirce, 'The First Rule of Logic,' in *The Essential Peirce: Selected Philosophical Writings, Volume 2 (1893–1913)*, ed. The Peirce Edition Project (Bloomington and Indianapolis: Indiana University Press, 1898/1998), pp. 42–56.
49. This guidance occurs both in the process of discovery (namely through the paradigm and specific research program), and in the process of transmission and reformation (in the forms of educational practices and institutions).
50. Sterelny draws on Peter Godfrey-Smith's 'helpful distinction between philosophy of science and philosophy of nature. The intellectual target of philosophy of science is science itself...The intellectual target of philosophy of nature is nature itself; the world in which we live (which, of course, includes humans and their practices, including science)' (*The Evolved Apprentice*, p. xi). Godfrey-Smith writes, 'When we export a picture of the world from the immediate context of science into a broader discussion, the features of scientific description that have their origin in these practicalities become potentially misleading...Work of this sort will also often aim at synthesizing the results of a number of different scientific fields, working out how they fit together – or fail to fit – into a coherent package,' before concluding 'So philosophy of nature refines, clarifies, and makes explicit the picture that science is giving us of the natural world and our place in it. Calling it "philosophy" does not mean that only philosophers can do it. Many scientists...undertake this kind of work. But it is a different kind of activity from science itself.' (*Darwinian Populations and Natural Selection*. New York: Oxford University Press, 2009, p. 3). He also notes that the philosophy of nature is 'an old term' – indeed, it was just what pragmatists like Dewey were doing (Dewey's influence here on Godfrey-Smith should not be underestimated as many of the latter's writings make use of the former's ideas).
51. Sterelny, *The Evolved Apprentice*.
52. Bill Bywater, 'Neuropragmatism's Pedagogy,' Presentation at the Annual Meeting of the Society for the Advancement of American Philosophy, March 15–17, 2012, Fordham University, New York City. See also Bywater and Piso's chapter in this volume for further elaboration along these lines, using Sterelny and data regarding mirror neuron systems to critique Kahneman's dual system approach.
53. See Franks, *Neurosociology*, as well as his chapter in this volume that elaborates further the contribution of the Chicago Pragmatists to neurosociology.
54. See also the new field of neuroanthropology: Daniel H. Lende and Greg Downey, *The Encultured Brain: An Introduction to Neuroanthropology* (Cambridge, MA: MIT Press, 2012).
55. C. P. Snow, *The Two Cultures* (New York: Cambridge University Press, 1959/1993).

56. Patricia S. Churchland, *Neurophilosophy: Toward a Unified Science of the Mind/Brain* (Cambridge, MA: MIT Press, 1986), pp. 302–303; and *Brain-Wise: Studies in Neurophilosophy* (Cambridge, MA: MIT Press, 2002), pp. 107–112.
57. Paul M. Churchland, and Patricia S. Churchland, *On the Contrary: Critical Essays, 1987–1997* (Cambridge, MA: MIT Press, 1998), pp. 25ff, and 4ff.
58. See Paul M. Churchland, *Plato's Camera: How the Physical Brain Captures a Landscape of Abstract Universals* (Cambridge, MA: MIT Press, 2012), pp. 128ff; see W. Teed Rockwell, 'Beyond Eliminative Materialism: Some Unnoticed Implications of Churchland's Pragmatic Pluralism,' in *Contemporary Pragmatism*, 8(1) (2011): 173–190, for a strong treatment of Churchland's previous pragmatist leanings. See also Rockwell's chapter in this volume, which continues this line of thought.
59. Owen Flanagan, *The Really Hard Problem: Meaning in a Material World* (Cambridge, MA: MIT Press, 2007), pp. 5ff.
60. See Dennett, 'How to Protect Human Dignity from Science,' in *Human Dignity and Bioethics: Essays Commissioned by the President's Council on Bioethics*, 2008, available at: http://bioethicsprint.bioethics.gove/reports/human_dignity/chapter3.html. Last accessed 21 May 2008. See also Dennett's 'Manifest Image and Scientific Image' in *Intuition Pumps and Other Tools for Thinking* (New York: W.W. Norton, 2013), pp. 69–72; and his 'Aching Voids and Making Voids: A Review of *Incomplete Nature: How Mind Emerged from Matter* by Terrence W. Deacon,' *The Quarterly Review of Biology* 88(4) (2013): 321–324.
61. Wilfrid Sellars, *Science, Perception and Reality* (London: Routledge and Kegan Paul, 1963), p. 25.
62. Rockwell, *Neither Brain Nor Ghost*.
63. Hilary Putnam, *The Collapse of the Fact/Value Dichotomy and Other Essays* (Cambridge, MA: Harvard University Press, 2002).
64. Dewey, *Experience and Nature*, p. 12.
65. Dewey, 'Experience, Knowledge and Value: A Rejoinder,' in *The Philosophy of John Dewey*, ed. Paul A. Schilpp (New York: Tudor, 1939), p. 531.
66. Dewey, *Logic: The Theory of Inquiry*, in *The Later Works of John Dewey, Volume 12*, ed. Jo Ann Bodyston (Carbondale: Southern Illinois University Press, 1938/1986), pp 71–72 and 69.
67. Ibid., p. 72.
68. Ibid.
69. We hasten to add the role of computer and information sciences both in advancing our understanding of biology and neuroscience and in significantly modifying our everyday lives. Without the shared questions about the nature of mentation, we would never have had the insights raised by the Turing Test, nor the application of those insights to biological phenomena. Furthermore, the further application of computer and information sciences to everyday life have, unfortunately, brought about a rise in disease that comes with a more sedentary lifestyle made possible by greater ease of communication.
70. John Dewey, 'The Influence of Darwinism on Philosophy,' in *The Influence of Darwin on Philosophy and Other Essays in Contemporary Thought*, ed. Larry A. Hickman (Carbondale, IL: Southern Illinois University Press, 1910/2007), p. 7.

2
Keeping the Pragmatism in Neuropragmatism

Mark Johnson

Whenever I hear the term 'neuropragmatism,' I am reminded of J. L. Austin's opening words in his famous article 'Performative Utterances,' where he says, 'You are more than entitled not to know what the word "performative" means. It is a new word and an ugly word, and perhaps it does not mean anything very much.'[1] Likewise, you are more than entitled not to know what "neuropragmatism" means. It is, indeed, a *new* word, and it is perhaps an *ugly* word, but I daresay that it is not an inconsequential word. Therefore, the first question to ask is what the term *might* mean. The second and more important question is why we ought to care about neuropragmatism. I shall argue that, although we have good reasons for thinking that cognitive neuroscience has a great deal to offer toward a psychologically sophisticated view of mind, experience, thought, and language, our enthusiasm for cognitive neuroscience should always be tempered by a critical and more comprehensive perspective supplied by pragmatism. What pragmatist philosophy has to offer is the broader philosophical context necessary for understanding the grounding assumptions of cognitive neuroscience, its fundamental limitations, and its place in a more expansive pragmatist framework for approaching both philosophy and our basic life problems. In short, pragmatism without neuroscience is (partially) empty, but neuroscience without pragmatism is (partially) blind.

I propose to address both of these questions about the meaning and importance of cognitive neuroscience, but first a preliminary word of caution is in order. Good pragmatists are well known for their tendency to eschew the 'pragmatist' label, insofar as the hardening of philosophical dispositions and practices into abstract rigid frameworks is one of the chief diseases classical pragmatism was trying to cure. If you then proceed to add the qualifier 'neuro' onto 'pragmatism,' there can be an unfortunate tendency to narrow and restrict the scope of your philosophical endeavors even more drastically. The risk, all too often realized, is the tendency to think that the work of neurophilosophy, or at least its important work, is done by the neuroscience alone, thereby denying any serious role for philosophical reflection. This would be a big mistake. Therefore, I shall be arguing that we need to keep the *pragmatism* in 'neuropragmatism,' which means that cognitive neuroscience needs a more comprehensive philosophical framework that can be supplied by a pragmatist orientation.

What is neuropragmatism?

As Tibor Solymosi[2] has suggested, 'neuropragmatism' is what you get when you plug 'pragmatism' into the 'philosophy' slot of neurophilosophy. As far as I am aware, the first person to use the term 'neurophilosophy' was Patricia Churchland, who argued for a constructive dialogue between the emerging empirical discoveries of the biological and neurosciences, on the one hand, and philosophy and other disciplines that have traditionally carried the burden of explaining 'mind,' on the other. Churchland described her project as follows:

> The sustaining conviction of this book is that top-down strategies (as characteristic of philosophy, cognitive psychology, and artificial intelligence research) and bottom-up strategies (as characteristic of the neurosciences) for solving the mysteries of mind-brain function should not be pursued in icy isolation from one another. What is envisaged instead is a rich interanimation between the two, which can be expected to provoke a fruitful co-evolution of theories, models, and methods, where each informs, corrects, and inspires the other.[3]

'Interanimation' and 'co-evolution' are the right kinds of terms to capture the complex non-reductive relation among various theories, models, and modes of explanation that are needed to understand human cognition. Churchland's enthusiasm for this relatively new perspective stems primarily from her sense that recent advancements in brain imagining and the cognitive sciences have given us shiny new tools for exploring how we perceive, feel, experience meaning, conceptualize, value, communicate, and act in the world. The insights that come from the neurosciences are both critical and constructive – critical insofar as they reveal the inadequacies of certain philosophical methods and theories, and constructive insofar as they suggest new ways of explaining various phenomena of human experience and cognition. Scientists are developing substantial bodies of solid empirical research that give us empirically responsible ways to explore the operations of 'mind,' and their various methods can either critique, replace, or supplement the typical armchair theorizing that plagues so much contemporary philosophy. Notice that in none of this does the neuroscience claim to make the philosophical dimension unnecessary. On the contrary, the philosophical framework is essential for understanding the limits and uses of the neuroscience.

Proponents of neuropragmatism claim that many of the insights that recommend pragmatism as a general philosophical orientation are supported by well-established results of empirical research in the neurosciences. In other words, pragmatist philosophy is much more compatible with our best current neuroscience than are many philosophical methods that set themselves over against the sciences and claim critical and reflective supremacy over empirical disciplines. In other places, I have praised the contributions of *both* the cognitive and neurosciences, because neuroscience is obviously not the whole of cognitive science, but here I shall restrict my focus to the contribution of the neurosciences alone, leaving out other cognitive science research that would be necessary to tell a more comprehensive, more nuanced story.[4]

Key shared themes of pragmatism and neuroscience

What follows is a brief list of some of the key insights of what is loosely known as 'pragmatist' philosophy that are strongly supported by some of the more robust findings of recent neuroscience.

1. *Organism–Environment Transaction as the Locus of Activity* – Dewey argued throughout his mature works that the locus of all experience, meaning, thought, and action is a living organism in ongoing interaction with environments that are complex and multidimensional, including at least physical, interpersonal, and culture dimensions. He repeatedly criticized the dualistic errors of placing all of the cognitive, epistemic, or ontological weight on either the organism (with its 'inner' functions) or the environment (regarded as wholly independent of the organism). In *Art As Experience* Dewey expressed the intimate organism–environment transaction as follows:

> The first great consideration is that life goes on in an environment; not merely *in* but because of it, through interaction with it. No creature lives merely under its skin; its subcutaneous organs are means of

connection with what lies beyond its bodily frame, and to which, in order to live, it must adjust itself, by accommodation and defense but also by conquest.... The career and destiny of a living being are bound up with its interchanges with its environment.[5]

Neuroscientist Antonio Damasio observes that the very continuation of life requires an organism to maintain a permeable boundary within which some measure of systemic equilibrium is possible: 'Life is carried out inside a boundary that defines a body. Life and the life urge exist inside a boundary, the selectively permeable wall that separates the internal environment from the external environment.'[6] The key word here is 'permeable,' because life cannot be sustained without an ongoing exchange between the internal and external environments. Damasio's claim is that the life of an organism requires what Walter B. Cannon called *homeostasis:* 'the coordinated physiological reactions which maintain most of the steady states of the body... and which are so peculiar to the living organism.'[7] Jay Schulkin has recently proposed the term *allostasis* (*allo*: change; *stasis*: stable/same) to emphasize that the equilibrium required for life need not be merely a return to some pre-set stable state, but rather involves the dynamic generation of new set-points for equilibrium, as the organism engages new aspects of its environment.[8]

It is a commonplace of the biological sciences today that any organism exists only in and through its ongoing engagement with its surroundings, in a continuous give-and-take of activity and receptivity. No environment, no organism. Moreover, boundaries between organisms and environments, however real, must be somewhat permeable for the organism to sustain life. 'Inner' and 'outer' are thus just convenient names for salient aspects of the unified processes of certain types of living beings, but they do not mark intrinsic ontological distinctions, and they cannot be defined independently of each other.

Because some neuroscientists assume such an inner/outer split, they have been justly criticized for treating what goes on *in* the brain as though it were somehow independent of the environing conditions that make cognition possible. One problematic result is a tendency to speak as if different types of cognition occur strictly *within* specified brain regions. This is the error of attributing some activity, relation, or property to a particular part, when the proper locus should be the whole multilevel and multidimensional organism as it is situated within its environments. M. R. Bennett and P. M. S. Hacker[9] are therefore rightly critical of familiar tendencies in the neurosciences to attribute various cognitive capacities and activities to selected parts of the brain, when it is actually the whole organism (typically a *person*) that has the capacities and performs the action, and not just their brain parts. One should add that it is not just the whole person, but the person engaging their environments, who performs the action, executes a particular function, or has an experience. Bennett and Hacker cite scores of passages from neuroscientists who slip into the error of attributing complex enactments to specific brain regions, as when one claims something like 'the amygdala experiences fear,' or 'vision occurs in the various regions of the visual cortex.'

The mistake here is just that of mistaking the part for the whole. It is also the error of conceiving the environing conditions of specific kinds of cognitive activity as externalities not intrinsically part of the cognition. That said, it certainly does *not* follow that neuroscience is intrinsically reductive in that unwanted sense. For example, Bennett and Hacker attribute to Antonio Damasio the view that 'emotions (are) essentially ensembles of somatic changes caused by thoughts (mental images).'[10] Although Damasio certainly does hold a view of this sort, he takes great pains to show that emotions require and manifest ongoing relations with both 'external' and 'internal' environments. Only a willful misreading of Damasio could overlook his constant insistence that emotions are the way in which certain advanced complex organisms take the measure of their interactions with their environment, and this involves monitoring changes in body-states, followed by adjustments to the internal milieu to regain some measure of dynamic equilibrium in the organism. Indeed, Damasio's theory of emotions depends upon the existence of what he calls an 'emotionally competent stimulus,'[11] which is some aspect of the organism or environment that evokes changes in the body-states of the organism and requires subsequent adjustments within the body to restore some measure of allostasis. Emotions are therefore not locked up within organisms or minds; instead, they implicate the world inhabited by the organism. It was for this reason that Dewey famously insisted that the locus of an emotion is not an isolated subject or experiencing organism, but instead the entire 'situation' in which the organism lives and moves.[12]

However much neuroscientists occasionally lapse into mistakenly speaking as if specific brain regions perform specific types of cognitive operations, the vast majority of them realize that *the mind is NOT the brain*, and that cognition implicates the environment just as much as it does the makeup of the organism.

2. *The Continuity of Experience* – There are at least two important and related notions of continuity central to pragmatist accounts of experience. The first has to do with the integrity and unity of experience, which cannot be captured by any set of polar oppositions. One of the distinctive contributions of the pragmatism of William James and John Dewey is their careful attention to the actual character of any given experience. Both argued that unless we begin with an adequate sense of the breadth, depth, and nuanced richness of experience, our philosophical theories will lose any relevance and transformative potential for our lives. Dewey called this encompassing world of experience *a situation*: 'What is designated by the word "situation" is *not* a single object or event or set of objects and events. For we never experience nor form judgments about objects and events in isolation, but only in connection with a contextual whole. This latter is what is called a "situation."'[13]

Both James and Dewey repeatedly argued that philosophy and psychology far too often begin with some part of a situation, such as an object, event, or quality, or, even worse, some reduction of experience to an alleged set of discrete sensations. Then, either by reducing experience to sense inputs supposedly somehow associated in the mind, or else by treating experience as if it consisted of 'objects of knowledge' known through various types of mental judgment, we miss what is really going on in a situation that gives meaning and value.

In his extraordinary two-volume *Principles of Psychology* (1890), James gave a remarkable phenomenological description of experience as a continuous 'flow' or 'stream' of thought-feelings, out of which we subsequently identify objects and events, based on our values, interests, and purposes: 'Consciousness, then, does not appear to itself chopped up in bits. Such words as "chain" or "train" do not describe it fitly as it presents itself in the first instance. It is nothing jointed; it flows. A "river" or a "stream" are the metaphors by which it is most naturally described. *In talking of it hereafter, let us call it the stream of thought, of consciousness, or of subjective life.*'[14]

Dewey is famous – or some would say infamous – for stressing the idea that every complete experience involves a situation unified by what he called a *pervasive quality*: 'The pervasively qualitative is not only that which binds all constituents into a whole but it is also unique; it constitutes in each situation an *individual* situation, indivisible and unduplicable. Distinctions and relations are instituted *within* a situation; *they* are recurrent and repeatable in different situations.'[15]

Dewey's point here is that (1) the integrity of a given situation depends on the unique pervasive unifying quality that marks it out from the stream of thought as *this experience here and now,* and (2) experience is thus meaningful and constituted for us not merely as *known* (which is a reductive and selective take on an experience), but rather as felt and encountered in all its cognitive, affective, and evaluative dimensions. Consequently, one important sense of the continuity of experience is the refusal to allow the fragmentation of any integral experience into discrete sensations, feelings, objects, actions, or events, without recognizing the operative power of the whole situation of which they are a part. Obviously, we can and do make such discriminations for various purposes, but we should never pretend that these selective processes constitute the fullness of an integrated experience. Rather, they are merely useful abstractions from what James called 'the much-at-onceness' of experience – selections we make because of our bodily capacities, our habits of perception and conception, or our interests and values.

A second important notion of continuity that is central to most pragmatist philosophy is the idea that in our explanations of any human actions, behaviors, or capacities, we treat them as emergent from 'lower-level' animal capacities and activities. Dewey regarded this methodological imperative as so important that he christened it the 'Principle of Continuity':

> The primary postulate of a naturalistic theory of logic is continuity of the lower (less complex) and the higher (more complex) activities and forms. The idea of continuity is not self-explanatory. But its meaning excludes complete rupture on one side and mere repetition of identities on the other; it precludes reduction of the "higher" to the "lower" just as it precludes complete breaks and gaps.[16]

The principle of continuity appears to be a fundamental postulate of most contemporary neuroscience, insofar as 'higher' cognitive operations are explained via the emergence of functional capacities, processes, and states on the basis of increasing complexity of 'lower' systems. What is denied in all naturalistic approaches is the sudden appearance on the scene of new metaphysical entities, forces, or activities as a way of explaining cognitive phenomena. As neuroscientist and psychologist Don Tucker explains,

> Primary process cognition is driven by the visceral controls at the core of the brain, and secondary process cognition [our more conscious

cognitive operations] must fit the somatic constraints of the sensory and motor interface with the world. This is an interpretation that fits all mammalian brains, and yet it must produce the most complex abstractions of the human mind. Complex psychological functions must arise from bodily structures. There is no other source for them.... There are no brain parts for abstract faculties of the mind – faculties such as volition, insight, or even conceptualization – that are separate from the brain parts that evolved to mediate between visceral and somatic processes. [17] [brackets added]

3. *Anti-dualism* – As conceived in both pragmatist philosophy and cognitive neuroscience, one of the most significant implications of the continuity of experience and of close attention to the intertwining of organism and environment is a rejection of dualism, both epistemic and ontological. Neuroscience provides abundant evidence that our preferred dualisms are not intrinsic to experience but are rather after-the-fact selections we make, for various purposes, from the ongoing flow of our experience. To cite the most obvious case, there can be no 'mind' without a living brain in a living body. We speak of the 'mental' versus the 'physical,' yet there is no mental process or experience that does not require processes of bodily activation and transformation. Damasio says that 'the apparatus of rationality, traditionally assumed to be *neo*cortical, does not seem to work without that of biological regulation, traditionally presumed to be *sub*cortical. Nature appears to have built the apparatus of rationality not just on top of the apparatus of biological regulation, but also *from* it and *with* it.'[18]

Although Damasio focuses here on rationality, one might extend this basic anti-dualism more broadly to encompass the traditional mind/body, mental/physical split. In other words, our capacities for what we call conceptualization and reasoning recruit structures and processes evolved for processing various bodily perceptual and motor activities. Reason (and more broadly, mind) is not separable from our other bodily capacities and operations, and this undercuts any strict separation of supposed cognitive faculties according to a dualistic metaphysics of mind. This assertion of the non-dualistic embodiment of mind is not an a priori truth, nor is it capable of any formal proof. Rather, it is a fundamental methodological assumption of naturalism without which neuroscience cannot get off the ground. It is, in other words, a working hypothesis and a methodological promissory note that neural (and other bodily) correlates will eventually be found for any so-called 'mental' operations of conceptualization, reasoning, and willing. It is a hypothesis that recognizes our various projected ontological and epistemic dualisms for what they are – not metaphysical realities, but convenient markers of dimensions or aspects of situations that we find it useful to notice.

There are dualistic neuroscientists (such as Crick), but cognitive neuroscience is, for the most part, predicated on continuity and the emergence of 'higher' functions from 'lower' ones via increasing complexity of organism–environment organization. In short, take virtually any received dualism – mind/body, subject/object, reason/emotion, knowledge/imagination, thought/feeling, fact/value – and you can find extensive empirical evidence undermining the postulation of these strict dualisms and suggesting that we use these terms, at best, only to mark aspects or dimensions of dynamic processes of experience that we find it methodologically useful to identify. I am not suggesting the draconian reconstructive project of attempting to give up all dichotomies and refrain from using terms like 'mental/physical,' 'mind/body,' 'reason/emotion,' and so forth. That would be impossible and unnecessary. Rather, we simply need to take great care to avoid treating our commonplace selections of various dichotomous aspects of our experience as if they constituted absolute ontological characteristics of particular beings.

I trust that it is hardly necessary to recapitulate any of the numerous arguments against epistemic and ontological dualisms that populate virtually all of Dewey's major works and that have their counterparts in William James. Suffice it to say that Dewey spent hundreds of pages showing how and why certain classic, yet historically contingent, dualisms arose in our language and conceptual systems. He then showed why hardening those dualities into absolute ontological or epistemic divisions leads us to dramatically mis-describe our experience and thereby to ruin our entire philosophical analysis by reading intellectual distinctions back onto our primary lived experience in a way that distorts its significance and causes us to ignore the complexity and richness of what is going on.

4. *The Intertwining of Reason and Emotion* – One of our most deeply rooted and troubling dichotomies that underlies so many traditional philosophical systems is the alleged split between our 'higher' faculty of reasoning and our 'lower' (bodily) capacity for feeling and emotion. On this view, reason is taken to exist independently of

feelings and to function properly only by rising above the influence of the emotions.

The denial of any rigid reason/emotion or thought/feeling dichotomy has been a staple of pragmatist psychology, epistemology, and ontology. Even as early as the late 19th century, James (in the 'Stream of Thought' chapter of his *Principles of Psychology*) had notoriously argued for a 'feeling of thinking,' and he often used the hyphenated term 'thought-feeling' to remind us that every thought has its corresponding feeling tone. James claimed that reasoning involves the 'transitive' movement from one thought to another, and he argued that there would always be a felt sense of the urgency, direction, and connection within the movement from one thought to another:

> If there be such things as feelings at all, *then so surely as relations between objects exist in rerum natura, so surely, and more surely, do feelings exist to which those relations are known.* There is not a conjunction or preposition, and hardly an adverbial phrase, syntactic form, or inflection of voice, in human speech, that does not express some shading or other of relation which we at some moment actually feel to exist between the larger objects of our thought. ... We ought to say a feeling of *and,* a feeling of *if,* a feeling of *but,* and a feeling of *by,* quite as readily as we say a feeling of *blue* or a feeling of *cold*.[19]

In this Jamesian framework, patterns of thought are metaphorical motions along paths to some metaphorical destination.[20] Thought *moves* (metaphorically) from one thought-location to another, and we feel the *tendency* and *direction* of any particular movement of thought.

James's focus was primarily on the feeling dimensions of thought (thinking), whereas Damasio (1994, 2003, 2010) focuses on the role of emotions in certain types of reasoning. He is known for demolishing this reason/emotion dualism, not in James's way, but by stressing the fact the reason relies on emotion. What he shows, using neuroimaging and lesion studies in clinical settings, is the necessity of an intact emotional system for proper social and practical reasoning:

> I propose that human reason depends on several brain systems, working in concert across many levels of neuronal organization, rather than on a single brain center. Both "high-level" and "low-level" brain centers, from the prefrontal cortices to the hypothalamus and brain

stem, cooperate in the making of reason. The lower levels of the neural edifice of reason are the same ones that regulate the processing of emotions and feelings, along with the body functions necessary for an organism's survival. In turn, these lower levels maintain direct and mutual relationships with virtually every bodily organ, thus placing the body directly within the chain of operations that generate the highest reaches of reasoning, decision making, and, by extension, social behavior and creativity.[21]

James and Dewey stand out among the classical American pragmatists for their recognition of the intertwining of thought, quality, feeling, and emotion. What contemporary neuroscience brings to the discussion are neuro-imaging technologies that allow us to move beyond the self-reflective phenomenology of our experience that was the hallmark of earlier pragmatist inquiry. We are now in a position to begin to see *that* and *how* parts of the brain responsible for emotional response patterns and our occasional feeling of those states are co-activated with, and partially constitute, our patterns of reasoning. In normal human beings, reasoning is felt, and it is made possible by emotional responses to our bodily monitoring of how our ongoing experience is developing, relative to our needs and desires.

5. *Reductionism That Involves Multiple Levels of Explanation* – One of the classic objections to any reliance on neuroscience in constructing a theory of mind, thought, and value, is the fear that all science is intrinsically reductionist in its methods and cannot therefore engage the depth and richness of the phenomena to be explained. The pragmatist response to this anxiety observes that, if there is genuine emergence of new levels of complex organization as we evolve and grow, then there may be forms of explanation fitted to a particular level of functional organization that are not necessary or appropriate for 'lower' levels of functional explanation. Different levels of explanation (for example, explanations in terms of processes at the cellular level versus processes of the autonomic nervous system versus complex interactions among subgroups within a society) are each likely to require different methods of inquiry whose key concepts are not reducible to those of a different level. William Bechtel argues that, although mechanistic explanations in the sciences are inherently reductive, it does not follow from this that we can always reduce 'higher' levels of organization to 'lower' ones:

A final feature of mechanistic explanation ... is that, insofar as it emphasizes the contributions made by parts of a mechanism to its operation, a mechanistic analysis is, in an important sense, reductionistic. However, insofar as it also recognizes the importance of the organization in which the parts are embedded and the context in which the whole mechanism is functioning, it not only sanctions looking to lower levels but also upward to higher levels.[22]

Bechtel insightfully observes that the entrenched fear of reductionism stems partly from the mistaken view that all explanations in the sciences fit the deductive-nomological model, in which a particular phenomenon is regarded as an instance of a causal law of some sort. However, deduction from so-called universal covering laws plays only a very small part in any of the cognitive sciences, where mechanistic explanations of what makes a certain cognitive activity possible are the norm.

Since, as just noted, mechanistic explanations are not just about the component parts of a mechanism (hence, reductionistic), but also about the organization of the parts into a functioning whole operating within a typically complex and multidimensional environment, there is always a need for multiple levels of explanation, each with their own terms and modes of explanation:

Mechanistic reduction only proposes to explain the response of an entity to the causal factors impinging on it in terms of its lower-level constituents. It does not try to explain the causal factors impinging on the mechanism. Understanding these requires inquiry focused on the mechanism as a whole, and often on yet higher-level systems in which the mechanism is embedded. Independence stems from the fact that inquiry at each level provides information additional to that which can be secured at other levels, and generally does so using different tools of inquiry.[23]

The typical need for multiple levels of explanation in accounting for all things human is one basis for the recognition that the mind is not the brain. Good neuroscience does not claim that we can explain all of the phenomena of mind by the firing of neurons in connected or co-activated clusters (even though there can be no mental activity without that). It recognizes, instead, that experience is a process of organism–environment interaction, and that our environments are at once physical, interpersonal, social, and cultural. Over decades and even centuries, we have developed a plurality of methods for inquiring into the various ways in which aspects of our bodily make-up and our complex environments shape our experience, thought, action, and communicative interactions. This methodological pluralism is a good and necessary thing, since we need multiple methods of inquiry that carve up the phenomena differently, indicate the complexity of underlying mechanisms, and operate at different explanatory levels in order to construct a suitably rich account of any cognitive process, action, or event.

Why neuropragmatism needs pragmatism

As the previous highly selective list of insights shared by neuroscience and pragmatist philosophy indicates, cognitive neuroscience has a great deal to contribute to a pragmatist naturalized approach that values empirical research as relevant to our understanding of concepts, meaning, thought, feeling, knowledge, values, action, communication, problem articulation, methods of inquiry, and so forth. Some of the discoveries and advances of cognitive neuroscience are so stunning, and are coming so quickly, that it is easy to be seduced into the mistaken view that, if we were just to let neuroscience research run its course for the next 50 or 100 years, we would be well on the way to a complete explanation of mind, thought, language, and nearly everything human. This, however, would be a serious mistake, because *neuropragmatism is not just neuroscience on steroids – neuropragmatism is the recruitment of neuroscience in the service of a pragmatist reconstruction of philosophy.*

There are three main reasons why we cannot simply replace certain fields and types of philosophy with neuroscience: (1) different scientific disciplines typically do not have sufficient reflective insight into the limitations of their own assumptions, values, goals, and methods of investigation, (2) no single scientific perspective can accommodate all of the levels of explanation necessary to account for the multidimensionality and complexity of the human organism in its cultural setting, and (3) a broader philosophical framework is necessary to see how the various levels interact and fit together.

Virtually all of the key points one might make regarding these three tenets follow from the main ideas outlined above. In particular, the traditional pragmatist insistence on beginning and ending with a sufficiently rich and nuanced account of experience gives important guidance into how we ought to appropriate the results of cognitive neuroscience. A multidimensional account of experience gives us a basis for critical reflection on the founding assumptions of various scientific methodologies, and it also reveals the need for a comprehensive account of the significance of neuroscience for our lives.

The first point concerns the grounding methodological and epistemic presuppositions of particular bodies of scientific research, which reveals the highly selective and biased nature of any scientific approach. One of the principal results of the past 50 years of philosophy of science is an understanding of the partial and perspectival nature of any scientific methodology. *Every* method must make assumptions about (a) what the relevant phenomena to be explained are, (b) what counts as relevant data, and (c) what constitutes an acceptable form of explanation of the phenomena in question. These three aspects of any scientific method are not self-evident or necessary truths about how any particular science must be conducted. They are tentative conclusions drawn from numerous studies of scientific practice. Contrary to a centuries-old commonplace of metaphysical realism, the world does not come to us with its basic categories of phenomena specified in some mind-independent, objective fashion.[24] Rather, what we take the phenomena to be – how we circumscribe and discriminate them – depends on our values, interests, technologies, and history. The methods we employ emerge in response to various historical circumstances that include cultural values, existing technologies, and institutional constraints of the conduct of science.[25]

A second significant contribution of pragmatist philosophy is the disclosure in various theoretical orientations of what William James coined the 'Psychologist's Fallacy': 'The great snare of the psychologist is the *confusion of his own standpoint with that of the mental fact about which he is making his report.*'[26] In other words, the psychologist or philosopher or scientist comes to regard the basic conceptual distinctions that they have found useful in constructing their theories as actually constituting the experience of ordinary people. The psychologist or philosopher reads their own theoretical (reflective, abstractive, selective) distinctions back onto the experience under examination, and then assumes that people must actually experience those distinctions. To cite one representative case of this type of error, recall how Enlightenment philosophers often assumed that experience had to consist of unorganized, disconnected 'givens' in the form of discrete perceptual sensations, which the mind must then organize into meaningful perceptions of objects. But this is clearly *NOT* how our world appears to us. We do not experience atomistic sense impressions that have to be combined by concepts we bring to our experience. Instead, we experience a flow of objects, persons, and events, and only later, if at all, and for very specific purposes, do we separate out a perceptual from a conceptual component of any fully developed perception.

A good example of pragmatist critique of this sensation/concept dichotomy is Richard Rorty's *Philosophy and the Mirror of Nature* (1979), where he showed the historical contingency of the way Kant fatefully shaped so much subsequent philosophy through his insistence that experience presented itself to us in terms of 'intuitions' that are 'given' through our perceptual capacities which are then structured into meaningful experience and thought via concepts that are 'brought to' experience by the mind. This fateful distinction led Kant to formulate the fundamental problem of empirical knowledge as being how these two radically different types of thing (matter vs. form, given vs. made) could ever come together in experience. Hence Kant's infamous project of setting out the nature and conditions of certain types of judgments (and experiences), given the separation of intuitions from concepts and the a priori from the a posteriori.

Dewey also identified a number of common fallacies in the sciences and philosophy that represent basic versions of James's 'psychologist's fallacy.' One of the worst is the still rampant error of treating all experience as if it were fundamentally a form of knowing. Because the inquirer so often seeks *knowledge* of the nature and character of experience (and experiences), she is tempted to regard *all* experience as a mode of knowing. Dewey explains the intellectualist mistake:

> By "intellectualism" as an indictment is meant the theory that all experience is a mode of knowing, and that all subject-matter, all nature, is, in principle, to be reduced and transformed till it is defined in terms identical with the characteristics presented by refined objects of science as such. The assumption of "intellectualism" goes contrary to the facts of what is primarily experienced. For things are objects to be treated, used, acted upon and with, enjoyed and endured, even before they are things to be known. They are things *had* before they are things cognized.[27]

Following Dewey's lead, Richard Rorty tracked several major historical moves by which dominant philosophical traditions were developed on the assumption of what he derisively calls 'philosophy-as-epistemology' – the reduction of all experience to acts of knowing and the obsessive focus in so much philosophy on issues of epistemic justification of knowledge claims. However, even Rorty, with his ingeniously clever deconstructions of foundationalist tendencies in philosophy, somewhat ironically never completely separates himself from many of the founding assumptions of analytic philosophy of language, in particular his exclusive emphasis on language over experience.

Another example of lack of sufficient reflective self-understanding can be found in some of the cognitive sciences. Certain early versions of cognitive science – which Lakoff and I[28] have called 'first-generation cognitive science' – are fraught with tendencies to read back onto experience various distinctions the theorist finds it useful to make in trying to explain certain cognitive phenomena. Thus, what was known as 'information-processing psychology' assumed something like a faculty psychology (with distinct cognitive operations performed by distinct faculties [perhaps parts of the brain]), brought to bear on pre-structured perceptual inputs, manipulated and 'processed' by internal mechanisms, to produce outputs such as behaviors like speech or bodily movement. These abstract models got a boost from early artificial intelligence research, which conceived the 'mind' as basically a computational program operating on meaningless symbols according to rigidly defined and automated processing rules. Mind was the software; the brain was the hardware (or 'wetware').

Burgeoning pragmatists like Hilary Putnam were some of the first to criticize these functionalist models of mind and the assumptions of the cognitive science that appeared to support them. Putnam had the courage and flexibility to criticize his own earlier commitment to a specific form of functionalism. Even without using emerging neuroscience research, Putnam was able to raise deep concerns about the adequacy of such information-processing models, given, for example, the role of the body in how we make and experience meaning.[29] Only later did the emergence of sophisticated neuro-imaging make it possible to show how perceptual and motor capacities, for instance, are crucial in carrying out many of our 'higher' cognitive functions associated with understanding, reasoning, and communicating.[30]

The serious oversimplifications and selectional abstractions to which neuroscience, and sciences in general, are prone are certainly understandable in light of the extreme difficulty of getting any reliable, reproducible results from experiments. Without idealizations and partial models, there would be no growth of science. But our appreciation of how hard the experimental science is to carry off successfully must not blind us to the damage that can be done by unavoidable assumptions that underlie specific methods of inquiry. The scientific work, in other words, simply cannot stand on its own, for it cannot criticize its own limiting assumptions and still go about getting its experimental work accomplished. Something more is needed beyond the science itself – a broader, more encompassing philosophical reflection on how the many different bodies of research, with their many different assumptions and methods, can fit together in a more comprehensive picture. This continuing critical reflection on the assumptions and limits of any particular scientific enterprise or orientation is thus the second great contribution of pragmatist reflection on scientific research.

One of the most important contributions to science from pragmatist thought has been its steadfast insistence on the fullest articulation of the breadth, depth, and multidimensionality of the experience that is the subject of our inquiries. Most of the errors of philosophy are due to failures to respect experience, not as we theorize it to be, but as we inhabit and carry it forward in our lives. Dewey nicely sums up the two prongs of this emphasis of the fullness of experience:

> Reference to the primacy and ultimacy of the material of ordinary experience protects us, in the first place, from creating artificial problems which deflect the energy and attention of philosophers from the real problems that arise out of actual subject-matter. In the second place, it provides a check or test for the conclusions of philosophic inquiry; it is a constant reminder that we must replace them, as secondary reflective products, in the experience out of which they arose, ... In the third place, in seeing how they thus function in further experience, the philosophical results themselves acquire empirical value; they are what they contribute to the common experience of man, instead of being curiosities to be deposited, with appropriate labels, in a metaphysical museum.[31]

In short, we need what Lakoff and I (1999) have called an 'empirically responsible' philosophy, that is, a philosophy that, in Patricia Churchland's words, 'co-evolves' with the work of the various sciences.

The third major contribution of pragmatist philosophy is that, since any given scientific approach will typically focus on but one level of explanation, we will always need a broader reflective take on how the various levels of explanation appropriate for different available scientific methodologies might be related to one another. In a few cases, there might be a relation of reduction from a 'higher' to a 'lower' level, but this will be rare in the sciences of mind.[32] More often, we will need to try to discern how the multiple methodological perspectives might possibly fit together to give us an at least partial picture of the complexity of human experience, thought, and behavior. Bechtel summarizes the ultimately non-reductive inter-level nature of human experience and behavior:

> Mechanistic reduction only proposes to explain the response of an entity to the causal factors impinging on it in terms of its lower-level constituents. It does not try to explain the causal factors impinging on the mechanism. Understanding these requires inquiry focused on the mechanism as a whole, and often on yet higher-level systems in which the mechanism is embedded. Independence stems from the fact that inquiry at each level provides information additional to that which can be secured at other levels, and general does so using different tools of inquiry.[33]

There are a number of distinguished cognitive neuroscientists who bring an appropriate philosophical perspective to the relevant research in their particular fields of expertise. There are also a number of philosophers of a pragmatist orientation who engage the developing empirical research in mind science. The former group would include neuroscientists such as Antonio Damasio, Gerald Edelman, Vittorio Gallese, Don Tucker, and Jay Schulkin, to name but a few. The latter group includes philosophers like Owen Flanagan, Paul Thagard, Alva Noë, Shaun Gallagher, Evan Thompson, Paul Churchland, Patricia Churchland, Larry Hickman, John Shook, Tibor Solymosi, William Casebeer, and, I would hope, myself. Not all of these philosophers explicitly espouse a pragmatist orientation, nor do they all cite the work of major pragmatists, but I see in much of their work significant parallels with pragmatist views of experience, mind, meaning, thought, reason, and value.

In closing, let us return to our ugly neologism 'neuropragmatism.' What are the prospects for its future? Pretty bright, on the whole, and for two reasons. First, nothing is about to stop the ever-increasing tsunami of cognitive neuroscience as it sweeps across the contemporary scientific, philosophical, and common sense cultural landscape. There is too much excellent and highly transformative research being done to ever doubt that cognitive neuroscience is now, and will remain into the distant future, *the* principal mode of insight into mind, thought, language, and values. Second, since the sciences themselves will always need critical analysis of their assumptions, exploration of their relations to other disciplines, and reflection on their significance for human existence, there will always be a pressing need for an empirically responsible, pragmatically oriented philosophy. 'Neuropragmatism' is as good a name as any for this blended approach for inquiring into the human condition. Its mantra, one more time: Pragmatism without neuroscience is (partially) empty; neuroscience without pragmatism is (partially) blind.

Notes

1. John Austin, 'Performative Utterances,' in Austin, *Philosophical Papers* (Oxford: Oxford University Press, 1970), p. 233.
2. Tibor Solymosi, 'Neuropragmatism, Old and New,' *Phenomenology and the Cognitive Sciences* 10(3) (2011), p. 348.
3. Patricia Churchland, *Neurophilosophy: Toward a Unified Science of the Mind-Brain* (Cambridge: Cambridge University Press, 1986), p. 3.
4. George Lakoff and I (George Lakoff and Mark Johnson, *Philosophy in the Flesh: The Embodied Mind and Its Challenge to Western Thought* (New York: Basic Books, 1999)) have surveyed some of broader cognitive science research that supports an embodied cognition approach to mind, thought, and language.
5. John Dewey, *Art as Experience,* in *The Later Works,* Vol. 10 ed. Jo Ann Boydston (Carbondale: Southern Illinois University Press, 1987), p. 13.
6. Antonio Damasio, *The Feeling of What Happens: Body and Emotion in the Making of Consciousness* (New York: Harcourt Brace, 1999), p. 137.
7. Walter B. Cannon, *The Wisdom of the Body* (New York: W. W. Norton, 1932).
8. Jay Schulkin, *Adaptation and Well-Being: Social Allostasis* (Cambridge: Cambridge University Press, 2011), p. 5.
9. M. R. Bennett and P. M. S. Hacker, *Philosophical Foundations of Neuroscience* (Oxford: Blackwell Publishing, 2003).
10. Ibid., p. 213.
11. Antonio Damasio, *Looking for Spinoza: Joy, Sorrow, and the Feeling Brain* (New York: Harcourt, Inc., 2003), p. 53.
12. John Dewey, "Qualitative Thought," in *The Later Works,* vol. 5, ed. Jo Ann Boydston (Carbondale: Southern Illinois University Press, 1930/1988), p. 248.
13. John Dewey, *Logic: The Theory of Inquiry* in *The Later Works,* vol. 12, ed. Jo Ann Boydston (Carbondale: Southern Illinois University Press, 1938/1991), p. 72.
14. William James, *The Principles of Psychology,* Vol. I (New York: Dover 1890/1950), p. 239.
15. Dewey 1938/LW12, p. 74.
16. Ibid., p. 31.
17. Don Tucker, *Mind from Body: Experience from Neural Structure* (Oxford: Oxford University Press, 2007), p. 218.
18. Antonio Damasio, *The Feeling of What Happens: Body and Emotion in the Making of Consciousness* (New York: Harcourt Brace, 1999), p. 128.
19. William James, *Principles of Psychology,* Vol. I, p. 245, italics in original.
20. An analysis of the basic metaphors for thinking, including a detailed treatment of the metaphor of Thinking As Moving, can be found in Lakoff and Johnson, *Philosophy in the Flesh*, Chapter 12.
21. Antonio Damasio, *The Feeling of What Happens,* p. xiii.
22. William Bechtel, *Mental Mechanisms: Philosophical Perspectives on Cognitive Neuroscience* (New York: Taylor & Francis, 2008), p. 21.

23. Ibid., p. 157.
24. Hilary Putnam, *Reason, Truth, and History* (Cambridge: Cambridge University Press, 1981), Chapter 3, provided an important early critique of what he calls 'metaphysical realism,' which he contrasts with his own 'internal realism.'
25. In *The Structure of Scientific Revolutions* (Chicago: University of Chicago Press, 1962), Thomas Kuhn famously argued for this historicized view of science, which has subsequently been followed up by many other parallel critiques coming from a diverse range of philosophical traditions.
26. James, *Principles of Psychology*, p. 196.
27. John Dewey, *Experience and Nature* in *The Later Works*, vol. 1, ed. Jo Ann Boydston (Carbondale: Southern Illinois University Press, 1925/1981), p. 21.
28. Lakoff and Johnson, *Philosophy in the Flesh*.
29. This criticism began as early as 'Brains in a Vat' and other key essays in *Reason, Truth, and History*, which drew hardly at all from the sciences of the mind and focused at first mostly on thought experiments that traded on allegedly shared intuitions about various mental states and processes.
30. See, for instance, Jerome Feldman, *From Molecule to Metaphor: A Neural Theory of Language* (Cambridge, MA: MIT Press, 2006); Don Tucker, *Mind from Body: Experience from Neural Structure* (Oxford: Oxford University Press, 2007); Benjamin Bergen, *Louder Than Words: The New Science of How the Mind Makes Meaning* (New York: Basic Books, 2012); and all of Antonio Damasio's books cited above.
31. John Dewey, *Experience and Nature*, pp. 18–19.
32. William Bechtel, *Mental Mechanisms*.
33. Ibid., p. 157.

3
How Computational Neuroscience Revealed that the Pragmatists Were Right

W. Teed Rockwell

Paul Churchland has recently attempted to distance himself from pragmatism to some degree, even though he still refers to himself as a 'closet pragmatist':

> That fringe account of classical truth occasionally advanced by the pragmatists...attempt(s) to define any representation as an instance of genuine *knowledge* just in case, when deployed, it produces successful behavior or navigation...crudely, a true proposition is *one that works*...This tempts me hardly at all...if we *define* or *identify* what counts as truth, or as knowledge, in terms of the behavioral successes it produces, then we will not be able to give a nontrivial *explanation* of those behavioral successes.[1]

This quote from Churchland expresses a common objection to pragmatism. It seems at first glance to be a fair paraphrase of this famous quote from Peirce.

> Consider what effects, which might conceivably have practical bearings, we conceive the object of our conception to have. Then, our conception of those effects is the whole of our conception of the object.[2]

Churchland seems to be interpreting this Peirce passage to imply that the only possible definition of truth is a list of dispositional properties: that is, that a true theory is nothing more than the sum total of individual true statements it confirms. It is possible that there were moments when the pragmatists really did think this. It is also understandable that

Churchland would consider this to be the essential principle of pragmatism. James quotes and extrapolates from this Peirce passage in his canonical popularization, *Pragmatism: A New Name for Some Old Ways of Thinking*, and his concept of 'Cash Value' was inspired by this view of Peirce's.[3] Nevertheless, many of us feel that this slogan leaves out most of what makes pragmatism important for contemporary neuroscience.

Contemporary pragmatists, both neo- and neuro-, are strongly divided on which aspects of pragmatism are the most important for our time. As Rorty puts it:

> James and Dewey, alas, never made up their minds whether they wanted just to forget about epistemology or whether they wanted to devise a new improved epistemology of their own. In my view they should have opted for forgetting.[4]

I think, however, that part of pragmatist ambivalence on this issue came from a dilemma that need not constrain us today. The pragmatists believed that epistemology (and all other philosophy) could not be autonomously independent from the physical sciences. Unfortunately, the limited neuroscience of the past made it impossible for anyone – rationalist, empiricist, or pragmatist – to provide anything more than abstract functional descriptions of cognitive systems. That is basically what phenomenology and epistemology have traditionally been: A description of the functional patterns of thought and experience as they appeared to those of us doing the thinking and experiencing. Without an awareness of the appropriate biological facts, this abstract thinking easily drifts up into the metaphysical stratosphere. Weariness with the resulting unresolvable controversies naturally tempted the behaviorists, the positivists, and sometimes even the pragmatists to refrain from theorizing and stick to cataloging dispositions. However, if Peirce and James thought this was a good strategy, they were wrong. Churchland correctly points out that post-Aristotelian science requires us to explain every potentiality by redescribing it in actual terms. Opium's dormitive powers, dynamite's explosiveness, glass's brittleness, and salt's solubility must be explicable in terms of their actual chemical microstructure. Similarly, the success of a cognitive system must also be accounted for in terms of the actual physical structure of the system.

James's *Principles of Psychology* contained some of the first attempts to give those functional descriptions actual biological flesh, but 19th century neuroscience could not settle the pragmatist's disputes with earlier traditions. The pragmatists mostly had to claim, without much empirical support, that their theories account for both phenomenology and cognitive skill at least as effectively as did the rationalists and empiricists. Today, however, we know much more about the upper and lower levels of biological cognition, and it appears that by and large the pragmatists were right. Classical GOFAI computers, which were modeled on the rationalist/empiricist view of cognition, have been shown to be only a brittle metaphor that is significantly inaccurate in important areas. Pragmatism informed by neuroscience (that is to say, neuropragmatism), points the way towards explaining both the accuracies and inaccuracies of that metaphor. The rest of this chapter will outline the basic components of the pragmatist theory of knowledge, and the neuroscience that supports it.

Knowing-how and knowing-that

The traditional assumption has been that knowing-that is the most fundamental kind of knowing, and that knowing-how is constituted by interrelating various sentences containing knowing-that knowledge.

> It can be argued that anything which can be properly called 'knowing how to do something' presupposes a body of knowledge-that; or to put it differently knowledge of truth or facts. If this were so, then the statement that 'ducks know how to swim' would be as metaphorical as the statement that they know that water supports them.[5]

Fodor both ridicules and endorses this view of 'knowing-how' knowledge when he says that there is a little man in our head with a set of books. When we tie our shoes, the little man gets down the book titled, *How To Tie Your Shoes*, reads each of the instructions, and then follows them in sequence. Fodor claims that this, minus the anthropomorphism, is a description of a computer with a series of branching subroutines (that is, the books on the inner shelf), which is basically how our minds actually work.[6] The pragmatists, however, considered this description to be exactly backwards. Dewey rejected the idea that 'knowledge is derived from a higher source than practical activity,'[7] and insisted that 'There is no such thing as genuine knowledge and fruitful understanding except as the offspring of *doing*.'[8] In other words, abstract knowledge (knowing-that) is dependent on practical activity (knowing-how), not the other way around. This was a very strange idea in 1916, and Dewey probably did not know exactly what he meant by it. However, in the artificial connectionist systems that have partially simulated knowing-that knowledge, this ontological relationship is reversed in pretty much the way Dewey described.

Instead of creating knowing-how abilities by writing knowing-that subroutines, these connectionist systems create knowing-that knowledge by a specialized truncation of knowing-how knowledge. Most of the first connectionist systems (hereinafter 'networks') attempted to simulate linguistic abilities because those were the ground rules of the puzzles set by the logic-based AI systems. These networks did this by positing a meaning for a particular configuration of outputs, and then training the network to produce that output as a way of identifying a particular input. For example, a network designed to distinguish mines from rocks with two output nodes might be trained to activate the left node when it sees rocks, and the right node when it sees mines. Like all linguistic symbols, this designation is arbitrary, and the reverse could have been posited with equal effectiveness. Networks can be trained to perform these linguistic tasks, but only by ignoring certain aspects of their output. A 'rock' signal is rarely 1/0; it is more likely to be .9/.02 or .8/.3 etc. Networks with linguistic aspirations are thus forced into accepting the law of the excluded middle, which does not really apply to them. The variations from a 1/0 output are seen as noise which obscures but does not completely obliterate the signal.

However, in a network which is part of a living organism, these variations are usually not noise. Even in this well-known rock/mine detector, these variations contain information about which rocks are more similar to mines. This is not crucial information if one is trying to avoid being blown up, but it is in principle available to the curious. More importantly, in an organism, these variations are used to control complex motor patterns. Neurophilosophers often depict information stored in networks by dividing up a multidimensional space into various regions, and then labeling each region verbally, as if each region were a very complicated symbol referring to a word. It would be more accurate to depict networks by showing relationships between two different spaces, one a perceptual space and the other a behavioral space. This is the kind of cognitive structure that enables someone to track an incoming fastball, and then position the bat to knock the ball out of the park. This is also the kind of structure that is needed to account for knowing-that abilities in actual human beings. Linguistic assertions are always parts of speech acts that involve multiple kinds of cognitive processing, such as judging context, modulating voice tones for emphasis, making similar judgments about the speech acts of the person you are conversing with, and so on. In the laboratory, we can isolate knowing how to recognize a word or sentence from all these various other knowing-how abilities. But the basic structure of word recognition is the same as for all of these other non-verbal knowing-how abilities. It is a neurological fact that knowing-that is only one variation on the vector transformation structure that makes all forms of knowing-how possible. Dewey was correct when he said, 'the brain and nervous system are primarily organs of action-undergoing...cognitive experience must originate within that of a non-cognitive sort.'[9]

Once we acknowledge that vector transformations, rather than language, are the fundamental structures of the mind, many of the other principles of pragmatist epistemology and metaphysics inevitably follow.

Continuity between truth and error

Pragmatism has gained many enemies by rejecting the allegedly undeniable truth that there is only one reality, and one truth that accurately describes it. William James's pragmatic pluralism entails accepting the possibility that there is no 'absolute totality,' and that 'a disseminated, distributed, or incompletely unified appearance is the only form that reality may yet have achieved.'[10] Trim away the qualifiers in this sentence and what we get is the apparently paradoxical claim that 'appearance is...reality.' This is an inevitable consequence of the pragmatist slogan that truth is whatever works. No theory works perfectly, and every theory that anyone has seriously adopted works at least part of the time. Even Ptolemaic astronomy worked as a basis for navigation, despite its numerous other weaknesses. Consequently, we end up with a continuum theory of truth, in which epistemic virtue admits of degrees, rather than being either completely present or completely absent. Ideas do not divide neatly into the true ones, which refer to reality, and the false ones, which refer to appearances, because the line between appearance and reality is blurry.

As counterintuitive as this may sound, this is the only kind of epistemic virtue possessed by connectionist networks. When a network is learning, Churchland argues, its response to its inputs is changing 'in a fashion that systematically reduces the error messages to a trickle.'[11] As Churchland continues,

> Nothing guarantees that there exists a possible configuration of weights that would reduce the error message to zero...nothing guarantees that there is only one global minimum...perhaps there will in

general be many quite different minima, all of them equally low in error, all of them carving up the world in quite different ways ... these considerations seem to remove the goal itself – a unique truth – as well as any sure means of getting there.[12]

Thus, Churchland claims that although there is no such thing as 'the truth' from a neurocomputational point of view, there is such a thing as learning. Learning in a neural network is defined not as eliminating error but as decreasing it. It is possible in principle that someday someone's networks may be so perfectly tuned that they will never make any errors again. Acknowledging this possibility is the pragmatist's way of escaping the charge of inconsistency in the claim 'it is absolutely true that there is no absolute truth.' The pragmatist does not need to prove that such a network is impossible to point out that there is no cash value in defining that possible network as the truth. We do not need an epistemology that only creates honorifics for a theory we have never seen. We need an epistemology that enables us to tell the difference between our best theories and our not-so-good theories. A theory based on connectionist vector transformations seems to be our best bet for that more modest goal.

Knowledge and experience

Dewey rejected the view 'that emotion as well as sense is but confused thought, which when it becomes clear and definite or reaches its goal becomes cognition.'[13] This view was later expressed in the Sellarsian slogan that 'All awareness is a linguistic affair.'[14] Dewey claimed instead that knowledge and experience were fundamentally different: 'To be a smell is one thing, to be known as a smell, another; to be a "feeling" one thing, to be known as a "feeling" another.'[15]

Some of Dewey's critics have thought that this distinction between knowledge and experience implies a kind of dualism between (subjective) experience and (objective) knowledge, of the sort defended by Frank Jackson in his famous 'Mary the Color-blind Neuroscientist' argument. This quasi-dualism is, however, saved from Cartesian substance dualism if we combine it with Churchland's theory of Semantic State Spaces (SSS). According to Churchland, the reason that Mary's subjective experiences of color cannot be communicated is not because they are non-physical but rather because 'the brain uses more modes and media of representation than the simple storage of sentences.'[16] Churchland later explains in greater detail why these modes of representation are both subjective and physical.[17] They are physical because they are embodied by the neural structures that semantically relate brain states to the world. Nevertheless, these neural structures are still subjective because they cannot be socially communicated the way linguistic ideas can be socially communicated. This distinction between language and the Semantic State Space that grounds it is a neurologically sophisticated version of Dewey's distinction between knowledge and experience. For both Churchland and Dewey, experience/SSS is capable of sophisticated cognitive functions without needing language. This is why we can sometimes have what is called tacit knowledge that is, knowing-how abilities that cannot be translated into knowing-that knowledge. However, language is empty and meaningless unless it is semantically grounded in experience/SSS. This is why Dewey insisted in *Democracy and Education* that memorizing facts was useless unless those facts were employed in the service of some embodied goal-directed activity. This kind of activity creates relationships between perceptual state spaces and behavioral state spaces and thus semantically connects our knowledge to the world. This distinction seemed vague and mystical in Dewey's time. Thanks to Churchland's neurophilosophy, we now have a version of Dewey's distinction with firm scientific support.

Radical empiricism

James's 'radical empiricism' is distinguished from what Dewey called 'sensationalistic empiricism' because of the alleged relationship between parts and wholes. Sensationalistic empiricism held that experience consisted of discrete sense data, and the mind's job is to assemble those sense data into unified experiences. Figuring out how the mind does this is called the 'binding problem' in psychology. Note also that sensationalistic empiricism is a position shared by both sides of the rationalist/empiricist debate. Both Kant and Hume are sensationalistic empiricists in this sense, their only disagreement being how much the mind adds to these bits of experience once they come in. James and Dewey dissolved this debate, and the binding problem that emerged from its presuppositions, by arguing that our experience is a fundamentally unified 'blooming buzzing confusion,' and that it is analysis, not synthesis, which is the fundamental cognitive achievement. James argues that this fact is revealed by careful phenomenology, once we free ourselves from the prejudices of sensationalistic empiricism. I will argue that seeing the mind as a 'connectionist network' provides an even stronger support for radical empiricism.

Let us consider our archetypical network that can distinguish mines from rocks. It works by setting the borders of a possibility space, then dividing that space into two parts, one of which triggers the 'mine' output and the other the 'rock' output. Suppose that we needed this network to make an additional distinction, say between rocks and corral? We would do this by training the network to make a third response, thus dividing the perceptual space of the network into 3 parts instead of 2. Suppose we needed to make a distinction between two different kinds of mines, such as metal and plastic? This would require dividing the perceptual space into 4 parts and so on. We can thus see that a network acquires greater cognitive sophistication by starting with a fundamentally unified input space, then dividing it into smaller spaces. An infant makes essentially no distinction between itself and everything else and acquires consciousness by dividing reality. The first such division is between self and other, an awareness also possessed by any cognitive creature with enough smarts to avoid eating itself. In a human infant, this distinction is the beginning of a long journey to cognitive sophistication, followed by divisions like 'mama' and 'everything else,' or 'hunger' and 'not hunger.'

Simple creatures remain permanently in a similarly simple state, being hardwired to divide perceptual space with a handful of very simple distinctions necessary for their survival and reproduction. A frog, for example, can only see things that move, and it can only make a rough distinction between big things that move and little things that move. This is enough for it to perform its simple behavioral repertoire: If it is little and moves, eat it, and if it is big and moves, flee from it. The difference between the frog and us is that we divide our input space into a much more complicated range of categories, each of which requires a different behavioral response depending on context and combinations. We do not get this cognitive sophistication by taking in discrete sense data and then assembling them into more complex patterns. There is nothing in our brains that does anything remotely like this, so once again the pragmatists have neuroscience on their side. Our minds do not begin as empty boxes into which we dump sense data. The ignorant mind is more like the 'uncarved block' of Taoist philosophy, which acquires knowledge by dividing and interrelating regions of perceptual and behavioral space.

Pragmatism and extended cognition

In each case, behavioral success is obviously dependent on a variety of factors. Consider Michael Jordan's success as a basketball player. Some of the factors that contribute to this success are clearly cognitive, such as his ability to negotiate relationships between behavioral space and perceptual space, due to his phenomenal hand to eye coordination. Others are clearly non-cognitive, such as his height. Is the pragmatist forced to ignore this distinction, because both of these factors are equally essential to Jordan attaining success as a basketball player? This is what Churchland is concerned about when he mentions 'the problem of how to evaluate candidates, when only *collectively* do they play a role in directing behavior.'[18] It is also Russell's objection to Dewey's theory of inquiry, which Russell said was so vague that Dewey would have to say that bricklayers were inquiring into the bricks when they built a house out of them.

It is of course true that there are many components of any behavioral system which are clearly cognitive, and others that are clearly non-cognitive. Nevertheless, the vast region in the middle is full of ambiguous hard cases, which get ignored when we assume that every factor that increases success must be unambiguously cognitive or non-cognitive. One of the main inspirations for the 'hypothesis of extended cognition' (HEC) is the fact that it is very difficult to make a principled distinction between cognitive and non-cognitive processes if all the processes under consideration are equally necessary for the performance of a successful behavior. When everyone assumed without question that whatever happened in the brain was cognitive, and whatever happened outside the brain was non-cognitive, there was only one puzzle cognitive scientists had to solve: 'What is happening in the brain at the time successful behavior is being performed?' This puzzle has been an effective research paradigm because, whenever cognitively sophisticated behavior occurs, there are always a great many of causally necessary things going on inside the brain. Recent research about the relationships between the brain and the body/world, however, frequently reveal isomorphic dynamic patterns outside the brain that also seem intuitively cognitive to many of us. This research is the primary empirical content of recent works by Wilson, Clark, Hutto, Menary, Noë, Rockwell, and others. Others (primarily Rupert, Adams and Aizawa) have argued that the externalist arguments in these works require additional explanation as to why these extra-cranial processes are actually cognitive. If you did not provide that additional explanation you were guilty of what Adams and Aizawa called the 'coupling/constitution fallacy.'[19] My reply to this[20] is that the burden of proof is actually on the critics of HEC, because traditional internalism cannot exist unless we can draw a principled distinction between Coupling and Constitution and/or Causation and Embodiment. My brand of HEC sees the line between self and world as both pragmatic and dynamic, which means you cannot draw a line between the two that is either principled or enduring. Consequently, cognition must be seen as extended, because there is no longer any principled reason to keep it inside the brain.

This deliberate blurring of the line between self and world is an important part of the pragmatist project, especially for Dewey.[21] One of the factors that separates Churchland from the pragmatists is his attempts to use computational neuroscience to draw that line afresh.[22] Churchland thinks that there is something uniquely cognitive about brain dynamics that cannot be found in other dynamic patterns present in the body and world. He may be right. Those of us who defend HEC have always insisted the question is an empirical one. However, the main thought experiment he uses to make this point could actually be turned against him, if we use the same techniques that he has taught us to use on so many others. Churchland asks us to consider a GPS (global positioning system) of the sort that are available in iPhones or 'upscale rental cars.'[23] He describes these devices as containing 'a stored street map,' and as receiving signals from satellites which then align the map up with the position of the car. Churchland claims that all of the cognitive work of locating the car position is done by the information and/or hardware that is stored in the car itself. He also says that 'the GPS link...is utterly inessential to its maphood. The system would work just as well if it were keyed to bar-coded magnetic beacons embedded in the roads every fifty feet, or to a video system on the hood.'[24]

Here I feel like borrowing one of the Churchland's picket signs with a crossed-out armchair on it. Churchland gives us no footnotes to any computer science texts justifying his claim that all of the cognitively relevant functions are stored in the car itself. It seems more likely to me that the map in this system is actually stored in some kind of online virtual computer cloud, and that the only cognitive machinery in the car itself is the receptor device that downloads pages as needed from the map website. If my guess is right, Churchland is mistaken when he claims that all of the cognition and representation is stored inside the GPS. Even if it turns out that Churchland is empirically correct about the structure of an actual GPS, this does not prove that biodynamic systems must operate that way. What little we know about human cognition is compatible with both the extended and brain-centered theories. If something like J. J. Gibson's theory of perception is correct, some of our cognitive processing could be uploaded into the array of perceptible light, just as the mapping procedures are uploaded onto the GPS website in my speculations about GPS functions.

It also strikes me as highly unlikely that this particular system would work every bit as well if all the machinery outside the car were changed, and the internal circuitry in the car remains the same. It seems more plausible that reconnecting the internal circuitry in the car's computer to these alternative systems would require a complete rewiring of the entire system. Some electronic systems have the kind of modularity which makes it possible to think of them as possessing what Churchland calls 'narrow content.' Because you can plug either a CD player or a cassette player or an iPod into the same hi-fi set, it is natural to assume that the musical content is intrinsically present in each CD or tape. However, this kind of portability is an artificial contrivance created by economic forces. It exists only because we often like to update components of our audio systems without having to buy a whole new system. Consequently, there is no reason to assume that these carvable joints exist in biodynamic systems, or even in other artificial systems. If you try inputting the digital output of a CD player into an analog amplifier, or information from an IBM computer directly into a Macintosh, what you are going to get is meaningless hash. In such a situation, it is quite clear that the so-called content in the mismatched information is a relational, not an intrinsic or internal, property. It is possible that neat lines can be drawn between the representational and non-representational parts of a biodynamic system. It is far more likely, however, that 'cognitive' and 'representational' are prototypes, and the various parts of a brain/body/world system cluster at various distances in logical space from those prototypes. If so, the fact that pragmatism cannot make sharp distinctions where none exist should be seen as a virtue.

DST and the metaphysics of process

Even if we assume that connectionist networks operate by starting with fundamentally unified wholes, is not neuroscience itself atomistic? Is not the fundamental unit of neuroscience the neuron, and does not neuroscience progress by seeing how these fundamental units are joined together to form cognitive structures? Not if the most accurate conceptual system for describing human cognition is dynamic systems theory (DST). Dynamic neuroscientist Walter Freeman remarked that trying to understand the mind by studying individual neurons is rather like trying to understand hurricanes by studying water molecules.[25] Many of us have argued that the essential cognitive categories are not neurons but multi-stable attractor spaces.[26] Attractor spaces are not discrete items, but patterns produced by the interaction of various forces, rather like whirlpools or ocean waves. The fluctuations between these attractor spaces can often be accurately approximated by modular computer flow charts, just as geocentric theories of astronomy can accurately map the 'movements' of the stars across the night sky. However, just as geocentric astronomy cannot effectively account for the behavior of the planets, rigidly modular models of mind cannot account for the fact that biological organisms are much more flexible than their computerized approximations. That is why many of us believe that we should see the mind as a fluctuating field rather than as a computer made out of meat.

This was the view that Dewey was prophesying when he described mind/world interactions as 'a change in the system of tensions.'[27] This view supports the pragmatist's extended cognition theory, because it demotes the neurons to being mere media through which the behavioral field flows, rather than the essential components of cognition. In neuroscience, the mathematics of vector transformation is used to measure variations in neural voltages, but there is no reason that the same mathematics cannot be used to measure relationships between other dynamically related factors in the body or the world. Many researchers are doing exactly that.[28] If this research continues to discover structures which are prototypically cognitive, neuropragmatism will have transcended itself by revealing that cognition is not fundamentally neural after all. As Tibor Solymosi puts it, 'once we have gone into the brain we must readily "get out of our heads."' Neuropragmatism must see itself as 'continuously and dynamically going in and out.'[29]

Despite Churchland's apparent commitment to the assumption that only cranial neural structures are the embodiments of what he calls narrow content, there are signs that he is shifting towards something like a dynamic view of the mind. More recently, he stresses that recurrent nets must be seen as dynamic systems and frequently refers to prototypes with DST terms such as attractors,' 'attractor basins,' or 'prototype wells.'[30] Perhaps most importantly, his critique of functionalism[31] actually supports an extended dynamic view of the mind, if he consistently follows out its implications. He gives three examples of physical phenomena – sound, temperature, and magnetic fields – each of which can be multiply realized in a variety of physical substrates.[32] Sound waves can travel through air, water, or solid rock, yet they always follow the same physical laws that govern wave propagation. Physical substances with radically different chemical structures still have temperatures governed by the laws of statistical mechanics. And magnetic fields follow the same basic laws in different metals and metalloids.

However, what Churchland calls 'a successful reduction to some underlying and highly general physical laws'[33] is not really a reduction in the literal sense. These are translations of concepts from one physical science to another, but they are not reductions. The success of this translation is a victory for a kind of physicalism but not for reductionism. Reductionism is the belief, rejected by the pragmatists and defended by the positivists, that the behavior of wholes can be fully *reduced* to the interactions of their underlying parts. The physical laws used in this translation are not *underlying* but *overarching*, which is why Churchland correctly refers to the items in the translated science as a 'substrate.' What DST is advocating is that neuroscience should be treated as a substrate for the field-based laws of abstract physics, which are doing the real cognitive work that is only approximately described by talks about neurons, and could in principle be embodied by other substrates in the body and/or world. This is not a reduction; it is rather an affirmation for the holism that was the central metaphysical and epistemological position of pragmatism. If neuropragmatism transforms itself into a kind of post-neural dynamic pragmatism, we will have yet one more scientific proof that the pragmatists were right.

Notes

1. Paul Churchland, *Neurophilosophy at Work* (New York: Cambridge University Press, 2007), p. 103.
2. C. S. Peirce, 'How to Make Our Ideas Clear,' in *The Essential Peirce*, vol. 2, ed. the Peirce Edition Project (Bloomington: Indiana University Press, 1992/1999), p. 132.
3. William James, *Pragmatism: A New Name for Some Old Ways of Thinking*, (Cambridge, MA: Harvard University Press, 1907/1975).
4. Richard Rorty, 'Dewey between Hegel and Darwin' in *Modernist Impulses in the Human Sciences*, ed. D. Ross (Johns Hopkins, Baltimore, 1994), pp. 59–60.
5. Wilfrid Sellars, 'Philosophy and the Scientific Image of Man' in Sellars, *Science, Perception, and Reality* (Routledge and Kegan Paul, 1963), p. 1.
6. Jerry Fodor, *RePresentations* (Cambridge, MA: Bradford Books The MIT Press, 1981).
7. John Dewey, *Democracy and Education* (New York: Macmillan, 1916), p. 321; *Democracy and Education*, in *The Middle Works of John Dewey*, vol. 9, ed. Jo Ann Boydston (Carbondale: Southern Illinois University Press, 1980), p. 284.
8. Ibid., p. 275, italics in original.
9. Dewey, *Experience and Nature* (New York: Dover, 1929), p. 23.
10. James, *A Pluralistic Universe* (New York: Longmans, Green, and Company, 1909), p. 112.

11. Churchland, *A Neurocomputational Perspective* (Cambridge, MA: Bradford Books The MIT Press, 1989), p. 177.
12. Ibid., p. 194.
13. Dewey, *Experience and Nature*, p. 19.
14. Although this is a direct quote from Sellars, his position on this issue was actually far closer to Dewey's than is usually acknowledged (see Teed Rockwell 'Experience and Sensation: Sellars and Dewey on the Non-cognitive Aspects of Mental Life,' *Education and Culture: the Journal of the John Dewey Society*, XVII Winter (2001). Available at: www.cognitivequestions.org/sellarsdewey.html). However, the interpretation of Sellars in Rorty, *Philosophy and the Mirror of Nature*, (Basil: Blackwell, 1980), which denies that there are separate cognitive functions for knowledge and experience, is the most commonly accepted interpretation of this slogan.
15. Dewey, *The Influence of Darwin on Philosophy and Other Essays* (Amherst, N.Y.: Prometheus Press, 1910), p. 81.
16. Churchland, *A Neurocomputational Perspective*, p. 63.
17. See Churchland, *Neurophilosophy at Work*.
18. Churchland, *Neurophilosophy at Work*, p. 103.
19. F. Adams, and K. Aizawa, *The Bounds of Cognition* (Walden, MA: Blackwell, 2008).
20. Rockwell, 'Extended Cognition and Intrinsic Properties,' *Philosophical Psychology*, December (2010).
21. See Rockwell, *Neither Brain nor Ghost: A Non-Dualist Alternative to the Mind/Brain Identity Theory* (Cambridge, MA: Bradford Books, MIT Press, 2005).
22. Churchland, *Neurophilosophy at Work*, especially Chapters 6 and 8.
23. Churchland, *Neurophilosophy at Work*, p. 155.
24. Ibid.
25. Personal communication.
26. Rockwell, *Neither Brain nor Ghost*; and Anthony Chemero, *Radical Embodied Cognitive Science* (Cambridge, MA: Bradford Books The MIT Press, 2009).
27. Dewey, 'The Reflex Arc Concept in Psychology,' *Psychological Review*, III (July 1896), pp. 357–370. See http://psychclassics.yorku.ca/Dewey/reflex.htm., par. 15.
28. See Chapter 10 of Rockwell, *Neither Brain nor Ghost*.
29. Tibor Solymosi, 'Neuropragmatism, Old and New,' *Phenomenology and the Cognitive Sciences*, 10(3) (2011), p. 366.
30. Churchland, *Neurophilosophy at Work*, pp. 12–13.
31. Churchland, *Neurophilosophy at Work*, Chapter 2.
32. Ibid., pp. 25–27.
33. Ibid., p. 27.

4
Pragmatism, Cognitive Capacity and Brain Function
Jay Schulkin

Introduction

Classical pragmatism understood that human action is replete with meaning and purpose; problem solving is continuous with adaptation; philosophy is continuous with, but not reduced to, science, but self-corrective inquiry is at the heart of the naturalizing self-correction and problem solving.[1] An appreciation of nature, in addition to human meaning bound with human community, scientific and otherwise, reveals the importance of classical pragmatism. In addition, an understanding of common sense 'critical realism'[2] is key to classical pragmatism, meaning abstractions are not considered more real than the objects encountered, and staying anchored to objects was of great importance.

The classical pragmatists were rooted in common core themes. Classical pragmatism avoided dualisms of diverse sorts that included the separation of facts and values, cognition and action, perception and cognition, analytic and synthetic separation. Instead, what often mattered was elaborating a sense of experience beyond classical empiricism, with an emphasis on sense data or modern positivism and the validity of statements being lodged in sense datum and an impoverished sense of the human experience of knowing. 'Radical empiricism,' as James propounded,[3] is rich in action with self-control with cognitive systems embodied in human performance.

Experiments and experience are at the heart of human investigation. They cut across all avenues of human endeavor, such as the cultivation of intelligence, which is a lifelong quest. Hypothesis formation and correction are at the heart of inquiry and our evolution. An axiological framework is at the heart of the knowing process.

The classical pragmatists[4] understood something about action, inquiry, and the brain. Their goal, a modern one, was an understanding of cognitive adaptation along with what John Dewey called 'lived experience.' They understood these events in the context of evolution of the brain. The evolution of cortical function represented an expansion of human capability and growth and thus part of their conception of inquiry.

A core feature of the pragmatist tradition is the embodiment of problem solving in the mind/brain. The pragmatists undermined all forms of dualism, and instead focused on the integration of systems. Indeed, cognitive and behavioral capacities work together to engage the world, to compute probability, and to assess our surroundings (for example, friendly or non-friendly social events).

Indeed, we come prepared by neural circuits to respond to all kinds of objects.[5] Diverse cognitive systems (e.g., recognition of animate/inanimate objects, agents, senses of space, time, and statistical relationships) underlie what Peirce (see below) called 'abduction,' or the genesis of ideas and problem solving. These biologically derived cognitive systems are not divorced from action, perception, or our physical body, but are endemic to them and are distributed across neural networks in the brain.

In what follows, I first give brief descriptions of the evolutionary and social contexts of neural systems and the naturalization of intelligence, within a perspective of the cognitive/neural sciences followed by the instrumental sense that runs through cognitive systems and the brain and through effort and human action, but first an account of the importance of human experience and inquiry, something essential to pragmatism.

Experience and the study of mind

The study of mind and experience is the integrative link that binds all psychological inquiry together. The tradition of William James[6] and John Dewey represents a link to experience that is construed in broad functionalistic and experiential terms within psychology. Dewey,[7] for example, constantly referred to our adaptations to the world. He states that our problem solving is knotted to the search for stability and the experience of security. But both James and Dewey were responsive to the biological sciences and their methods, and this would remain constant throughout their lives.

For James and Dewey, experience is not simply about sensations, but rather the engagement with objects one encounters in the world while trying to cope and adapt. Experience for pragmatists like James is, as he put it, 'radical.' Experience is a means to interact with the world and not something that cuts us off from the world. As James said, 'knowledge of sensible realities come to life inside the tissue of experience. It is made and made by relations that unroll in time.'[8] As Dewey stated, 'Experience is not a veil that shuts man off from nature, it is a means of penetrating continually further into the heart of nature.'[9] I suggest that any account of mind and experience, in either the neural or cognitive sciences, needs to recapture this tradition.

For pragmatists like James or Dewey, mind and experience are tied to function and mechanism.[10] Experience is replete with cognition and embodied in central states of the brain. Experience is active, and not just passive. Experience for pragmatists is linked to action as well as the resolution of problems. An 'enactive' orientation of understanding human adaptation and knowing ties action and perception,[11] something dear to the heart of pragmatists such as Dewey.[12]

The overwhelming dominance of cognitive science in the study of memory, perception, emotion, social behavior, language, and learning indicates the power of the cognitive turn. These events are construed as often unconscious.[13] This cognitive orientation did not return to James's notion that psychology, in addition to the study of specific mental events, also tries to captures the richness of experience. The experiences are rooted with others, in understanding the social context of being others, surviving with others; there is no Cartesian devolution of function into isolated self-certainty. The orientation is into expanding human experience and its core embodiments as we forage for understanding.

The cognitive revolution needs anchoring with an emphasis on the expression of experience and the biological context in which the mind evolved. Dewey challenged stimulus-response theories of behavior early on and introduced cognitivism to the understanding of how one copes with and adapts to one's surroundings.[14] But both Dewey and James wanted to capture the richness of experience, a world replete with causal efficacy, problem solving linked to function, biological adaptation, and behavioral functions in appetitive and consummatory behaviors.[15] Experience is the conduit to get outside one's head and be responsive to the world. Any theory that attempts to capture humans within the cognitive and neural sciences needs to take that into account, in spite of the difficulty of doing so.

The tradition of James's and Dewey's American pragmatism united psychology and philosophy and placed them in the context of biology and experience, a foothold in the humanities and the sciences.[16] The task is to uncover the features of experience and its relationship to biology and culture. This is the rich tradition established by James and the classical pragmatists. James, Peirce, and Dewey envisioned psychology as responsive to biology and culture. Psychology's task is to integrate across disciplines and to ask questions about the underlying mechanisms that account for our experience.

The James-Deweyian view of philosophy and psychology looks to discern the affordances of the environment in which we evolved[17] and our underlying psychobiology with the abilities that we have. Underlying this integration is the acknowledgement of our social evolution and the utter profundity of our social discourse determining how we think.[18]

Two senses of functionalism are therefore reinforced by one another: the first sense is the one the pragmatists held – mind adapting to niches and solving problems. The second is the idea of internal computational mechanisms[19] or what Peirce meant by 'logical machines.'[20] Neither of these senses is reducible to the neural sciences, and yet links with brain events are part of what we want to understand (for instance, in physiological psychology). Both senses are tied to experience while recognizing that experience is not the same as consciousness or learning. Functionalism is not about an abstract code but about real enactive embodied experiences.[21] When discerning the features of mind, one needs to co-inhabit without pernicious reduction the features of experience with that of an underlying mechanism, be that cognitive or neural.

We are neither Cartesian machines thinking in a vacuum, nor inductive machines, nor empirical blank slates. We bring with us diverse forms of cognitive devices that underlie what Dewey used to call 'lived experiences' or what others have called 'embodied cognition.' Cognitive systems, as Dewey notes, are endemic to the organization of action, to one's sense of self and of others, and to pragmatism's emphasis on human purpose and well-being.[22] Our sense of self and meaning is rooted in our life histories and our trajectory of movement through space and experience. 'The stories are journey's metaphors,' with metaphorical reasoning being an important part of our cognitive architecture, grounded in action and rooted in human meaning and purpose.[23]

Thinking is understood by the classical pragmatist, and certainly by me, in the context of action and transacting with others. This is quite close to a pragmatist position, where cognitive systems are thought of as embedded in the organization of action. The emphasis is on embodied and expanded cognitive systems in which the sensorimotor systems are themselves knotted to cephalic machinations across all regions of the

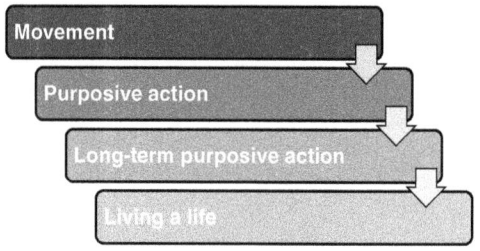

Figure 4.1 Action and purpose in living a life
Source: Adapted from Mark Johnson, *Moral Imagination* (Chicago: University of Chicago Press, 1993).

brain. In other words, cognitive systems are not just a cortical affair but are endemic to cephalic function.

Evolution, pragmatism cortex and social contact

Naturalistic pragmatists are rooted in biology. 'Naturalization of intelligence,' as Dewey certainly understood it, places problem solving and inquiry in a psychobiological context, in a context of behavioral adaptation. The development of the theory of evolution had an impact on a wide area of thought. In particular, Chauncey Wright understood the importance of Darwin and the concepts of evolution, wrote about the logic of evolutionary theory in the context of problem solving and our sense of the world and our experiences, and would help introduce American thinkers to this paradigm shift in thought.[24]

Evolutionary themes are grounding perspectives, and Dewey never lost sight of the integration of this theoretical base. In Dewey's article, 'The Influence of Darwinism on Philosophy,' Darwin's theory was considered an 'epoch in the development of the natural sciences and for the layman.'[25] Dewey maintains the importance of understanding that biology is essential to social beings, while still resisting the urge to reduce all behaviors purely to the biological.

Figure 4.2 depicted here is an endocast of the frontal region of a putative *Homo* around 2 million years ago and a representation of (A) chimpanzee, (B) orangutan, (C) gorilla, and (D) human frontal plane (Falk, 1983). Frontal cortical expansion is a primary feature of each illustration.

Indeed, we now know that diverse forms of hominids competed and perhaps interbred[26] while living during the same time period.[27] *Homo*

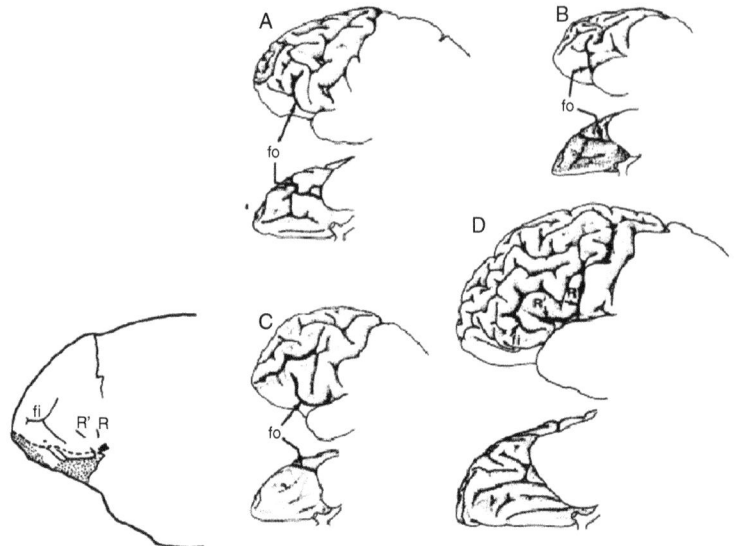

Figure 4.2 Comparisons of primate brains

sapiens came to dominate the landscape, as other human-like primates became extinct.

Some 30,000 years ago, there is evidence to suggest Neanderthals and *Homo sapiens* co-inhabited different but overlapping geographical locations.[28] One description of the evolution of Homo sapiens is depicted below.[29] Some core features in the origins of the genus *Homo*: longer gestational period, longer life spans, forward locomotion with heel and hind limb dominance, dominance of stereoscopic vision and forward movement of the eyes, and expanding use of the hands.[30]

In addition to these features, there is the often overlooked feature that contributed to the ascent of our genus: our increased social skills. An elaborate set of neural structures widely distributed throughout the brain is linked to keeping track of others, watching what they do, and getting a foothold in the world of approachable and avoidable events. For primates, a wide variety of evidence links the degree of social interaction with neocortical expansion. Diverse models of group size have been linked to neocortical enlargement and cognitive competence across distributed cognitive systems. The greater the degree of social contact and social organization experienced by diverse primates, including us, the greater the trend toward cortical expansion.

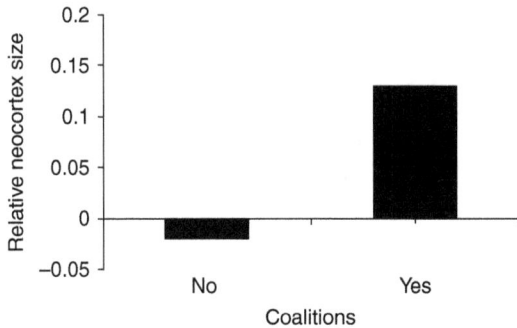

Figure 4.3 Social contact in primates is consistently linked to neocortical expansion
Source: R. I. M. Dunbar, and S. Shultz, 'Evolution in the Social Brain,' *Science* 317 (2007): 1344–1347.

In addition, a broad-based set of findings in non-primates has been the link between social complexity and larger brain size.[31] The metabolic investment of larger brained animals is expensive; a brain is a high-energy organ and neural tissue expands while other tissue does not, or at least not to the same degree. Interesting correlations have been suggested between neocortical size and social cognitive skills. The pressure on coming into touch with others, creating alliances, and tracking lineages no doubt required more cortical mass. The metabolic investment of larger brained animals in expensive neural tissue indicates that this high-energy organ provides compensating social benefits, as seen in the changes in longevity.

We are pedagogical animals, though core cephalic capabilities and the social context set the conditions for our continued social evolution.[32] As social animals, we are oriented toward diverse expressions of our conspecifics that root us in the social world. This social world is one of acceptance and rejection, of approach and avoidance of others, and of social and ecological objects rich with significance and meaning.

A diverse set of cognitive functions underlies the great range of behavioral capacities that we display. In addition, biological cortical functions contribute to the expression of social and communicative cognitive-behavioral systems. In other words, cognitive capacity and expression in primate evolution is knotted to neocortical expansion and the flexibility that marks our species' behavioral repertoire is the result of corticalization of function.

This cognitive competence also reflects rapid brain growth during critical periods in our evolution, which perhaps figured into the dramatic expression of our social intelligence. Cultural variation in decision-making, for instance, is anchored to core cephalic predilections. Perhaps this resulted in a brain oriented to change, rather than stability.[33] The world is social for our species, and the self is embedded in the larger linguistic/social community. Communicative competence in this realm is essential for pragmatists because it allows one to transform high ideals into practical outcomes and consequences that matter.

Dewey's theory of problem solving and its expansion into every aspect of the human condition goes a long way to explain the extension of the adaptive use of our cognitive resources.[34] As our cognitive powers became greater, humans began applying those cognitive abilities to different and more complex problems. This not only continued to change our knowledge base but also perpetuated the cycle of cortical expansion. By our efforts to ameliorate devolved function, using a toolbox of cognitive capabilities as resources, we developed an expanding sense of memory and culture, thereby setting the conditions for further human advances.

Objects and cephalic competence

One of the anchoring points of pragmatism is that a sense of nature should not be mythologized as a panacea. Instead, good sense, sound consequences, and instrumental capabilities allow us to understand and appreciate the organic features of nature.

The orientation of a human child to a physical domain of objects can appear quite similar on some tasks to the common chimpanzee or orangutan in the first few years in development. In particular, children perform similarly to chimpanzees and orangutans when given problems concerning objects in space, quantities or drawing inferences.[35] What becomes quite evident early on in ontogeny, however, is the link to the vastness of the social world in which the neonate is trying to get a foothold for action.

Indeed, we come into life prepared to interpret our surroundings as defined by the social milieu, and the degree to which we succeed in this task determines to a great extent our success in coping, achieving, and thriving. The fact that we come prepared to recognize others and learn from their experiences is thus a fundamental social behavioral adaptation. Dewey emphasized the development of our capabilities, yet he always remained grounded in the social fabric around us. The human world is one of shared meaning and of practices that bind us together; it is in the diverse shared experiences, in the transitions between events in which the enrichment of experiences emerges[36] and also the transactions with others.[37] We, perhaps, take these practices and links for granted, yet they form the fundamental glue in human behavior.

We come prepared with an evolved brain and set of cognitive predilections that are situated towards context, flexibility, and perceptual embodiment about objects that are conceptually rich and vital to behavioral adaptation of action, perception, and the brain. Core orientations to events and to kinds of objects underlie cephalic capabilities and behavioral coherence.

We come with a toolbox of core orientations that prepare us to learn, inquire, and theorize. The lenses through which we interpret the world and the inferences we make are based on an orientation to events. An orientation to objects is part of the adaptive specialization that has prepared us for action to recognize animate from animate objects, to locate objects in space and to anticipate them in time.

The range of hypotheses one can create are constrained by what Peirce called 'abduction' or 'retroduction' in hitting on the right idea about an event or object; where seeing an object as something already presupposed some context of understanding, amidst diverse contexts of

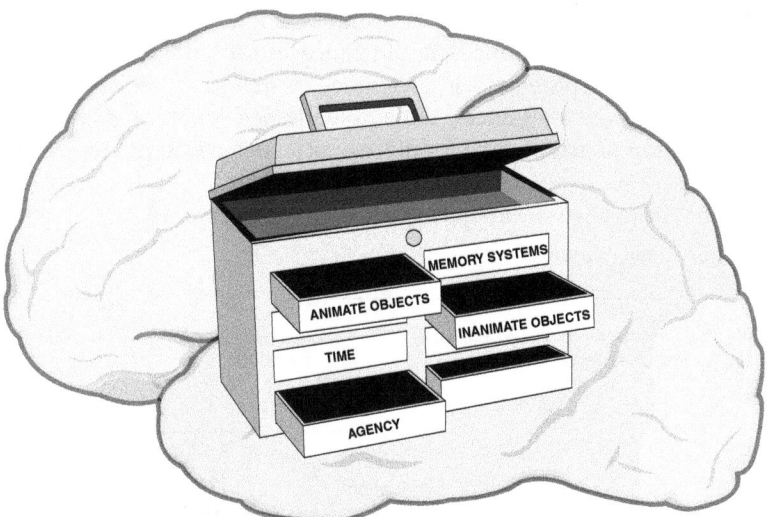

Figure 4.4 Toolbox of cognitive functions

practices that we presuppose and in which highlighting central tendencies about object relations, for instance, and inductive inferences are placed in a warranted context.

Abduction (that is, the genesis of an idea) is always constrained by context and ecology: foraging for coherent expression with others in a musical moment is constrained by cognitive capacity, individual competence, and that moment of pulling things together in the creation of something new amidst the resources available.

Our sensory capacity is keen to detect objects that afford sustenance or harm.[38] Fast forms for detecting information reflect diverse heuristics;[39] fast ways to solve problems, both specific and general, developed in the evolution of our brain. We come prepared to associate a number of events linked by causal building blocks in cephalic structures with worldly events. Ecological sensibility and rationality describe readily available heuristics well grounded in successful decision making.

Classical pragmatists understood this to mean that ecological rationality places decision making and the use of statistical features within cephalic predilections about numbers and representations of frequencies in real contexts.[40] Moreover, we lean less on memory being strictly in the head by contextual cues to which we scaffold our cephalic capabilities.

Part of Peirce's purpose was to establish warranted beliefs and legitimated habits of action, routines towards goals that serve diverse purposes. Science, like conduct, is established by expectations. These expectations are not reduced to simple sensations that legitimate the action; detecting noticeable difference in sensation is one thing, grounding all of epistemological legitimacy is quite another. And action is rich in cognitive resources; the automatic perception of events, the orchestration of action, vital for diverse social behaviors, has long been

Table 4.1 Peirce on inference and hypothesis

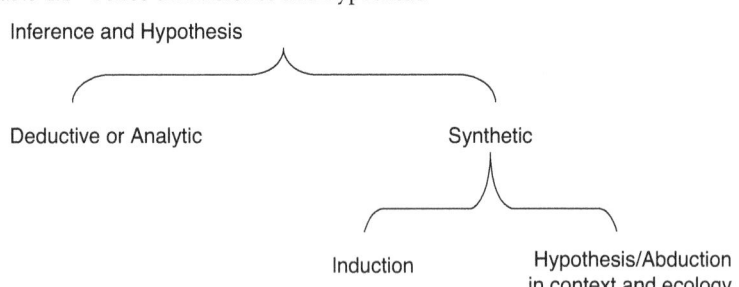

Source: Adapted from Peirce, 'Questions Concerning Certain Faculties Claimed for Man,' in *The Essential Peirce*, Vol. I, ed. Nathan Houser and Christian Kloesel (Bloomington, IN: Indiana University Press, 1868/1992).

noted. One exception is that we have re-envisioned the motor regions below the neocortex with regard to the codification of action as we have expanded our notion of cognitive systems; there are many diverse cognitive systems that underlie the organization of action.

Semiotics, Peirce thought, are never as precise as one might want; vagueness and generality are ingredients in the perceptual availability of everyday transactions. Signs that serve as icons, indices, and symbols permeate the cephalic social space. These are rich bearers of information in which the knower and the known are in contact. This is certainly true in music. There is no Cartesian space of separation in a fabricated rationalistic space, only engaged activity, the signs rich in scaffolding towards conceptual sanity and expansion.[41]

Forging links between events is one essential feature of our cognitive efforts. Semiotics is the connective glue that helps us to understand the links between an image or symbol and meaning.[42] Semiotics occur in an individual (for example, learning to read music) but are also transmitted culturally and play an important role in our ability to cooperate socially (for example, the meaning of traffic lights). A sense of semiotics, Peirce thought, is something 'virtual' in the bounds of information and human connectedness, building into a larger public space of meaning.[43] This is true whether we are talking about understanding, our sense of music, or our sense of inquiry about other species.

Thus, our inferences are constrained by an orientation to events and the kinds of objects that are detected. Linking mammals and finding what seem like counterexamples (such as the platypus, an egg-laying mammal) first requires a background of linking diverse kinds of events and knowledge, which may (perceptual) or may not (conceptual) have clear common properties. The taxonomic and thematic conditions may be simple or complex, but there is always a background condition. In diverse human societies, these cognitive events are apparent in taxonomic organization about basic objects (for example, plants, animals, and so forth).

In this context, representations are strengthened with what Dewey called 'funded knowledge.' While the term 'representations' is unfortunately often thought of as disembodied or divorced from the object, Dewey's notion is that a representation is an essential part of the object and our understanding of it. The representation becomes part of the fabric of the epistemological engines and coherence that underlie the way we engage the world. The notion of representation is not one thing, just like mind is not one thing as James long ago noted. Pragmatists look for functional relationships between mind rooted in action and a respect for human experience.

Thus, for instance, representations of others such as those that we care about, do not divide us from others. Instead, these internal representations guide the organization of cognition, which underlies the organization of action. This is important when we consider others, their beliefs, their desires, the way they are oriented, to what they are oriented, the tools they use, etc. These internal representations give us our understanding of the world around us and inform our interactions with the world.

Indeed, diverse cognitive adaptations, including our ability to predict the behaviors of others, are a function of the fact that we tag our fellow humans in terms of their beliefs and desires. These 'tags' are based on our observations and the integration of those observations into an internal representation of a person. This of course is a higher order cognitive function, and we use that adaptation in part to predict what other human beings do in our social world, as well as to predict their intentions.

An elaborate neural set of structures widely distributed throughout the brain is linked to keeping track of others, watching what they do, and getting a foothold in the world of approachable and avoidable events. Our brains are prepared to recognize animated objects, motion, and action. The detection of motion and our sense of being a causal agent are embedded in the concept of agency. Moreover, the detection of movement, knotted to intentionality, is an important means of discrimination in understanding others. The brain comes prepared to discern, or at least to try to discern, such relationships.

Pragmatism, experience and cognitive science

The modern era has ushered in an older perspective held by most of the classical pragmatists: bodily sensibility is an essential way in which we explore the world. Within the pragmatists' theoretical approach, there is the assumption that, through cephalic systems, a mind inhabits a body and the two work together. This outlook includes no Cartesian fallacies but rather says that cognitive systems work with our bodily sensations to interpret and understand the world around us. A core feature of the pragmatist tradition is the embodiment of problem solving in the mind/brain. This link shows us how our interactions with the external world are processed internally in order to problem solve. While also maintaining a healthy respect for the individual experiences that we have, pragmatism focuses on the way that those experiences, both internal and external, interact and create a deeper sense of understanding.

Like Leibnitz, Peirce was searching to understand 'thinking machines' in his original work in logic and mathematics.[44] His idea of cerebration and the brain and his tie to simple thinking machines are reminiscent of the first Turing machine. To move from Alan Turing and his envisioning of machine knowledge – in reality very low level computational devices – to the tiny micro chip-driven devices that we carry in our pockets in such a short time period is astonishing, but we are in some ways no wiser than either Peirce[45] or Turing,[46] the great British logician 20th century logician.

These machines needed to be embedded in the larger world of semiotics, of social embeddedness.[47] Inferences to better explanations must follow something like Peirce's deduction, induction, and hypothesis.[48] Peirce held that 'settlement of opinion is the sole object of inquiry.'[49] Of course, that is misleading, since inquiry is broader, but ideally, as a normative goal, finding a level of agreement about the investigation holds for most forms of human activity (for example, law). Peirce was in a perpetual state of modifying his list of categories[50] that like Kant[51] set the conditions for human inquiry.

Peirce understood that 'thinking is a species of the brain and cerebration is a species of the nervous action.'[52] He always noted that there are no precognitive events; for Peirce, degrees of cognitive systems underlie perception, attention, and action and hypothesis testing.

The theory of inquiry, of hypothesis testing, is rooted in this cognitive perspective; fixation of belief[53] is rooted in the organization of action. The orientation is not simply reactive, but anticipatory – and responsive to discrepancy with expectations. German cognitivism was part of Peirce's depiction of what he called 'Scottish common sense realism.' Indeed, Max Fisch, a scholar of Peirce, has noted that Immanuel Kant and Alexander Bain were particularly influential with regard to the origins of pragmatism.[54] Bain, less well known, had written an influential book in the early 19th century on the will, linking beliefs to a tendency and strength of action.

Indeed, we know that human organization is replete with anticipatory cognitive systems, most of which encompass the vast cognitive unconscious.[55] Action sequences are well orchestrated, embedded in successful survival for both short- and longer-term expression. When Peirce noted correctly that 'it was impossible to know intuitively that a given cognition is not determined by a previous one,'[56] he was close to recognizing that cognitive resources figure in the organization of action, something that Dewey would note in his critique on the 'reflex arc' some 30 years later.[57]

Of course, action is often habitual routine. Memory, attention, and other cognitive resources are minimized during diverse routines;

cognitive capabilities are then recruited elsewhere in the ongoing action. It is the breakdown that helps generate further action and cognitive resources to learn, attend, and construe new resolutions and new forms of adaptations, and part of the clarity to determine the diverse consequences of the new set of hypotheses determinant in human action in the long term.

Epistemology is rooted in nature. Since the categories are constrained by nature, the range of hypothesizes has an instinctual component – they are not infinite but finite. Peirce pulled up reason by linking it to action for consequences and instinct for some basic responses. It is interesting, again, that Peirce mixed writing these philosophical articles with items for wide-audience journals; he was rooted equally in the practical world, the rarified world of his father, and his roots in the Metaphysical Club at Cambridge.[58]

Keeping track of events is expanded by the scaffolding of the age of information and the cognitive revolution. This expansion is clearly strengthened by the same cataloging and quantification of pockets of information that underlie musical competence and composition.

Action and perspective

The ability of 'perspective taking' is an evolved central state. It is an active state in which we consider of the experiences of others, and it is tied to communicative competence, which is essential to human bonds and well-being. The information processing that entails this ability is a cognitive and behavioral achievement. While perception and action are represented in similar regions of the brain, they are not identical. When we learn of another person's experience, we are able to reflect upon our own experiences in order to make sense of another's internal perceptions.

Embedded in this is the commonsense assumption of something shared. Whether it is bodily experience or cognitive states embedded in the sensorimotor experiences, there is something innately human and understandable about these internal experiences. One could suggest quite reasonably that the imitative processes seen in neonates is an elementary form of perspective-taking made manifest and expanded from following others, to engaging others, learning from others, and eventually challenging others, generating hypotheses in an expanding, self-corrective process.[59]

Joint attention or gaze following is a fundamental adaptation demonstrated in a number of primates (like macaques). Diverse regions of the brain in primates, including neocortical regions and amygdala, are linked to gaze following. The temporal cortex, temporal pole, and amygdala are importantly involved in social contact and discernment of social meaning.[60] Gaze following may therefore be a precursor to perspective-taking, in which we observe another's outward interactions with the world in order to gain insight into their internal state.[61]

At a basic level, imitation is a vital way in which we get to the world through learning about others. Starting in neonates, imitation is easily incorporated into how we become connected to others. It is part of the roots of social bonding. Under the right conditions, human imitation presumably can become connected to prosocial sensibilities. Imitation is richly expressed by neocortical activation including frontal and parietal cortex as well as to striate and amygdala activation.

A cognitive resource is this ability to track others by what we think they desire and believe. This understanding is based on our tracking and integration of many behaviors that are simpler. For instance, we may observe what another person is looking at by joining eye contact on a common object. This roots people together in a coordinated fashion and is at the heart of pedagogy but also gives us information about the internal state of the person that we are observing.

We look at each other and closely watch what others do; facial expression – to mention just one form of appraisal and cephalic information processing – is endlessly rich. In our species, this trend contributed to our ability to make joint contact and mutual understanding. This ability allows us to combine the actions of two individuals into coherent social organization.

Figure 4.5 depicts some of the visual pictures from Darwin's book and a modern view of some of the anatomical sites involved in information processing of the face and social context from visual cortex to amygdala and visceral information (Emery and Amaral, 2000).

A very early evolutionary trend towards stereoscopic binocular vision is linked to the expansion and design of the visual system, and the evolution of the primate brain, which integrates visual information into almost all other systems. This visual system makes our object interactions possible and helps to shape our understanding of the world around us.

The visual cortex is essential for human beings, and projections to a region of the amygdala, linked to the formation of social attachment and social aversion, are key neural connections.[62] This link allows us to process information that we obtain visually directly to its implications for our social interactions. Our cephalic systems are bulging with

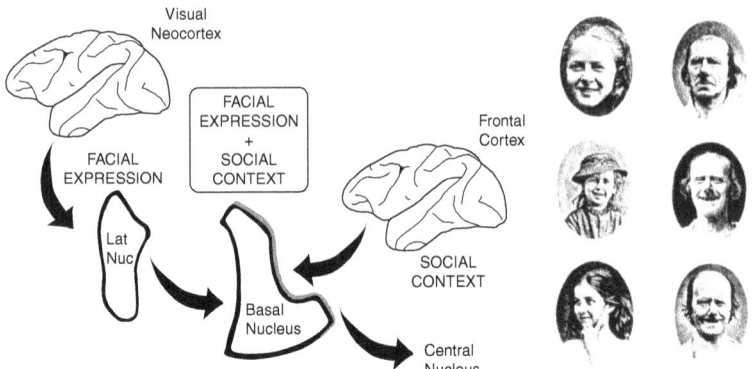

Figure 4.5 Facial recognition and the cortex

visual input, and it is transformed into vital pieces in the organization of action meaning.[63]

Facial expressions are vastly informative and rich in emotional content. Pragmatists like Dewey understood the informative nature of the emotions and linked them to learning, and learning is what we do best. The pedagogical predilection is vast – a human core capability. Moreover, Darwin,[64] like Dewey,[65] did not separate emotions and cognition. Neither have I. Emotional systems are forms of adaptation; consider, for instance, the importance of the immediate detection of facial expression for survival. The diverse forms of its bodily expression are strengthened with cognitive capability, rich in information transfer and processing.

Darwin, and certainly Dewey, understood that emotions are rich in cognitive functions and appraisal processes. There are diverse forms of appraisals and some of these, like music and faces, are affectively opulent. Moreover, function and cognitive processing are tied to an appreciation of nature.

Cognitive resources are rich in the generative processes within the expectations that surround, for instance, facial expression reflecting cognitive architecture, the generative processes, the diverse variation and embodiment of human meaning within almost all spheres of human expression.[66] The human face is a very informative piece of visual information and is, of course, just one amongst many others. But human meaning is tied to social contact:[67] the making of it, participating in it, remembering it, and sharing diverse forms of experience in vectors of meaning.

Cognitive systems and the brain

We now know that cognitive systems are not simply cortical but traverse all regions of the brain, something Dewey suggested in his 1896 paper 'The Reflex Arc Concept in Psychology'.[68] Others would also suggest that cognitive expression is endemic at all levels of the neural axis.[69]

Diverse forms of information processing are portrayed and are endemic to bodily functions. As we have come to understand, the brain is, in part, a cognitive organ. It deciphers information and organizes behavioral responses, projecting future possibilities and expressing a wide variety of behavioral options, all varying with the species and the environmental context. From bug detectors in the brainstem to semantic networks linked to the neocortex,[70] the brain is in the business of processing, utilizing, and organizing behavioral responses.

Earlier theories in the field identified thought with the cortex, with the idea that cognition or thought is non-reflexive. But cognition can surely be reflexive (syntax, spatial perception, probability judgment) in the same way that bodily responses can. Moreover, the brain is essentially involved in both lower-level and higher-level appraisal and information processing systems.[71]

Identifying cognition with the cortex is the traditional trapping of the mind/body split. Structures on the bottom (to speak metaphorically) of the brain are low level, reflexive, and brute; structures on the top of the brain are free and non-reflexive. The mistake inherent in this approach lies in the dominant view of bodily sensibility as dumb and inert. This error is perhaps best expressed in the context of the emotions, which are often characterized as only bodily changes.

This misleading and mistaken view held for James and was expanded by a number of investigators. James, it is well known, asserted that 'my heart beats fast; thus I know I am afraid.' According to James, the brain must interpret one's bodily state in order to make sense. For instance, when confronted by a bear, the bear is perceived, and my body acts accordingly by issuing responses that my mind interprets as fear. In a coordinated fashion, my mind then feeds information back to my body to ensure, under most conditions, that I can stay out of harm's way. I either stay immobilized or run fast.

Cognition, mistakenly, was placed on the mantle of the brain. In a beautiful, anatomically rich vision of the neural organization of basic behavioral functions, cognition is associated with the cortex.[72] This is shorthand, and cognition is also often treated as synonymous with consciousness. The routine assumption is that consciousness, cortex, and cognition are knotted together. But the thirst that generates a conscious cognitive problem-solving orientation is driven more from regions below the cortex,[73] and many of the routine appraisal systems (for example, where water is located, what time of day it might appear, what the water is associated with, in addition to the detection of water needs) are below the cortex. Appraisal systems in the brain that allow detection of signals (such as snake, harm, smiling faces, safety) are not necessarily accessible to consciousness. Moreover, the conceptions that give structure to the detection system can be incredibly low-grade; recognizing one taste over another (salt or sweet), this facial expression over that (happy or sad), and so on. The mechanisms are not conscious, they are mediated by brainstem structures, and they certainly can be reflexive.[74]

Cognitive systems reflect an orientation to events; they can be reflexive, as with face detection and imitation in neonates.[75] One feature of this evolution and successful adaptation is the panoply of information processing in the brain and the various lenses that we have for seeing the object.[76] Thus, information processing system networks, whether naturally or culturally derived, are richly coded in frameworks that allow inferences to be made easily. Again, it is important to be rid of the old pernicious split appearing between the mind (free and distant) and the body (reflexive and obligatory); reflexes do not render something non-cognitive.

Pragmatists like Peirce, James, Dewey, or Mead also noted, and importantly, that cognitive systems are endemic to motor control. Our sense of effort is tied to motor systems and to the cognitive systems that underlie the organization of action. Re-envisioning the motor system with regard to a consideration of, for example, attention and effort would suggest that there is no absolute separation of the motor system from the cognitive systems in the brain.

Cognitive systems cross the neural axis and appear to be more distributed than strictly hierarchical.[77] Labile systems and distributed neural networks have been redefined in terms of strict hierarchical levels.[78] But the forebrain is essential in the initiation of diverse forms of motivational behaviors[79] and particularly social behaviors. The brainstem coordinates the basic reflexes, and the forebrain orchestrates competing signals of importance in the organization of action.[80]

The idea of embodied cognitive systems requires two concepts: a sense of animacy and a sense of agency. These are key cognitive categories that underlie our sense of each other, ourselves, and our understanding of human history. The origins of our psychology are perhaps found in the fundamental distinction between animate and inanimate objects. Core knowledge, such as which of these two categories objects fall into, permeate the cognitive architecture including concepts about physical objects, causation, and orientation toward others based on their experiences, expectations, and intentions. The classificatory distinction of the living from non-living is a fundamental cognitive adaptation much expanded and developed. Categorical attributions of animacy and agency are dominant early but not totally unconstrained, and they are intertwined.[81] Both categories matter in determining the world around you. The cognitive architecture is linked to making sense of our world, of participating with others in practice and expertise.[82]

Lakoff and Johnson have nicely depicted such relationships between perception and action with such characterizations in the table below (Table 4.2).[83]

Cognitive adaptation is in the doing of things for coherence of action in complex social environments, from ecological conditions to social communicative functions. Thus the traditional view assumed that the cortex was the only cognitive part of the brain. Importantly, cognitive systems run across the brain, and from cortex to brainstem, the central nervous system is knotted to cognitively rich information processing resources. Diverse regions of the cortex are tied to motor function. There is an expanding notion of motor regions as rich with cortical functions. The broad repertoire of social perception and action is codified across diverse regions of the brain.

As Dewey noted in his critique of the reflex arc,[84] and as I have indicated already in my view, cognitive systems are pervasive in the organization of action and more generally linked to contexts in which the orchestration of the behavioral responses and enaction of human action are not strictly in the head.[85]

Within cephalic systems, human action and meaning are embedded in diverse regions of the brain, which include the basal ganglia, cerebellum, and motor cortex and the function of these chunks of organized

Table 4.2 Lakoff and Johnson on perception and action

Thinking is perceiving	Imagining as moving
Knowing is seeing	Attempting to gain knowledge is searching
Representing as doing	Becoming aware is noticing
Communicating as showing	Impediments to knowledge are impediments to vision
Searching as knowing	Knowing from a 'perspective' is seeing from a point of view

action expanded by suitable environments and well-orchestrated practices.[86] Moreover, the perceptual apparatus is linked to preprogrammed movements. The range of the underlying action codes for movement or actions that are expanded in use and action. They become coupled to habits, the cement of everyday actions for us. Sustaining acts are at the heart of agency.[87]

The brain generates movements and information processing systems underlie motor programs in the brain.[88] Endogenous motor groups are extended in capacity as more of the brain became involved in self-generated contextual motor programs, in which representations of planning and the control of action predominate central nervous function.[89]

While sensory systems are indeed separate from cognitive systems, they reflect cognitive systems in the brain. The child's world, while narrow in scope, is replete with information processing systems that are essential for its taking possession of its world, for organizing itself in relation to the objects that it might encounter. The child's world is not a 'blooming buzzing confusion,' as James suggested,[90] nor is the adult's. The scope of sensorimotor development is learning how to cope and how to adapt.

Imagined action and the brain

Modern brain imaging studies have revealed some interesting connections between action and perception and imagination. For instance, when subjects were asked to imagine grasping objects, significant activation of regions of the brain concerned with movement occurred. In further studies that used neuromagnetic methods to measure cortical activity, the primary motor cortex was active both when subjects observed simple movements and when the subjects performed them.[91] Of course, the motor cortex is activated in a wide array of human cognitive/motor activities. Importantly, motor imagery is replete with cognitive structure and is reflected in the activation of neural circuitry, so auditory imagery is reflected in different regions of the brain, including anticipatory musical imagery.[92]

In another study focusing specifically on sensory events in a functional magnetic resonance imaging scanner (fMRI), subjects were presented with spoken words via headphones. Then, in a second experiment, the same individuals were asked to identify the words with silent lip-reading.[93] Not surprisingly, many of the same cortical regions were activated.

In other words, hearing sounds is like imagining them. Both tasks recruit many of the same brain regions.[94] Across a number of perceptual experiences, imagining entails the neural systems involved in seeing them, hearing them, or touching them. Imagining a visual rotation, for example, versus actually looking at a rotating object, takes a similar time period. Moreover, very similar neural circuits are also activated when the object is imagined or viewed.[95] Imagining is the process of creating brain stimulation internally that is similar to what would be created by external stimulation. In other words, the neural structures that are active in imagining objects appear similar to those structures that are active when looking at the objects.

Not surprisingly, hearing music activates many of the regions linked to auditory perception and to music. Similarly, regions of the auditory cortex are also activated when subjects are asked to imagine music or other auditory stimuli (Figure 4.6).

Thus, despite the difficulty of, in the end, not knowing what people are actually imagining, one can dissociate hearing something from seeing it through diverse regions of the brain. Of course, the inverted spectrum, always humbling, reminds us that it is difficult to discern what we see and hear, as well as the corresponding relationships to what we report and observe behaviorally.

Of course, this is as it should be: why should there be an extra imaginative site? Action is generated by context mediated by the brain, but the brain is not in a vacuum. Central command units are not isolated but are informed by other units in the organization of action.

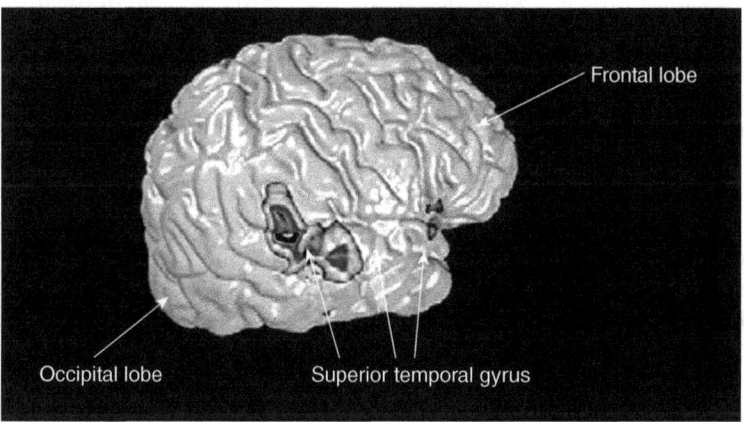

Figure 4.6 A scan revealing that in even in silence the auditory cortex is activated

Source: R. J. Zatorre, A. R. and Halpern, 'Mental Concerts: Musical Imagery and Auditory Cortex,' *Neuron* 47 (2005): 9–12.

Representations of goals are endemic in the motor systems, just as representations of action are endemic to the motor cortex.[96] In one experiment with macaques sitting in a chair, motor neurons are active when the macaque is shown movement but more so in selective neurons when the movements are organized in terms of goals. Grasping for an object or watching someone else grasp the object resulted in the activation of corresponding regions of the motor cortex.[97]

The region of the frontal lobe has been construed as one site for the organization of action, where goals and the perception of others are integrated. A semantic storehouse of actions is coded in the region. In humans, for example, researchers have demonstrated that this region is active when I watch you perform a well-rehearsed, intentional, goal-oriented action and also when I perform the action myself.[98]

Perhaps one the most interesting findings that has emerged in the organization of action is that sets of neurons in the motor cortex that are active when I watch you perform an action and when I perform the same action myself.[99]

Other regions of the brain, including the inferior parietal lobe, are active in the representation of action patterns, for example, in electrophysiological studies in macaques.[100] These neurons are also active both in viewing, and in anticipation of, the action. Thus neurons in both the premotor region and regions of the parietal lobe are active in action and observation.

Regions of the temporal lobe are responsive to detection of the direction of movement in another, whether the movement was intentional or not. Central representations of the movement, or the form of something with movement, is integrated along pathways essential in visual and auditory information processing.[101] Deciphering action in others, in addition to the organization of one's own actions, is a fundamental function of the brain and, for humans, our visual system.[102]

Thus neural networks designed to orchestrate the organization of action are recruited when observing and deciphering the actions of others. Regions of the brain are active when the child observes you perform an action and when the child imitates the action that she just observed you perform.[103]

Distinct sets of neurons in diverse regions of the cortex are active when one performs an action and when one watches others do so; this is pristinely shown in studies in macaques. That does not mean that there is not overlap in neurons that fire to mirroring others and in performing the action; it just is so that we come prepared to respond to others.

Tools, kinds, and cortex: neuropragmatism

Dewey[104] in particular, understood tool use as just an expansion of normal problem solving, part of the broader natural condition.[105] Part of what he had in mind is the fact that cephalic expansion set the stage for technological creations, expanding our sensory systems. Seeing by magnification became an evolving theme as our capacities were extended, and we turned from managing nature toward understanding nature.

Tool use and tool making helped make this shift possible. Tool making must have taken a dramatic leap in human primate ancestors around 2.5 million years ago. This was no trivial event; fine motor control and an expanded and extended use of the motor cortex and sub-motor cortical areas no doubt figured importantly in this evolutionary development.[106]

Cephalic expansion set the stage for technological creations, expanding our sensory systems and mnemonic capacities. Many species use diverse tools in adapting to their environment, and this reflects the evolution of cortical and subcortical systems that are important to tool use, tool making, and tool recognition. Tool use is an expression of an expanding cortical motor system in which cognitive systems are endemic to motor systems. This is a view of modern evolutionary trends and classical pragmatists such as Peirce and Dewey.[107]

Moreover, regions of the brain are prepared to recognize differences between different kinds of objects, one of which are mechanical tools. Importantly, frontal motor regions have been linked to the motor features of tool use.

Thus an expanded cognitive/motor system with diverse cognitive capacities was no doubt pivotal in our evolutionary ascent. It is not only the evolution of the cortex or brain that is knotted to social function; tool use and other diverse abilities are also important to our ability to thrive and to be social. The diverse and expanding cognitive systems are rich in motor regions of the brain and are fundamental to our evolutionary ascent. Tool use is an expression of an expanding cortical motor system in which cognitive systems are endemic to motor systems.

We are, of course, rooted in diverse contexts about social objects. The automatic perception of events, the orchestration of action vital for diverse social behaviors, has long been noted. Cognitive systems are often bodily in nature: struggling to learn something, persevering to acquire something, knowing as a contact sport with others, getting linked to others, enjoying the solitude of one's self enclosure amidst the safety of others, or despite others, forming boundaries of protective parlance.

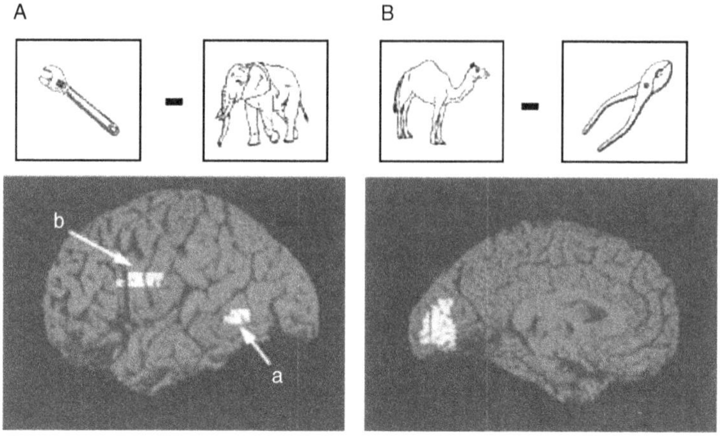

Figure 4.7 Kinds of objects and corresponding brain activation
Source: K. R. Gibson, and T. Ingold, ed., *Tools, Language and Cognition in Human Evolution* (Cambridge, UK: Cambridge University Press, 1993).

Neuropragmatism emphasizes anticipatory predictive responses that enhance evolutionary human capability; prediction of events[108] in active enactive and embodied systems easily sample events and forage for coherence in the organization of experience and action.

Tool use requires an evolving motor system that runs the gamut of the brain and enhances the cognitive/motor abilities that underlie the use of tools in perception and action, and enhance brain function. What emerged in us, ultimately, were improved methods of passing information to others, including future generations, something understood by an instrumental sense of human action and understanding.

Modern neuroscience has caught up with some of insights of the classical pragmatist: that brain action and perception are always linked, whether the action is real or imagined; that there is no no separation between action and perception, whether real or imagined; and that action and perception recruit the same brain regions whether real or imagined. Information processing, foraging for coherence, and sampling within a context of self-correction is at the heart of inquiry, as was envisioned by pragmatists like Peirce, James, Dewey, and Mead.

Neuropragmatism emphasizes no separation of cognition from motor expression, in which anticipatory and feed-forward circuits are predominant.

Conclusion

The classical pragmatists understood something about action, inquiry, and the brain; their goal, a modern one, was to understand cognitive adaptation, along with what John Dewey called 'lived experience.' They understood these events in the context of evolution of the brain. The evolution of cortical function represented an expansion of human capability and growth and thus part of their conception of inquiry.

We now know that cognitive systems are not simply cortical but traverse all regions of the brain, something Dewey suggested in his 1896 paper, 'The Reflex Arc Concept in Psychology,' and something that a number of others would suggest in terms of cognitive expression being endemic at all levels of the neural axis.

Any biologically grounded view of brain function understands that bodily sensibility is essential for the diverse ways in which we explore the world. Dewey, for instance, understood that cephalic (brain/body) capability – and not simply computational/representational systems that divorced one from the world – underlie our explorations of objects. We are oriented to kinds of objects in a world that we are trying to understand. Minds inhabit bodies in cephalic systems that explore the world – no Cartesian fallacies; cognitive systems are embodied in real life events – the stuff of adaptation, long noted by Darwin and James and incorporated by Dewey.

A core feature of the pragmatist tradition is the embodiment of problem solving in the mind/brain. The pragmatists undermine all forms of dualism. Indeed, cognitive/behavioral capacities are ways to engage the world, to compute probability and assess, for example, friendly or nonfriendly social events. We come prepared by neural circuits to respond to kinds of objects. Diverse cognitive systems (such as recognition of animate/inanimate objects, agents, senses of space, time and statistical relationships) underlie what Peirce called 'abduction,' or the genesis of ideas and problem solving. These biologically derived cognitive systems are not divorced from action or perception but are endemic to them and are distributed across neural networks in the brain.

In our transactions with others, we uncover theory and practice in everyday life in the sciences. These events are both existential and historical and can be tied to self-corrective methods of inquiry and hypothesis testing that inform a broad spectrum of human experience.

A root metaphor for pragmatists like Peirce is having a 'laboratory frame of mind,' which is at the heart of this endeavor. Peirce, an essential pragmatist thinker, established a philosophy of self-correction and set up the first experimental laboratory in psychophysics in America. He understood that all experience is embedded in practices that are rich with frames of reference. As we perceive the world around us, we presuppose background sets of inferences.

Our evolution is knotted to social groups working in unison across diverse terrains. Key abilities include discerning the wants and the desires of others (a core feature of our adaptations), along with cognitive adaptations such as recognizing the kinds of objects that are useful, affordable, or avoidable, coupled with a wide array of inhibitory capacities that contribute to social cooperative behaviors.

Moreover, long-term social bonds, plasticity of expression, and corticalization of function evolved in our species. And as our cortical visual functions increased dramatically, an evolutionary expansion of standing up, looking around, and making eye contact began in many primates. Human social contact, representation of objects, and use of objects are core cognitive capacities. Technology is an extension of ourselves, expanding what we are able to explore.

Stripping away the absolute abyss between cognition and motor systems and speaking in terms of degrees without absolute separation, the concept of will figures in our lexicon of understanding. In explaining people's wants and their capacity to satisfy their wants and beliefs, the concept of agency with that of the will emerges. The degree to which cognitive systems are part of the hardwiring of the motor systems embedded in the brain is apparent. Thus I suggest that a number of cognitive systems in the brain are not separate from motor programs that underlie behavioral adaptation. Rather, cognitive systems evolved as part of the organization of action,[109] but they are endlessly and effortlessly embedded in extended shared practices that bind us together.[110] Shared experiences and practices – what Peirce called 'frozen mind' – pervade our sense of the world as a shared one.[111] Such experiences and practices effortlessly scaffold our extended minds to which we tie action and perception in our embodied cephalic systems.[112]

But 'no ghosts' does not translate into the devolution of the voluntary into the involuntary. There is always a level of ambiguity inherent in these concepts. Ambiguity does not translate as not usable, not valid, and not empirical. We can distinguish contexts in which the motor systems orchestrate movements that are goal-directed, in which representations of goals permeate the motor systems in the organization of action. Looking at others and doing the same actions recruit many of the same circuits in the brain. Perhaps this cognitive capacity plays a role in anticipating the level of effort that the action requires.

One pragmatist who continually wrote about cooperative behaviors, an essential feature of our evolution and our intelligence, was John Dewey. Social behavior is rooted in our psychobiology, anchored to an individual who is social in nature. Moral experience is pervasive in our social nature. From it develops an ethics that sets the condition for a participatory democracy, which emerges with our capacity to work through endless conflicts. Cooperative behaviors were linked to a sense of human progress, grounded in a new sense of biology and human possibilities. Indeed we know that our evolutionary success is linked to social cooperative behaviors, something emphasized by both classical pragmatists and critical realists.[113]

Pragmatism is a consequential approach, up front and personal. A decision matters and is judged by consequences. Those consequences can be quite immediate and narrow, while consideration of generalities are perhaps longer term and have more encompassing consequences. The dilemmas of determinism are brought down to earth, to what bears directly on a radical empiricist. James's pragmatism, as he understood it, was broadly humanistic, with a tinge of effort to conserve and embrace a form of naturalism.

Dewey, as most of us pragmatists want to do, aimed to anchor a rich sense of human experience to the social context of civilized action, and also to anchor it to cephalic (that is, whole-brained) propensity embedded in a sense of objects. We are rooted in objects, as he often put it.[114] We are forged in communicative social contexts of adaptation while coping with diverse forms of precarious experience. One key element is to anchor the prosocial sensibility into adaptive, culturally, and socially bonded individuals. Diluting differences, encouraging engagement, and, as Dewey often noted, 'all human experience' have a social component.[115]

Notes

1. John Dewey, *Experience and Nature* (New York: Dover Press, 1925/1989).
2. Roy Wood Sellars, *Evolutionary Naturalism* (Chicago: Open Court Press, 1922).
3. William James, *Essays in Radical Empiricism* (New York: Longman, Green and Co., 1912/1958).
4. For example, James, *The Principles of Psychology* (New York: Dover Press, 1890/1952); Charles Sanders Peirce, 'Reasoning and the Logic of Things,' ed. K. L. Ketner and H. Putnam (Cambridge: Harvard University Press, 1899/1992); Dewey, *Experience and Nature*; George Herbert Mead, *Mind, Self, and Society: From the Standpoint of a Social Behaviorist* (Chicago: University of Chicago Press, 1934/1972).

5. S. Carey, *Conceptual Change in Childhood* (Cambridge, MA: MIT Press, 1985, 1987); F. Keil, *Semantic and Conceptual Development: An Ontological Perspective* (Cambridge, MA: Harvard University Press, 1979).
6. James, *The Principles of Psychology* and *Essays in Radical Empiricism*.
7. Dewey, *The Influence of Darwin on Philosophy* (Bloomington, IN: Indiana University Press, 1910/1965) and *Experience and Nature*.
8. James, *Essays in Radical Empiricism*, p. 57
9. Dewey, *Experience and Nature*, from the Preface.
10. For example, James, *Pragmatism: A New Name for Some Old Ways of Thinking* (New York: Longmans, Green and Co. 1907), and Dewey, *The Influence of Darwin on Philosophy*.
11. M. Wheeler, and A. Clark, 'Culture, Embodiment, and Genes: Unraveling the Triple Helix,' *Philosophical Transactions of the Royal Society B*, 373 (2008): 3563–3575; Alva Noë, *Action in Perception* (Cambridge, MA: MIT Press, 2004).
12. Dewey, *Experience and Nature*.
13. P. Rozin, 'The Evolution of Intelligence and Access to the Cognitive Unconscious,' in *Progress in Psychobiology and Physiological Psychology*, ed. J. Sprague and A. N. Epstein (New York: Academic Press, 1976); and Rozin, 'Evolution and Development of Brains and Cultures: Some Basic Principles and Interactions,' in *Brain and Mind: Evolutionary Perspectives*, ed. M. S. Gazzaniga and J. S. Altman (Strassbourg: Human Frontiers Science Program, 1998).
14. Dewey, 'The Reflex Arc Concept in Psychology,' *Psychological Review* 3 (1896): 357–370.
15. Dewey, *Experience and Nature*.
16. Jay Schulkin, *The Pursuit of Inquiry* (Albany, NY: SUNY Press, 1992).
17. J. J. Gibson, *The Senses Considered as Perceptual Systems* (New York: Houghton-Mifflin, 1966).
18. Mead, *Mind, Self, and Society*; J. Sabini, and J. Schulkin, 'Biological Realism and Social Constructivism,' *Journal for the Theory of Social Behavior* 24 (1994): 207–217.
19. G. W. Parrott, and J. Schulkin, 'Neuropsychology and the Cognitive Nature of Emotions,' *Cognition and Emotion* 7 (1993): 43–59.
20. Peirce, 'Deduction, Induction and Hypothesis,' *Popular Science Monthly*, 13 (1878): 470–482.
21. Dewey, *Experience and Nature*; and Noë, *Action in Perception*.
22. J. E. Smith, *Themes in American Philosophy* (New York: Harper and Row, 1970); and R. C. Neville, *The Cosmology of Freedom* (New Haven: Yale University Press, 1974).
23. Mark Johnson, *The Body in the Mind* (Chicago: University of Chicago Press, 1987/1990).
24. Chauncey Wright, 'Evolution of Self-Consciousness,' *The North American Review* 116 (1873): 245–310.
25. Dewey, *The Influence of Darwin on Philosophy*, p. 214.
26. P. Mellars, 'Why Did Modern Human Populations Disperse from Africa 60,000 Years Ago?' *PNAS* 103 (2006): 9381–9386.
27. For example, R. Foley, 'The Emergence of Culture in the Context of Hominin Evolutionary Patterns,' in *Evolution and Culture*, ed. S. C. Levinson and P. Jaisson (Cambridge: MIT Press, 2006).

28. Mellars, 'Why Did Modern Human Populations Disperse from Africa 60,000 Years Ago?'
29. H. M. McHenry, 'Human Evolution,' in *Evolution: The First Four Billion Years*, ed. M. Ruse and J. Travis (Cambridge: Harvard University Press, 2009); and R. Foley, 'In the Shadow of the Modern Synthesis?' *Evolutionary Anthropology* 10 (2001): 5–14.
30. S. Robson, and B. Wood, 'Hominin Life History: Reconstruction and Evolution,' *Journal of Anatomy* 212 (2008): 394–425.
31. R. I. M. Dunbar, and S. Shultz, 'Evolution in the Social Brain,' *Science* 317 (2007): 1344–1347.
32. Dewey, *Experience and Nature*.
33. Dewey, *Experience and Nature*; Jay Schulkin, *Cognitive Adaptation: A Pragmatist Perspective* (Cambridge: Cambridge University Press, 2009).
34. Dewey, *The Influence of Darwin on Philosophy*; Dewey, *Essays in Experimental Logic* (New York: Dover Press, 1916).
35. E. Herman, J. Call, M. V. Hernadez-Lioreda, B. Hare, M. and Tomasello, 'Humans Have Evolved Specialized Skills of Social Cognition,' *Science* 317 (2007): 1360–1366.
36. James, *The Principles of Psychology*; R. Rorty, *Philosophy and Social Hope* (New York: Penguin Books, 1999); C. Koopman, *Pragmatism as Transition* (New York: Columbia University Press, 2009).
37. Dewey, *Experience and Nature*.
38. Gibson, *The Senses Considered as Perceptual Systems*.
39. G. Gigerenzer, *Adaptive Thinking: Rationality in the Real World* (New York: Oxford University Press, 2000); Gigerenzer, *Gut Feelings* (New York: Viking Press, 2007).
40. Gigerenzer, *Adaptive Thinking*; Andy Clark, *Being There* (Cambridge: MIT Press, 1997).
41. M. Donald, *Origins of the Modern Mind* (Cambridge: Harvard University Press, 1991); Noë, *Action in Perception*; Noë, *Varieties of Presence* (Cambridge: Harvard University Press, 2012); M. Wheeler, *Reconstructing the Cognitive World* (Cambridge: MIT Press, 2005); and Clark, *Being There*.
42. C. Morris, *The Pragmatic Movement in American Philosophy* (New York: George Braziller, 1970).
43. A. Schütz, and T. Luckmann, *The Structures of the Life-World* (Evanston, IL: Northwestern University Press, 1973).
44. See M. H. Fisch, 'Evolution in American Philosophy,' in *Peirce, Semiotic and Pragmatism* ed. K. L. Ketner and J. W. Kloesel (Bloomington, IN: Indiana University Press, 1986).
45. Peirce, 'Deduction, Induction and Hypothesis.'
46. Alan Turing, *The Essential Turing* (Oxford: Clarendon Press, 2004).
47. See also Clark, *Being There*; Noë, *Action in Perception*; Donald, *Origins of the Modern Mind*.
48. Peirce, 'The Fixation of Belief,' *Popular Science Monthly* 12 (1877): 1–15; Peirce, 'Logic, Chapter 1,' in *Writings of C. S. Peirce*, Vol. 4 (Bloomington, IN: Indiana University Press, 1880).
49. Peirce, 'Questions Concerning Certain Faculties Claimed for Man.'
50. J. Buchler, *Charles Peirce's Empiricism* (London: Routledge, 1939/2000).

51. Immanuel Kant, *Critique of Practical Reason*, trans. L. W. Beck (New York: Bobbs Merrill Company, Inc., 1788/1956).
52. Peirce, 'Logic, Chapter 1,' p. 352.
53. Peirce, 'The Fixation of Belief.'
54. M. A. Fisch, 'Alexander Bain and the Genealogy of Pragmatism, *Journal of the History of Ideas* 15 (1954): 413–444; and Fish, 'Evolution in American Philosophy.'
55. Rozin, 'The Evolution of Intelligence and Access to the Cognitive Unconscious,' C. R. Gallistel, *The Organization of Learning* (Cambridge: MIT Press, 1993).
56. Peirce, 'Questions Concerning Certain Faculties Claimed for Man.'
57. Dewey, 'The Reflex Arc Concept in Psychology.'
58. J. Brent, *Charles Sanders Peirce* (Bloomington, IN: Indiana University Press, 1993); Louis Menand, *The Metaphysical Club* (New York: Farrar, Straus and Giroux, 2001).
59. A. Gopnik, and A. N. Meltzoff, *Words, Thoughts, and Theories* (Cambridge, MA: MIT Press, 1997).
60. N. J. Emery, and D. G. Amaral, 'The Role of the Amygdala in Primate Social Cognition,' in *Cognitive Neuroscience of Emotion* ed. R. D. and L. Nadel (New York: Oxford University Press, 2000).
61. M. Tomasello, *The Cultural Origins of Human Cognition*) Cambridge, MA: Harvard University Press, 1999).
62. D. C. Van Essen, 'Corticocortical and Thalamocortical Information Flow in the Primate Visual System,' *Progress in Brain Research* 149 (2005): 173–181.
63. Emery and Amaral, 'The Role of the Amygdala in Primate Social Cognition'; and Van Essen, 'Corticocortical and Thalamocortical Information Flow in the Primate Visual System.'
64. Charles Darwin, *The Expression of Emotions in Man and Animals* (Chicago: University of Chicago Press, 1872/1965).
65. Dewey, 'The Theory of Emotion,' *The Psychological Review,* 2 (1895): 13–32.
66. Darwin, *The Expression of Emotions in Man and Animals*; P. Ekman, 'Universals and Cultural Differences in Facial Expressions of Emotion,' in *Nebraska Symposium on Motivation, 1971*, ed. J. Cole (Lincoln: University of Nebraska Press, 1972).
67. K. Jaspers, *Way to Wisdom* (New Haven: Yale University Press, 1951/1954).
68. Dewey, 'The Reflex Arc Concept in Psychology.'
69. R. A. Barton, 'Binocularity and Brain Evolution in Primates,' *PNAS* 101 (2004): 10113–10115; Schulkin, *Adaptation and Well-Being: Social Allostasis* (Cambridge: Cambridge University Press, 2011).
70. A. Martin, 'Organization of Semantic Knowledge and the Origin of Words in the Brain,' *The Origin of Diversification of Language,* 24 (1998): 69–87.
71. For example, M. S. Gazzaniga, *The New Cognitive Neurosciences* (Cambridge, MA: MIT Press, 1995/2000); L. W. Swanson, 'Cerebral Hemisphere Regulation of Motivated Behavior,' *Brain Research* 886 (2000): 113–164; Swanson, *Brain Architecture* (Oxford: Oxford University Press, 2003).
72. Swanson, 'Cerebral Hemisphere Regulation of Motivated Behavior,' and *Brain Architecture.*
73. D. Denton, *The Hunger for Salt* (Berlin: Springer-Verlag, 1982).

74. K. C. Berridge, 'Motivation Concepts in Behavioral Neuroscience,' *Physiology and Behavior* 81 (2004): 179–209.
75. A. N. Meltzoff, and M. K. Moore, 'Imitation of Facial and Manual Gestures by Human Neonates,' *Science* 198 (1977): 75–78.
76. N. R. Hanson, *Patterns of Discovery* (Cambridge: Cambridge University Press, 1958/1972); Gigerenzer, *Adaptive Thinking*.
77. See C. R. Gallistel, *The Organization of Action: A New Synthesis* (Hillsdale, NJ: Lawrence Erlbaum, 1980); Berridge, 'Motivation Concepts in Behavioral Neuroscience'; G. G. Berntson, and J. T. Cacioppo, 'From Homeostasis to Allodynamic Regulation,' in *Handbook of Psychopathology*, ed. J. T. Cacioppo, L. G. Tassinary, and G. G. Berntson (Cambridge, UK: Cambridge University Press, 2000).
78. Barton, 'Binocularity and Brain Evolution in Primates.'
79. Berridge, 'Motivation Concepts in Behavioral Neuroscience.'
80. Gallistel, *The Organization of Action*, and *The Organization of Learning*.
81. Carey, *Conceptual Change in Childhood*, and Schulkin, *Cognitive Adaptation*.
82. Noë, *Action in Perception*, and *Varieties of Presence*.
83. George Lakoff and Mark Johnson, *Philosophy in the Flesh* (New York: Basic Books, 1999).
84. Dewey, 'The Reflex Arc Concept in Psychology.'
85. Hilary Putnam, *Realism With a Human Face* (Cambridge, MA: Harvard University Press, 1990); Putnam, *The Collapse of the Fact/Value Distinction* (Cambridge, MA: Harvard University Press, 2000); Maurice Merleau-Ponty, *The Structure of Behavior* (Boston: Beacon Press, 1942/1967); and Noë, *Action in Perception*.
86. A. M. Graybiel, 'The Basal Ganglia and Chunking of Action Repertoires,' *Neurobiology of Learning and Memory* 70 (1998): 119–136; M. Donald, 'Hominid Enculturation and Cognitive Evolution,' in *The Development of the Mediated Mind* ed. J. M. Luraciello, J. A. Hudson, R. Fibush, and P. J. Baver (Mawash, NJ: Erlbaum Press, 2004).
87. Schulkin, *Action, Perception and the Brain* (London: Palgrave Macmillan, 2012).
88. R. R. Linas, *I of the Vortex* (Cambridge, MA: MIT Press, 2001).
89. Gallistel, *The Organization of Action*.
90. Jerome Kagan, *The Nature of the Child* (New York: Basic Books, 1984); and Carey, *Conceptual Change in Childhood*.
91. M. Jeannerod and V. Frak, 'Mental Imaging of Motor Activity in Humans,' *Current Opinion in Neurobiology* 9 (1999): 735–739.
92. G. Rizzolatti, and G. Luppino, 'The Cortical Motor System,' *Neuron* 31 (2001): 889–901.
93. G. A. Calvert, E. T. Bullmore, M. J. Brammer, R. Campbell, S. C. R. Williams, P. K. McQuire, P. W. R. Woodruff, S. D. Iverson, and A. S. David, 'Activation of Auditory Cortex During Silent Lip-Reading,' *Science* 276 (1997): 593–596.
94. R. J. Zatorre, 'Neural Specializations for Tonal Processing,' *Annals of the New York Academy of Sciences* 930 (2001): 193–210.
95. M. S. Cohen, S. M. Kooslyn, H. C. Breitter, G. J. DiGirolamo, W. L. Thompson, A. K. Anderson, S. Y. Bookheimer, B. R. Rosen, and J. W. Belliveau, 'Changes in Cortical Activity During Mental Rotation: A Mapping Study Using Functional MRI,' *Brain* 119 (1996): 89–100.

96. Rizzolatti and Luppino, 'The Cortical Motor System.'
97. Jeannerod and Frak, 'Mental Imaging of Motor Activity in Humans'; Rizolatti and Luppino, 'The Cortical Motor System.'
98. M. Jeannerod, *The Cognitive Neuroscience of Action* (Oxford: Blackwell Publishers, 1997); J. Decety, and P. W. Jackson, 'A Social Neuroscience Perspective on Empathy,' *Current Directions in Psychological Science* 15 (2006): 54–58.
99. Jeanearod, *The Cognitive Neuroscience of Action*.
100. D. Perrett, and A. Mistlin, 'Perception of Facial Characteristics by Monkeys,' in *Comparative Perception, vol 2: Complex Signals*, ed. W. Stebbins and M. Berkeley (New York: Wiley, 1990); V. Gallese, L. Fadiga, L. Fogassi, and G. Rizzolatti, Action Recognition in the Premotor Cortex, *Brain*, 119 (1996): 593–609.
101. R. Desimone, 'Neural Mechanisms for Visual Memory and Their Role in Attention,' *Proceedings of the National Academy of Sciences*, 93 (1996): 13494–13499.
102. D. Milner, and M. A. Goodale, *The Visual Brain in Action* (Oxford: Oxford University Press, 1995).
103. Decety and Jackson, 'A Social Neuroscience Perspective on Empathy'; A. N. Meltzoff, 'The Case for Developmental Cognitive Science: Theories of People and Things,' in *Theories of Infant Development*, ed. G. Bremmer and A. Slater (Oxford: Blackwell, 2004), pp. 145–173; A. N. Meltzoff, '"Like me:" A Foundation for Social Cognition,' *Developmental Science* 10 (2007): 126–134.
104. Dewey, *Experience and Nature*.
105. L. Hickman, *Philosophical Tools for Technological Culture* (Bloomington, IN: Indiana University Press, 2001).
106. Foley, 'The Emergence of Culture in the Context of Hominin Evolutionary Patterns.'
107. Ibid.
108. Peirce, 'Reasoning and the Logic of Things'; Clark, *Being There*.
109. Dewey, 'The Reflex Arc Concept in Psychology'; Gallistel, *The Organization of Action*.
110. Donald, *Origins of the Modern Mind*; Clark, *Being There*.
111. Dewey, *Experience and Nature*; Mead, *Mind, Self, and Society*; A. Schütz, *The Phenomenology of the Social World*, trans. G. Walsh and F. Lehnert (Evanston, IL: Northwestern University Press, 1932/1967).
112. Clark, *Being There*.
113. Dewey, *Experience and Nature*.
114. Ibid.
115. Dewey, *Experience and Education* (New York: Collier, 1938/1973).

Part II

Cognition, Emotion, and the World

5
The End of the Debate over Extended Cognition

Jeffrey B. Wagman and Anthony Chemero

Introduction

One of the more lively current debates in philosophy of cognitive science and philosophy of mind is over the possibility of extended cognition. Recently, though, this debate has hit a dead end, with proponents and opponents agreeing that whether cognition is sometimes extended is an empirical matter, but not knowing what an empirical demonstration of extended cognition would look like.

In very rough outline, which is all that is necessary for current purposes, the philosophical debate over extended cognition has gone as follows.[1]

PRO: Thought experiments, and some real experiments, show that cognitive systems are not in principle bound by skull and skin.[2]

CON: The inferences from the thought experiments and real experiments to the claim that cognition is extended are fallacious. Things outside the skin and skull are not of the right sort to be parts of cognitive systems.[3] These (thought) experiments merely show that cognitive systems are situated, embedded, and/or embodied.[4]

PRO: The arguments against extended cognition are mere armchair philosophizing, which make it impossible in principle for cognition to be extended. This sort of argument should not be taken as telling against a contingent empirical claim.[5]

CON: The arguments against extended cognition are not armchair philosophizing. Whether cognition is extended is a contingent, empirical claim. As a matter of contingent empirical fact, cognition is not extended.[6]

What this rough sketch reveals is that, although there is still open dispute over whether cognitive systems are extended, even the foes of extended cognition agree that whether there are extended cognitive systems is an empirical matter. This can be confirmed in these quotes from critics of extended cognition.

> In the end, empirical research should decide this question: we should commit resources to the framework of extended cognitive systems, apply the extended view in the study and the lab, and see whether doing so generates a flourishing research program in cognitive science. It is very difficult to predict the future of science; matters might work out in favor of extended systems. There are, however, reasons for pessimism.[7]
>
> Our view has always been that, as a matter of contingent empirical fact, pencils, papers, eyeglasses, Rolodexes, and so forth happen not to be parts of any currently existing cognitive economy. It is, in principle, possible that, say, a pencil or notebook could be so employed as to be a contributor to a cognitive process.[8]

This state of play, with everyone agreeing that the claim that whether cognition is extended is an empirical matter, would seem to leave open the possibility of a scientific resolution to the debate. That is, if experiments designed to detect the presence of extended cognition do in fact detect extended cognition, this would seem to constitute empirical confirmation of the existence of extended cognition. Should this happen, it would seem that the debate over extended cognition would simply be over.

In what follows, we will argue that recent experimental evidence demonstrates the existence of extended cognition. Doing so will require careful description of the methodological details of experiments that show that cognition is extended. Having seen this evidence and claiming that it is an empirical matter whether extended cognition is possible, opponents of extended cognition should relent. Their failure to do so would indicate that they do not in fact take this to be an empirical matter.

Ecological psychology and dynamical modeling

In the third section, we will present voluminous evidence that cognition is extended. Nearly all of this evidence is drawn from a loosely organized research program, what Chemero calls 'radical embodied cognitive science.'[9] In this section, we briefly describe this research program in order to say how participants in it understand cognition, and what they take it to mean for cognition to be extended.

The roots of radical embodied cognitive science go back to the beginning of psychology in the United States. William James, in his attempt to bring Darwinian thinking to psychology, took the mind to be a tool for adapting animals to their environments. This functionalist approach took the object of the psychological sciences to be the animal as a whole in its interaction with its environment. The culmination of this approach is seen in the radical empiricism of the later James, in which he argues that the act of perception includes both the perceiver and the perceived.[10] As far as we know, this is the first articulation of the basic extended cognition view. Although contemporary psychology has little to do with Jamesian ideas, especially the later ones, radical empiricism did influence E.B. Holt and, later, James Gibson.[11] Gibson developed this idea into a full-fledged theory of perception, action, and cognition. Gibson's ecological psychology[12] is the center of today's radical embodied cognitive science. According to ecological psychology, perception is direct in that it does not involve internal maps, images, or representations of the environment. Instead, perception is an unmediated relation between the animal and the environment. Gibson argues that this sort of unmediated relation to the environment is sufficient for guiding action because the environment contains affordances, or opportunities for behavior, and information sufficient to specify them.

This is a very strong claim, and one that contradicts the widely held belief in the poverty of the stimulus. For example, the traditional story goes, the retinal image of a large, distant car would be identical to that of a small, nearby car. To determine whether we can cross the street before the car gets to us, then, we have to use information from memory concerning the typical size of cars and from this, (unconsciously) compute how distant the car is. Then we would need to store the distance of the car, recalculate the distance some particular time later, use the difference between these two distances to calculate the car's velocity, and then use the distance and velocity to determine when the car will get to where we are. Gibson claims that it only seems that we need all this unconscious calculation because of our mistaken belief that vision begins with a retinal image. Seeing, according to Gibson, is an action, something that animals do, not just with their eyes, but also with their moving eyes on a face on a head on a neck on a torso on legs. Among the many, many possible motions that an animal can take as part of seeing, one is especially relevant here: swaying. As a human stands still, she sways slightly forward and backward. Swaying while looking at a car yields a regular visible change in the area of background that the car is occluding, and indeed this change is lawfully related to the distance of the car, with the change in occlusion of nearby objects being greater than the change in occlusion of distant objects. So for an animal that moves, even by more or less involuntarily swaying, there is information available that is lawfully related to the distance of objects. Moreover, in the case we are imagining, the car is also moving. David Lee and his colleagues[13] have shown that the optical variable tau, the ratio of the apparent size of an approaching object to the rate of change of the apparent size the object, is available in the light and is lawfully related to the time to contact with the approaching object. So, because animals and objects in the environment move during the course of perceiving, there is sufficient information in the light to specify both distance and time to contact, and an animal can simply see, without calculation, whether it can cross the street. The same goes for many, many other affordances. There is information available in the environment for perceiving and acting upon these affordances, and this information can be picked up and acted upon without internal computations.

Notice that picking up this information involves moving and change over time, which makes dynamical systems theory an appropriate modeling tool for modeling perception and action. And, indeed, dynamical modeling has been used by ecological psychologists for several decades.[14] In dynamical systems modeling, one uses the tools of calculus to explain the change over time of some system of variables.[15] Two features of dynamical modeling will be important in what follows. First, the locations of the quantities tracked by the variables in dynamical models are not constrained to be on just one side of the skin-environment boundary.[16] Second, as will be argued below, certain kinds of dynamical systems are non-decomposable, in that the equations that model them cannot be solved separately. This is best seen in a description by Beer.[17] The figure depicts a coupled agent-environment system, whose parts, the agent A and the environment E are modeled with the following equations:

$$X_A = A(X_A; S(X_E))$$
$$X_E = E(X_E; M(X_A))$$

Put in English, equation 1 says that the changes in the agent (dA/dt) are a function of the current state of the animal (A) and the agent's sensing of the environment (S(E)); equation 2 says that the changes in the environment (dE/dt) are a function of the current state of the environment (E) and movement of the animal (M(A)). These equations are coupled in that each has a variable that is a parameter in the other. Therefore, the equations cannot be solved separately. Beer takes this as indicating that the proper object of study in a system like this is U, the agent-environment system. Notice, of course, that Beer's equations here are not a real dynamical model of an actual system. However, similar claims about the non-decomposability of actual coupled agent-environment system have been made by many others.[18]

This non-decomposability is a key feature of extended cognition, which we will now define. A cognitive system is extended whenever it is in part constituted by things outside the biological body. To say that a cognitive system is in part constituted by things outside the biological body is to claim something stronger than that these things causally impact or enable cognition. De Jaegher, Di Paolo and Gallagher[19] make distinctions that are useful here. A *contextual factor* in some cognitive activity is some variation which causes variations in the cognitive activity; an *enabling factor* in cognition is something without which the cognitive system could not operate; a *constitutive element* in a cognitive system is actually part of the processes of the cognitive system. Words on a page or monitor are contextual factors, in that different words would lead to different cognitive processes; oxygen is an enabling factor, in that cognitive systems (at least cognitive systems like ours) could not operate in its absence. Neither of these are constitutive elements in cognitive systems. To claim that cognitive systems are extended is to say that things outside the body constitute them in part. In the next section, we pile on evidence that cognitive systems are sometimes extended in exactly this sense.

Empirical research on extended cognition

In this section, we discuss several cases of empirical research gathered by radical embodied cognitive scientists. Collectively, they provide substantial evidence that cognitive systems, sometimes at least, extend beyond the skin. The first set of results show that cognitive systems are sometimes extended from the perspective of a person controlling his or her action. The second set of results shows that cognitive systems are sometimes extended from the point of view of scientists mathematically modeling cognitive systems. The research described in the first and second sub-sections concerns affordances, a notion that we introduce here.

In order to successfully perform an intended behavior (for example, retrieving an object from a shelf), perceiver-actors must be able to perceive whether that behavior is possible, and (if so) how to control their movements such that this possibility is realized. Such possibilities for behavior are known as affordances and are determined by the fit between a perceiver-actor's action capabilities and behavior-relevant properties of the environment.[20] Therefore, perceiving affordances means perceiving the environment in terms of this relationship. Empirical work on perception of affordances has shown that animals do, in fact, perceive the environment in these terms. For example, perception of whether doorways (of varying widths) can be passed through is relative to the relationship between the width of the doorway and the perceiver-actor's ability to fit through the opening.[21] This ability is constrained, in part, by the perceiver-actor's widest horizontal body dimension. The perceptual boundary on this behavior (that is, defined as the smallest aperture that is perceived to be pass-through-able) occurs at a wider doorway width for perceivers with wide shoulders than for perceivers with narrow shoulders. However, the perceptual boundary occurs at the same doorway-width-to-shoulder-width ratio for both groups of perceivers (that is, a ratio of approximately 1.2).[22] Importantly, when humans and other animals actually attempt to pass through doorways, the behavioral boundary occurs at a similar ratio (Warren and Whang, 1987, see also Ingle and Cook, 1977).

Perception of affordances *for* the person-plus-object system

The action capabilities of any perceiver-actor are in continual flux. Action capabilities can change over long time scales (on the order of months or years) by means of developmental changes in strength, coordination, and flexibility (Adolph, 2008; Konczak, Meeuwsen, and Cress, 1992). Action capabilities can also change over short time scales (on the order of seconds or minutes) by means of a change in posture (e.g., from sitting to standing), locomotion speed (from walking to running), or by means of attaching an object to the body. Attaching an object to the body (such as a handheld tool, backpack, or wheelchair) creates an integrated person-plus-object system with action capabilities and control dynamics that are entirely different from those of the person without the attached object (Wagman and Taylor, 2004). Thus, controlling the person-plus object system is tantamount to manually controlling a novel dynamic system.[23] Perceiving what behaviors are possible for the person-plus-object system means perceiving the world in relation to this integrated, dynamic system.

In some cases, attachments to the body change a perceiver-actor's action capabilities by changing their geometric properties. For example, a large carried object hinders the ability to pass through narrow spaces. Wagman and Taylor[24] conducted a series of experiments investigating whether perception of affordances for carrying an object through a doorway was relative to (the width of) the person-plus-object system. In the first experiment in this series, participants held T-shaped objects ranging in width from 50 to 140 cm such that the branches of each object extended out to their sides. In addition, they wore specially designed goggles that occluded their view of the object but enabled them to see a 90 cm wide doorway that was several meters away. On a given trial, they held one of the objects at their waist and reported (yes or no) whether they would be able to carry that object through the doorway without rotating their shoulders or the object.[25] The perceptual boundary on this behavior (here defined as the largest ratio of doorway-width-to-object-width that was perceived to be pass-through-able) occurred at a ratio of approximately 1.0. In an additional experiment in this series, participants held pairs of rods at their sides – one in each hand – such that each rod pointed out to one side of their body. Again, they wore the specially designed goggles and reported whether they would be able to carry the objects through the doorway without turning their shoulders (or excessively rotating the objects). In contrast to the first experiment, in this experiment, both the shoulder width of the participant and the width of the objects contributed to the width of the person-plus-object system. Therefore, participants could be divided into wide and narrow shouldered groups, and the perceptual boundaries of each group of participants could be compared as in the earlier work of Warren and Whang.[26] As expected, the perceptual boundary on this behavior occurred at wider object widths for perceivers with narrow shoulders than for perceivers with wide shoulders but occurred at the same doorway-width-to-person-plus-object ratio for each group (that is, a ratio of approximately 1.0).

The work by Wagman and Taylor (2005a) shows that perception of whether an object can be carried through a doorway is relative to the width of the person-plus-object system both when the when the width of the person contributes to the width of the system and when it does not. Moreover, the perceptual boundaries on this behavior are analogous to those for perception of whether a doorway can be passed through when the perceiver-actor is not carrying an object.[27] Importantly, perceiver-actors also produce analogous patterns of locomotion (movement velocity, relative body position at center of the doorway) when they attempt to pass through a doorway while carrying an object as when they attempt to do so without doing so.[28] Thus, both in terms of perception and action, objects attached to the body that change a perceiver-actor's geometric properties seem to be experienced (and treated behaviorally) as extensions of the body.

Attachments to the body not only change action capabilities by changing the perceiver-actor's geometric properties, they also do so by changing the perceiver-actor's dynamic properties. In general, a heavy backpack hinders the ability to walk long distances or stand on an inclined surface. Accordingly, Bhalla and Proffitt[29] found that distances appear farther and hills appear steeper to participants who are wearing heavy backpacks than to those who are not. Such findings provide preliminary evidence that attachments to the body are perceived as part of the body. However, given that such perceptual experience is only valuable to a perceiver-actor to the extent to that it influences decisions about whether (and how) to perform a given behavior, such evidence would be strengthened by findings that such a manipulation also influences perception of affordances of inclined surfaces. Wagman and colleagues[30] investigated this hypothesis. Specifically, they investigated whether affordances for standing on an inclined surface are perceived in relation to dynamic capabilities of the person-plus-object system. A heavy backpack not only increases the mass of the perceiver-actor, it also changes the *mass distribution* of the perceiver-actor. Importantly, though, not all such changes in a perceiver-actor's mass distribution will hinder the ability to stand on an inclined surface. For a person facing (or ascending) an inclined surface, wearing a heavy backpack on the back (as one would normally do so) shifts the center of mass away from the surface (that is, it makes that person 'back heavy'). Such a change compounds the pull of gravity and serves as a destabilizing force for the perceiver-actor, making it harder to stand on the surface. However, wearing a backpack on the front shifts the center of mass toward the surface (that is, it makes that person 'front heavy'). Such a change *counters* the pull of gravity and thus serves as a *stabilizing* force for the perceiver-actor, making it easier to stand on the surface. Malek and Wagman[31] investigated whether perceiver-actors are sensitive to how such changes in their mass distribution would influence their ability to stand on an inclined surface. On a given trial, participants viewed a surface at one of seven different inclinations (15° to 45° in 5° increments) and reported (yes or no) whether they would be able to stand on that surface without bending their waist or knees. They performed this task while (a) wearing a backpack loaded with 15% of their body weight on their back, (b) wearing the same backpack on their back, or (c) not wearing a backpack. After the perceptual task, participants attempted to stand on a surface at increasing angles of inclination in each of these conditions.

The perceptual boundary (here defined as the steepest inclination that was perceived to be stand-on-able) occurred at a steeper angle of inclination when participants wore the backpack on their front (26.8°) than when they wore it on their back (23.9°). Importantly, behavioral boundaries (defined as the steepest inclination that was actually able to be stood on) also occurred at a steeper inclination when they wore the backpack on the front than when they wore it on their back.[32] In other words, perception of whether the inclined surface could be stood on reflected the action capabilities of the integrated person-plus-object system. Again, both in terms of perception and action, objects attached to the body that change a perceiver-actor's dynamic properties seem to be experienced (and treated behaviorally) as extensions of the body.

Such effects, of course, are not limited to objects that are merely transported by a perceiver-actor. They also occur with objects that are (or could be) used to accomplish an intended behavior, such as tools or other implements. When a person uses a handheld tool, for example, they must perceive and act on affordances of an integrated person-plus-tool-system. Successfully doing so requires prospectively adjusting postures and movements of the body based on both the controllability of this integrated system and the constraints of a given task. Bongers, Michaels, and Smitsman investigated this ability.[33] They asked participants to displace a target with the distal end of a handheld rod. The researchers manipulated the length, mass, and mass distribution of the rod as well as size of the target to be displaced (that is, the precision constraints of the task). Participants held the rod at a 45° angle as they walked toward the target. They were instructed to come to a complete stop when they felt that they could most comfortably displace the target with the tip of the rod, lower the rod, and then use tip of the rod to horizontally displace the target. The researchers measured a number of variables including the stopping distance from the target and the positions of the toe, ankle, knee, hip, shoulder, elbow, and wrist at the moment of displacement. The results showed that, together, properties of the tool and the task influenced both (a) the chosen stopping distance and (b) how the upper and lower limbs were used in performing the task. Specifically, the stopping distances chosen by participants anticipated the postures and movements required to control the person-plus-tool system given the precision constraints of the task. Again, both in terms of perception and action, objects attached to the body that change a perceiver-actor's geometric and dynamic properties seem to be experienced (and treated behaviorally) as extensions of the body.

The research reviewed so far shows that perception of affordances for an intended behavior reflects the changes to action capabilities that are brought on by (current) use of objects attached to the body. Importantly, research has also shown that perception of affordances for an intended behavior also reflects *impending* changes to action capabilities that would be brought by *future* use of an object attached to the body. Wagman and Morgan[34] investigated whether perception of maximum reaching height reflected impending changes in reaching ability brought on by how the person intended to perform the reaching task (for example, with or without a handheld implement). Participants stood several meters from an object suspended from the top of a vertical surface. They instructed an experimenter to adjust the height of that object until it was at their maximum vertical reaching height. All participants reported perceived maximum reaching height with three different intentions as to how they would perform the reaching task. In one condition, participants intended to walk across the lab and reach for the object with their fingertips of their preferred hand. In a second condition, they intended to walk across the lab, pick up a (visible) plastic stick, and use the distal tip of the stick to reach for the object. In the third condition, they intended to walk across the lab, step on a plastic step stool, and reach for the object with the fingertips of their preferred hand. The perceptual boundary occurred at taller object heights when participants intended to reach by means of using the stick or standing stepstool than when they intended to do so by means of standing on the floor. Importantly, however, the perceptual boundary occurred at the same ratio of perceived-to-actual-maximum-reaching height in each condition (that is, a ratio of approximately 0.90).[35]

In a similar study,[36] participants reported their maximum reaching height when they intended to reach by means of using a stick or by means of standing on tiptoe. Again, the perceptual boundary occurred at different object heights in the different conditions, but at the same ratio of perceived-to-actual-maximum-reaching height in each case. Together, such results suggest that perception of maximum reaching height is relative to reaching ability both when intended use of an object attached to the body would change reaching ability and when it would not. Moreover, changes in reaching ability that would be brought on by use of such objects are treated the same as changes in reaching ability that would be brought on by changes in body posture (that is, standing on tiptoe).

If a handheld tool is experienced as an extension of the body, then use of a tool that changes the action capabilities of a perceiver-actor may influence subsequent motor behaviors performed without the tool. This hypothesis was investigated by Cardinali, Frassinetti, Brozzoli, Urquizar, Roy, and Farne.[37] They conducted a series of studies in which participants grasped small objects of different sizes with their right thumb and index finger before and after a lengthy training session in which they used their right hand to perform this task with a trigger-operated pincer tool. The results showed that the kinematics of grasping movements performed without the tool changed from pre-test to post-test. In particular, reaching movements occurring after the tool use task (a) took longer to initiate, (b) took longer to complete, and (c) occurred at smaller peak velocities than those occurring before the tool use task. Moreover, such changes in movement kinematics generalized to pointing movements performed without the tool (even though such movements were never practiced with the tool).

A follow up experiment investigated whether changes in movement kinematics following tool use were likely due to changes in the perceived length of the arm brought on by experience using the tool in the training session. In this experiment, blindfolded participants reported the perceived location of tactile stimulation on their right arm before and after performing the same tool use training task as described above. The results showed that the distance between the perceived location of the stimulation on the tip of the middle finger and the perceived location of the stimulation on the elbow increased from pre-test to post-test, suggesting that the perceived length of the arm increased as a result of experience using the tool. So, just during the course of an experimental trial, using the tool altered the participants' perception of their own bodies so that later action without the tool was more difficult.

Perception of affordances *by means of* the person-plus-object system

Merleau-Ponty famously claimed that a blind man who is adept at using a cane to navigate does not perceive the cane, but the world at the end of the cane.[38] When he does so, the cane becomes part of an extended perceptual system. Objects attached to the body not only change a person's action capabilities but also their perceptual capabilities. The ability to perceive environmental properties by means of an object attached to the body is known as extended haptic perception.[39] A fairly well-worn example of this ability is that visually impaired individuals can perceive properties of the environment by means of a handheld cane, but there are many other examples of this ability as well. Amputees can perceive objects by means of prosthetic devices, and surgeons can perceive internal organs by means of laparoscopic tools. More mundanely, perceiver-actors can perceive properties of the ground surface through their footwear and can perceive properties of the road surface through a vehicle such as a car or bicycle. This ability is continuous with the ability of both human and non-human animals to perceive environmental properties by means of non-innervated appendages such as fingernails, claws, vibrissae, whiskers, antennae, horns, and quills.[40]

Just as affordances can be perceived by means of vision, affordances can be perceived by means of extended haptic perception.[41] For example, merely by exploring a surface with a handheld object, a perceiver-actor can determine whether a gap in that surface can be stepped across.[42] In many ways, viewing a surface and probing that surface with a handheld implement are very different tasks. They differ not only in the perceptual machinery involved and the energy array detected but also in the point of observation at which the perceiver-actor encounters that energy array.[43] Nonetheless, in many cases, perception is relative to the action capabilities of the perceiver-actor regardless of whether perception is by means of vision or extended haptic perception. In the study by Wagman and Malek[44] described above, a different group of participants also explored the inclined surface with a 1 m wooden dowel (while blindfolded) in each of the three mass distribution conditions. For these participants, perceptual boundaries also occurred at a steeper angle of inclination when they wore the backpack on their front than when they did so on their back. In fact, there were no differences in the perceptual boundaries for this group and the group that performed this task visually.[45] These results suggest that *both* the weighted backpack *and* the handheld wooden dowel are experienced as part of the body.

One intriguing possibility is that attachments to the body are experienced as part of the body because they are perceived in the same way as the body.[46] Oftentimes, awareness of objects attached to the body occurs by means of dynamic touch – the kind of touch used when an object is wielded by means of muscular effort.[47] Perception by dynamic touch is constrained in a lawful and predictable manner by how the wielded object resists movement about a rotation point, typically located in the wrist.[48] Furthermore, this stimulation pattern constrains perception of properties of objects attached to the limbs and perception of properties of the limbs themselves in the same way. For example, perceived orientation of the limb (or an object attached to a limb) is constrained by the orientation of the mass distribution of the limb (or limb-plus-object-system). Moreover, if a handheld object is weighted such that the mass distribution of the limb-plus-object system is shifted away from the longitudinal axis of the arm, the perceived orientation of the limb-plus-object is also shifted in the same direction.[49] If such an object is used as a perceptual tool, the perceived location of any surface touched by the limb-plus-object should also be shifted in that direction, and the perceived affordances of that surface should shift accordingly. Wagman and Taylor investigated this hypothesis.[50] In this study, blindfolded participants grasped a 1 m T-shaped object in a 'handshake' grip at the intersection of the stem and the crossbar, such that the crossbar extended vertically, and the stem extended out from the longitudinal axis of the arm. On a given trial, they used this object to explore a horizontal bar at seven different heights (42 to 133 cm in 13 cm increments) and reported (yes or no) whether they would be able to step over that bar (while keeping at least one foot on the floor at all times and without touching the bar). They performed this task when (a) the upper branch of the crossbar was weighted with a 160 g mass, (b) the lower branch of the crossbar was weighted with a 160 g mass, and (c) when each branch of the crossbar was weighted with an 80 g mass.

The perceptual boundary (here, defined as the tallest bar height that they perceived they could step over) occurred at a taller bar height for tall participants than for short participants but at the same ratio of bar-height-to-leg-length for each group (approximately 1.0). Such results are consistent with work showing that perception of affordances is relative to action capabilities. However, what is more important for present purposes is that perceptual boundaries occurred at *both* taller bar heights *and* larger ratios of bar-height-to-leg-length when the lower branch of the T-shaped object was weighted than when the upper branch of the T-shaped object was weighted. Such results show that perceptual boundaries on this behavior are influenced by the perceived orientation of the limb-plus-object-system. In particular, when one of the branches of the T-shaped object is weighted, the perceived orientation of the arm-plus-object system is shifted in the direction of the added mass. When the bar was touched or probed with this object, it was perceived to be higher or lower than it actually was, which shifted the perceptual boundary of whether the bar could be stepped over. Thus, perceiver-actors can perceive affordances by means of objects attached to the body, and this is likely because perception of objects attached to the body occurs by the same means as perception of the body itself. In short, objects attached to the body are perceived as extensions to the body both when using those objects to perform a behavioral task and when using those objects to perform a perceptual task.

1/f Scaling in Human-Machine Systems

Many natural phenomena exhibit what is called *1/f scaling*. 1/f scaling (sometimes called '1/f noise' or 'pink noise') is a type of variability in a time series that is neither random nor predictable. 1/f scaling occurs when the components of a system are so tightly integrated with one another that they cannot be understood independently.[51] Systems with this type of tight integration are often said to exhibit *interaction-dominant dynamics*. This technical term can be read quite literally: a system exhibits interaction-dominant dynamics when the interactions among the components dominate or override the dynamics that the components would exhibit separately. Systems with interaction-dominant dynamics (that is, interaction-dominant systems) are not modular. Over the last few decades, it has been demonstrated that many well-functioning physiological systems are interaction dominant. For example, 1/f scaling has been found in human heartbeats,[52] gait patterns,[53] and brain activity,[54] indicating that the chambers of the heart, the locomotory system, and parts of the human brain are interaction-dominant systems. More recently, it has been shown that 1/f scaling is ubiquitous in cognition, indicating that, in some cases at least, cognitive systems are interaction-dominant.[55] For example, van Orden, Holden and Turvey[56] use 1/f scaling to gather direct evidence showing that cognitive systems are not modular; rather, these systems are fully embodied, and include aspects that extend to the periphery of the organism. The question for current purposes is whether interaction-dominant systems extend beyond the body periphery, whether person-plus-tool systems can be shown to exhibit interaction-dominant dynamics. Demonstrating this is tantamount to demonstrating that human-plus-tool systems can be as tightly integrated as the components of a beating heart are.

Dotov et al.[57] have shown that cognitive systems can be made to extend beyond the periphery to include artifacts that are being used. Participants in their experiments play a simple video game, controlling an object on a monitor using a mouse. At some point during the one-minute trial, the connection between the mouse and the object it controls is disrupted temporarily before returning to normal. Dotov et al. found 1/f scaling at the hand-mouse interface while the mouse was operating normally but not during the disruption. As discussed above, this indicates that, during normal operation, the computer mouse is part of the interaction-dominant system engaged in the task; during the mouse perturbation, however, the 1/f scaling at the hand-mouse interface disappears temporarily, indicating that the mouse is no longer part of the extended interaction dominant system. That is, during smooth play of the video game, the mouse was a constituent in the cognitive system, but during the disruption, it became a mere contextual factor that the cognitive system was engaged with.

These experiments therefore were designed to detect, and did in fact detect, the presence of an extended cognitive system. The fact that such a mundane experimental setup (using a computer mouse to control an object on a monitor) generated an extended cognitive system suggests that extended cognitive systems are quite common.

Dealing with the evidence

We have provided multiple sources of evidence that cognitive systems are sometimes partly constituted by non-biological, external parts. That is, we have shown that cognitive systems are sometimes extended. As we noted in the first section, all participants in the debate over extended cognition agree that it is an empirical debate. So we should expect Adams, Aizawa, and Rupert to relent at this point. This does not mean that they will, alas. Rupert will argue that what matters is cognitive science as an enterprise.[58] These demonstrations are not sufficient to overturn the internalist understanding of cognition at the foundations of the discipline. Rupert might be convinced by many, many more similar demonstrations in the future. We find Rupert's theoretical conservatism to be a dreary but reasonable position, one in keeping with the claim that it really is an empirical matter whether cognitive systems are extended. It will be interesting to see how he feels in a decade. We predict two responses from Adams and Aizawa: one related to intrinsic content; the other related to the 'mark of the cognitive.' We address these in order.

Adams and Aizawa[59] have argued that to be a part of a cognitive system, an entity must have intrinsic content. To say that something has intrinsic content is to say that it has content in its own right, and not in virtue of a relationship to an interpreter. So, for example, the words 'metaphysical silliness' as they appear here have merely derived content in that they do not mean anything other than in the context of readers of English. In contrast, a mental representation (should there be such) of 'metaphysical silliness' has its meaning all on its own, without the need for an interpreter. Adams and Aizawa claim that, as far as they know, nothing non-biological can be the bearer of intrinsic content, so canes and computers could not be constituents of cognitive systems. In addition to finding intrinsic content to be a terrible idea, we think that it is at odds with the claim that the debate over extended cognition is empirical. Until there is some device to detect intrinsic content, its presence or absence cannot be an empirical matter. Thus intrinsic content has no place in a scientific debate.

The second response is over the nature of cognition. Proponents of extended cognition, Adams and Aizawa[60] claim, need to supply a definition of the term 'cognition,' especially if they reject Adams and Aizawa's suggestion that intrinsic content is the 'mark of the cognitive.' We deny that cognitive scientists need to provide a 'mark of the cognitive,' if that is taken to be a set of necessary and sufficient conditions on being a component of a cognitive system. That said, if no sense of what the term 'cognition' means is in place, claims that cognitive systems are extended can seem empty. We take it that cognition is what cognitive scientists study, and that includes both the subject matter of traditional, computational cognitive science and that of radical embodied cognitive science. In this we follow William James, who as always says things better than we could.

> *The Pursuance of future ends and the choice of means for their attainment, are thus the* mark and *criterion of the presence of mentality* in a phenomenon. We all use this test to discriminate between an intelligent and a mechanical performance. We impute no mentality to sticks and stones, because they never seem to move for *the sake of* anything, but always when pushed, and then indifferently and with no sign of choice. So we unhesitatingly call them senseless.[61]

We can ruin this by putting it in more modern language: cognition is the ongoing, active, purposeful maintenance of a robust animal–environment system, achieved by closely coordinated perception and action. Of course, this understanding of the nature of cognition is intended primarily to reflect the phenomena of dynamicist cognitive scientists in philosophy, psychology, AI, and artificial life – that is, perception-action. But notice that it also applies to learning, speaking, reasoning and other traditionally cognitive phenomena. This pursuance of ends is cognition, and we have shown that things larger than a biological body sometimes accomplish it.

Acknowledgment

Anthony Chemero's work on this paper was partly funded by the Office of the Provost at Franklin and Marshall College. Thanks to Ann Steiner for her generosity.

Notes

1. Notice that this ignores the rich history of the idea of extended cognition, which goes back at least to William James. It also ignores several side debates, such as whether evolutionary considerations favor extended cognition (A. Clark, *Being There* (Cambridge, MA: MIT Press, 1997); M. Rowlands, *The New Science of the Mind* (Cambridge, MA: MIT Press, 2010); L. Shapiro, 'James Bond and the Barking Dog: Evolution and Extended Cognition,' *Philosophy of Science*, 77 (2010): 400–418), or whether extended cognition implies or enables anti-representationalism (R. Beer, 'A Dynamical Systems Perspective on Agent-Environment Interactions,' *Artificial Intelligence* 72 (1995): 173–215; A. Chemero, *Radical Embodied Cognitive Science* (Cambridge: MIT Press, 2009).
2. Clark, *Being There*; A. Clark and D. Chalmers, 'The Extended Mind,' *Analysis* 58(1) (1998): 7–19.
3. F. Adams, and K. Aizawa, 'The Bounds of Cognition,' *Philosophical Psychology*, 14 (2001): 43–64.
4. Rupert, 'Challenges to the Hypothesis of Extended Cognition,' *Journal of Philosophy*, 101 (2004): 389–428.
5. Robert A. Wilson and Andy Clark, 'How to Situate Cognition: Letting Nature Take its Course,' in *The Cambridge Handbook of Situated Cognition*, ed. Murat Aydede and P. Robbins (Cambridge, UK: Cambridge University Press, 2009), pp. 55–77.
6. Adams and Aizawa, 'The Bounds of Cognition,' Adams, and Aizawa, *The Bounds of Cognition* (Malden, MA: Blackwell, 2008); Rupert, *Cognitive Systems and the Extended Mind* (New York: Oxford University Press, 2009).
7. Rupert, 'Challenges to the Hypothesis of Extended Cognition,' pp. 425–426.
8. Adams and Aizawa, 'The Bounds of Cognition,' pp. 128–129.
9. Chemero, *Radical Embodied Cognitive Science*, and Chemero, 'Radical Embodied Cognitive Science,' *Review of General Psychology* 17 (2013): 145–150.
10. William James, *Essays in Radical Empiricism* (New York: Longman Green and Co., 1912).
11. H. Heft, *Ecological Psychology in Context* (Mahwah, NJ: Erlbaum, 2001).
12. J. J. Gibson, *The Senses Considered as Perceptual Systems* (Boston: Houghton Mifflin, 1966); J. J. Gibson, *The Ecological Approach to Visual Perception* (Boston: Houghton Mifflin, 1979).
13. For example, D. N. Lee, and P. E. Reddish, 'Plummeting Gannets: A Paradigm of Ecological Optics,' *Nature* 293 (1981): 293–294.
14. P. N. Kugler, J. A. S. Kelso, and M. T. Turvey, 'Coordinative Structures as Dissipative Structures I, Theoretical Lines of Convergence,' in *Tutorials in Motor Behavior*, ed. G. E. Stelmach and J. Requin (Amsterdam: North Holland, 1980).

15. Saying that dynamical models in cognitive science explain, rather than merely describe, is a contentious claim. See T. van Gelder, 'What Might Cognition be if not Computation?' *Journal of Philosophy* 91 (1995): 345–381; Chemero, 'Anti-Representationalism and the Dynamical Stance,' *Philosophy of Science,* 67 (2000): 625–647, and *Radical Embodied Cognitive Science*; N. Stepp, A. Chemero, and M. Turvey 'Philosophy for the Rest of Cognitive Science,' *Topics in Cognitive Science* 3 (2011): 425–437, for the pro side. See D. Kaplan, and W. Bechtel, 'Dynamical Models: An Alternative or Complement to Mechanistic Explanations?' *Topics in Cognitive Science* 3 (2011): 438–444; David Michael Kaplan and Carl F. Craver, 'The Explanatory Force of Dynamical and Mathematical Models in Neuroscience: A Mechanistic Perspective,' *Philosophy of Science* 78(4) (2011): 601–627 for the con side. Nothing in this paper rides on this debate.
16. van Gelder, 'What Might Cognition be if not Computation?'
17. Beer, 'A Dynamical Systems Perspective on Agent-Environment Interactions,' see Figure 1.
18. G. Van Orden, J. Holden, and M. T. Turvey, 'Self-Organization of Cognitive Performance,' *Journal of Experimental Psychology: General,* 132 (2003): 331–351; J. Holden, G. Van Orden, and M. T. Turvey, 'Dispersion of Response Times Reveals Cognitive Dynamics,' *Psychological Review,* 116 (2009): 318–342; A. Chemero, and M. Silberstein, 'After the Philosophy of Mind,' *Philosophy of Science* 75 (2008): 1–27; D. Dotov, L. Nie, and A. Chemero, 'A Demonstration of the Transition from Readiness-to-Hand to Unreadiness-to-Hand,' *PLoSONE* 5 (2010): e9433; and C. Zednik, 'The Nature of Dynamical Explanation,' *Philosophy of Science* 78(2) (2011): 238–263, among others.
19. H. De Jaegher, E. A. Di Paolo, and S. Gallagher, 'Can Social Interaction Constitute Social Cognition?' *Trends in Cognitive Sciences* 14 (2010): 441–447.
20. Chemero, 'An Outline of a Theory of Affordances,' *Ecological Psychology* 15 (2003): 181–195; Gibson, *The Ecological Approach to Visual Perception*; M. T. Turvey, 'Affordances and Prospective Control: An Outline of the Ontology,' *Ecological Psychology* 4 (1992), 173–187.
21. D. Comalli, J. Franchak, A. Char, and K. Adolph, 'Ledge and Wedge: Younger and Older Adults' Perception of Action Possibilities,' *Experimental Brain Research* 228 (2013): 183–192.
22. W. H. Warren, and S. Whang, 'Visual Guidance of Walking through Apertures: Body-Scaled Information for Affordances,' *Journal of Experimental Psychology: Human Perception and Performance,* 13 (1987): 371–383.
23. R. J. Jagacinski, and J. M. Flach, *Control Theory for Humans: Quantitative Approaches to Modeling Performance* (Mahwah, NJ: Erlbaum, 2003).
24. J. B. Wagman, and K. R. Taylor, 'Perceiving Affordances for Aperture Crossing for the Person-Plus-Object System,' *Ecological Psychology* 17 (2005): 105–130.
25. Cf. Warren and Whang, 'Visual Guidance of Walking through Apertures.'
26. Ibid.
27. Ibid.
28. T. Higuchi, M. E. Cinelli, M. A. Greig, and A. E. Patla, 'Locomotion through Apertures when Wider Space for Locomotion is Necessary: Adaptation to Artificially Altered Bodily States,' *Experimental Brain Research* 175 (2006): 50–59; T. Higuchi, G. Murai, A. Kijima, Y. Seya, J. B. Wagman, and K. Imanaka,

'Athletic Experience Influences Shoulder Rotations when Running through Apertures,' *Human Movement Science* 30 (2011): 534–549.
29. M. Bhalla, and D. R. Proffitt, 'Visual-Motor Recalibration in Geographical Slant Perception,' *Journal of Experimental Psychology: Human Perception and Performance* 25 (1999): 1076–1096.
30. E. A. Malek, and J. B. Wagman, 'Kinetic Potential Influences Visual and Remote Haptic Perception of Affordances,' *Quarterly Journal of Experimental Psychology* 61 (2008): 1813–1826; T. Regia-Corte, and J. B. Wagman, 'Perception of Affordances for Standing on an Inclined Surface Depends on Height of Centre of Mass,' *Experimental Brain Research*, 191 (2008): 25–35.
31. Malek and Wagman, 'Kinetic Potential Influences Visual and Remote Haptic Perception of Affordances.'
32. See also Regia-Corte and Wagman, 'Perception of Affordances for Standing on an Inclined Surface Depends on Height of Centre of Mass.'
33. R. M. Bongers, C. F. Michaels, and A. W. Smitsman, 'Variations of Tool and Task Characteristics Reveal that Tool-User Postures are Anticipated,' *Journal of Motor Behavior*, 36 (2004): 305–315.
34. J. B. Wagman, and L. L. Morgan, 'Nested Prospectivity in Perception: Perceived Maximum Reaching Height Reflects Anticipated Changes in Reaching Ability,' *Psychonomic Bulletin and Review* 17 (2010): 905–909.
35. Wagman and Morgan, 'Nested Prospectivity in Perception'; cf. Warren and Whang, 'Visual Guidance of Walking through Apertures.'
36. J. B. Wagman, 'Perception of Maximum Reaching Height Reflects Impending Changes to Reaching Ability and Improvements Transfer to Unpracticed Reaching Tasks,' *Experimental Brain Research* 219 (2012); 467–476.
37. L. Cardinali, F. Frassinetti, C. Brozzoli, C. Urquizar, A. C. Roy, and A. Farne, 'Tool-Use Induces Morphological Updating of the Body Schema,' *Current Biology* 19 (2009): R478–R479.
38. Maurice Merleau-Ponty, 'Phenomenology of Perception,' trans. by Colin Smith (New York: Humanities Press, and London: Routledge and Kegan Paul, 1962), p. 142.
39. G. Burton, 'Nonvisual Judgment of the Crossability of Path Gaps,' *Journal of Experimental Psychology: Human Perception and Performance* 18 (1992): 698–713.
40. Burton, 'Nonvisual Judgment of the Crossability of Path Gaps,' and 'Non-Neural Extensions of Haptic Sensitivity,' *Ecological Psychology* 5 (1993): 105–124.
41. Burton, 'Nonvisual Judgment of the Crossability of Path Gaps,' and 'Non-Neural Extensions of Haptic Sensitivity'; P. Fitzpatrick, C. Carello, and M. T. Turvey, 'Eigenvalues of the Inertia Tensor and Exteroception by the "Muscular Sense",' *Neuroscience* 60 (1994): 551–568; Wagman and Taylor 'Perceiving Affordances for Aperture Crossing for the Person-Plus-Object System.'
42. Burton, 'Nonvisual Judgment of the Crossability of Path Gaps,' and 'Non-Neural Extensions of Haptic Sensitivity' and 'Non-Neural Extensions of Haptic Sensitivity.'
43. Regia-Corte and Wagman, 'Perception of Affordances for Standing on an Inclined Surface Depends on Height of Centre of Mass.'

44. Malek and Wagman, 'Kinetic Potential Influences Visual and Remote Haptic Perception of Affordances.'
45. See also Regia-Corte and Wagman, 'Perception of Affordances for Standing on an Inclined Surface Depends on Height of Centre of Mass.'
46. C. C. Pagano, and M. T. Turvey, 'Eigenvectors of the Inertia Tensor and Perceiving the Orientation of Limbs and Objects,' *Journal of Applied Biomechanics* 14 (1998): 331–359; C. C. Pagano, and M. T. Turvey, 'The Inertia Tensor as a Basis for the Perception of Limb Orientation,' *Journal of Experimental Psychology: Human Perception and Performance* 21 (1995): 1070–1087.
47. Gibson, *The Senses Considered as Perceptual Systems.*
48. M. T. Turvey, and C. Carello, 'Obtaining Information by Dynamic (Effortful) Touch,' *Philosophical Transactions of the Royal Society B*, 366 (2011): 3123–3132; C. Carello, and J. B. Wagman, 'Mutuality in the Perception of Affordances and the Control of Movement,' in *Progress in Motor Control: A Multidisciplinary Perspective*, ed. D. Sternad (New York: Springer, 2009), pp. 271–289.
49. Pagano and Turvey, 'The Inertia Tensor as a Basis for the Perception of Limb Orientation,' and 'Eigenvectors of the Inertia Tensor and Perceiving the Orientation of Limbs and Objects.'
50. J. B. Wagman, and K. R. Taylor, 'Perceived Arm Posture and Remote Haptic Perception of Whether an Object Can be Stepped Over,' *Journal of Motor Behavior* 37 (2005): 339–342.
51. P. Bak, C. Tang, K. Wiesenfeld, 'Self-Organized Criticality,' *Physical Review A* 38 (1988): 364–374; van Orden, Holden and Turvey, 'Self-Organization of Cognitive Performance.'
52. J. Hausdorff, C. Peng, Z. Ladin, J. Wei, and A. Goldberger, 'Is Walking a Random Walk? Evidence for Long-Range Correlations in Stride Interval of Human Gait,' *Journal of Applied Physiology* 78 (1995): 349–358.
53. Ibid.
54. Walter J. Freeman, 'Origin, Structure, and Role of Background EEG Activity, Part 4, Neural Frame Simulation,' *Clinical Neurophysiology* 117 (2006): 572–589.
55. M. Ding, Y. Chen, and J. A. S. Kelso, 'Statistical Analysis of Timing Errors,' *Brain and Cognition* 48 (2002): 98–106; M. Riley and M. Turvey, 'Variability of Determinism in Motor Behavior,' *Journal of Motor Behavior* 34 (2002): 99–125; van Orden, Holden and Turvey, 'Self-Organization of Cognitive Performance'; Holden, Van Orden, and Turvey, 'Dispersion of Response Times Reveals Cognitive Dynamics.'
56. van Orden, Holden and Turvey, 'Self-Organization of Cognitive Performance.'
57. Dotov et al. 'A Demonstration of the Transition from Readiness-to-Hand to Unreadiness-to-Hand.'
58. Rupert, *Cognitive Systems and the Extended Mind* (New York: Oxford University Press, 2009).
59. Adams and Aizawa, *The Bounds of Cognition.*
60. Adams and Aizawa, 'The Bounds of Cognition,' and *The Bounds of Cognition.*
61. James, *The Principles of Psychology* (New York: Henry Holt, 1890). Retrieved from http://psychclassics.yorku.ca/James/Principles/index.htm, p. 8.

6
Knowing and the Known: Brain Science and an Empirically Responsible Epistemology
David D. Franks

Introduction

This chapter will relate three subject matters: neurosociology, mirror neurons, and the pragmatic, social behaviorism of John Dewey and George Herbert Mead. Social behaviorism was basically initiated right after the turn of the last century in what has been called the Golden Age of the University of Chicago. As will be discussed below, it was an epistemology that avoided the dualism of the Enlightenment idealism and the British empiricists. Important for this chapter is the often neglected interest that Dewey and Mead had in the neuroscience of their times, as undeveloped as it was. Before coming to Chicago, Mead had worked in a neuroscience lab. The relevance of this for what follows is that whatever epistemology one chooses as most fruitful should include compatibility with the knowledge of the brain and mind that we have today. After all, it is the brain by which we know. Furthermore, as argued below, this brain is designed for sociality. It is social to the core. It is the social emotions as well as mirror neurons that drive and organize the brain. Any epistemology we choose must therefore be a social one. This fits in with Mead's social behaviorism and the epistemology that unfolds from it. This is the heart of a new sociologically oriented neuroscience referred to as neurosociology. Below I lay out why the brain is a social organism and why social behaviorism as an epistemology fits so well with it.

Dualism

One virtue of social behaviorism is that it avoids dualism. This either/ or thinking pits against each other the knower and the known, nature and nurture, individual and society, objective and subjective, theory and fact as if they are antithetically opposed to each other. Seeing them as antithetical means that one has to choose one over the other as the important one. This is what occurred in the fruitless quarrels between the Enlightenment empiricists and idealists. It ignores the fact that human experience consists equally of both. For example, for objectivity to exist as a preferred human norm we must embrace it as some kind of feeling that allows us to identify affectively with it. Assuming it exists independently from subjectivity leads ultimately to logical incoherence (as with objectivity in the dictionary sense as seen as independent of anything human). We will see below how impossible this really is. We can even *passionately* embrace the norm of objectivity that the social behaviorists argued we needed in order to understand how these dualities overlap and are inseparably involved with one another while nevertheless recognizing that there is always a possible tension between them. The formula was: cancel their total opposition, but do not merge them so much that you cannot retain their possible opposition and tension. This should be especially clear with the individual and society. Society makes an individual possible, but the individual can be in utter tension with particular aspects of it. We will return to this later when discussing theories of knowledge and mirror neurons that basically bring us back to the old pragmatic tradition and its rejection of dualisms.

Neurosociology

The term neurosociology started to gather momentum in 1999 with a collection of chapters edited by this author and Thomas Smith titled *Mind, Brain and Society*.[1] It was not until 2010 that I wrote the first monograph in this field titled *Neurosociology: The Nexus Between Neuroscience and Social Psychology*.[2] Also, this author and Jonathan Turner, a leading grand theorist of sociology, edited *The Handbook of Neurosociology*.[3] In other words, the field is young, but it is becoming firmly established.

One of the most interesting findings to a neurosociologist is that of mirror neurons first published by the Italian neuroscientists Marco Iacoboni in *Mirroring People*,[4] and Rizzolatti and Sinigaglia in *Mirrors in the Brain*.[5] It would be fair to say then that this chapter will bring together the new and the old in the sense that the recent findings about mirror neurons confirm in many ways the old social behaviorism of G. H. Mead and John Dewey. As will be seen below, a description of social behaviorism will involve a critique of empirical and Idealistic epistemologies since their social behaviorism was born in reaction to the dualism implicit in both.

American analytic philosophy: a critique from brain science

Currently, American analytic philosophers often embrace a disembodied version of mind. Though inconceivable to those knowledgeable about the brain, the link between the knower and the known involves the inherent reason of human-kind and the principled reasons that drive the non-human universe: that is, we understand the universe because it follows the same principles that drive our minds. Some might see this as 'highly convenient' and caution that much is yet to be known of such rational principles in higher theoretical physics where much is counterintuitive.

Reason itself is going through rough times. To paraphrase Robert Laughlin's *A Different Universe*,[6] nature does things, but whether we can understand these processes with human reason may be highly improbable. One possible implication of this is that nature may be still reasonable, but our human reasoning has limits that diminish the link between us and the reason guiding and explaining the universe.[7]

Other problems stem not from human reasoning processes but from threats to the sufficiency of reason itself. As early as 1958, Hannah Arendt in her *The Human Condition* expressed such doubts:

> The modern astrophysical world view, which began with Galileo, and its challenge to the adequacy of the senses to reveal reality, have left us with a universe of whose qualities we know no more than the way they affect our measuring instruments, and – in the words of Eddington – 'the former have as much resemblance to the latter as a telephone has to a subscriber.' Instead of objective qualities, in other words, we find instruments, and instead of nature or the universe – in the words of Heisenberg – man encounters only himself.[8]

In these words from our leading physicists, we see the fracture of any link at all and the immanent arrival of post-World War II nihilism and the 'century of the self.'

Reason and emotion

By the end of the last century, seeing reason as purged from emotion was also becoming untenable, even though philosophers like Ronald deSousa,[9] neuroscientists like Antonio Damasio,[10] and artificial intelligence workers have shown that decision making (surely at the heart of reasoning) is impossible without some basic emotional preferences that determine saliency or importance, setting the agenda for thought, and that are 'what we see the world in terms of.' Damasio's 'Prefrontal' patients, who had been traumatized in the lower region of the prefrontal lobes, which integrate cognition and emotion, were often quite intelligent in terms of IQ scores, but could not make the simplest decisions. Since the brain makes possible everything human and living, it is hard for those who take brain science seriously to argue with Lakoff and Johnson[11] when they insisted that we have an *empirically responsible philosophy* or that reason cannot be set apart from our bodily-driven brains. To them, the very structure of reason comes from the structure of our brains and bodies. Furthermore, reason is not 'universal' in any structural sense, and it is certainly not the structure of the universe. It is universal in the sense that we all share bodies whose motor behaviors respond to and are responded to by the world in uniformed and shared ways. This essentially pragmatic framework views reason as metaphorical – as stemming from our experience with our sensory motor activities.[12] We say, 'I was *knocked out*' or '*bowled over* by that idea.' My study of neurosociology has been a hard *journey across* disciplines; I was *drawn* to the social nature of the brain. Reason, then, in non-dualistic fashion includes input from our bodies as well as from the world acting on us. I will not bore you with the thousands of other examples.

Before leaving this topic I will add, as do Lakoff and Johnson, that reason is unconscious primarily and not as dispassionate as we think, since decision making is enabled by some kind of affective preference as we have seen above. In light of the cautions above, we may renew our interest in an old philosophy that seems to be confirmed by current inquiry into the brain, especially in the case of what we know about mirror neurons.

Neurosociology: an introduction to the social brain

As we have said, one of the newest fields of neuroscience is neurosociology, and it is within its development that mirror neurons have been related to philosophy. As said above, it seems reasonable to think that any epistemology that is useful today should be in line with our social brains and current neurosociology. We start with a broader meaning of the social that stretches back to the beginnings of pre-humans on earth.

In the beginning of hominid evolution, we were scavengers who had to fight away other animals attracted to rotting meat. Stone Age axes were hardly enough to make this possible, and much depended on our social abilities to communicate and organize others to help in the process. From the beginning, social skills were necessary for our survival. From what we can tell, this was an important part of the beginning of the social brain.

In my opinion, Leslie Brothers wrote the most thoroughgoing description of this brain.[13] In doing so, she also laid the foundation for what is now known as neurosociology. Most treatments of our social brains (for example, 'the social brain hypothesis') miss the point about how deeply social our brains are. There is no question, she argues, that our brains are structurally encased, well protected in our individual isolated brains. But in order for this brain to *function* emotionally and cognitively, other attentive brains must surround it. These brains must be further drawn together by shared symbols that allow for such individuals to be drawn together intersubjectively by a common language and culture. Last, but very important for sociology, these brains must influence each other mutually. Our basic unit of analysis, therefore, is not one person, but at least two people engaged in social interaction with each other. When, for whatever reason, infants are simply fed by a bottle and are without other social and affective interactions, they do not develop emotionally. Since emotion is a critical organizer of the brain, infants do not develop cognitively and become depressed and anxious. Such infants do not develop interest in, or even a capacity for, social activity. In its place is heartbreaking fear, anxiety, and depression. In many cases, ordinary infant diseases led to death.[14] Tredway et al. have shown the specific brain processes that make this happen.[15]

Brothers goes on to catalogue for us what nature supplies the infant in terms of social interests from the beginning of birth. Infants 9-minutes old were more likely to be interested in a lifelike picture of a stylized human face rather than an upside down one, or simply the shape of a face. There is a fixation on the mother's face that Brothers says is a brain stem imprinting of this important information. 36 hours after birth, babies imitate their caretakers opening their mouths and sticking out their tongues. Though eye-gaze is often used as an indicant of social interest and attention, infants tend to look at eye-regions (which show emotion) much more than the mouths even though its movement should be inherently more interesting. Nonetheless, 7–11 week old infants were more likely to gaze at regions around the eyes than mouths. In the study above about the socially neglected depressed children, nature had done its job by supplying them with these and other inherent social interests. It was the social environment, or the lack of it, that was tragically missing.

The social editor is Brothers's way of presenting the brain's social structuring. A critical structure here is the amygdala, mostly known as a very fast warning signal that moves us safely through our physical world. The brain however is not a perfect device, and the amygdala is a classical example of this. It operates at speeds of less than one hundred milliseconds, allowing us to know whether something is safe or not, so quickly that we also know whether an object is good or bad before we know what the object is. It will surprise no one, therefore, to know that at such speeds it is often wrong. But, when it is accurate, it can save our lives. Frequently, writers think that the story ends here. But they miss the amygdala's importance for social life. It also assesses others' emotions as to whether these others are safe and accommodating or dangerous and apathetic. Brothers cites a man stimulated electronically in the left amygdala who was made to feel unwelcomed and uneasy in social situations. In the early stages of our prehistory, when life hung in the balance and others of our kind were needed for safety in the numerous stages of gathering food, the quickness of the amygdala served us well. Even today, the speed of social interaction is so fast that we need such assessments – the faster, the better – even though life may or may not be hanging in prehistoric precariousness.

The importance of learning is as important as ever, and the amygdala is critical to socially related memories as is the hippocampus. A lesson learned but forgotten is not really a lesson learned in the larger and more relevant picture.

In the numerous cases of post-traumatic stress, the amygdala may be quick to learn, but it makes it hard to forget. The critical importance of the amygdala is that it enables so much, like fear, assessing facial expressions, attachment, unconscious learning, early emotional memory, and emotion across the life span. One should not therefore be surprised at its complexity, despite its small size.

Another critical part of the social editor is the often over-appreciated prefrontal cortex. It is unappreciated because it is treated as if it stands alone in royal fashion to produce our intellectual capacities. Damasio[16] as well as philosophers and even artificial intelligence workers have convincingly shown that without certain emotions, we are unable to make the simplest choices. Without emotion and its limbic system, the prefrontal cortex gives us nothing. Patients traumatized in the ventral medial parts of the prefrontal lobes were capable of anger but incapable of signals of apprehension that lead us away from bad social associations, especially untrustworthy business partners. On the other hand, they were not able to have normal emotions. While interviewers wept at their stories of their familial and business downfalls, they felt no such remorse, even though many such patients were quite intellectually capable.

Social behaviorism as an epistemology

Aspects of mirror neurons and social functioning draw one inevitably into philosophical issues. As a neurosociologist, I am particularly sensitive to the fact that because the brain is the organ by which we know, we automatically become amateur epistemologists. Because we all share bodily experiences, this thesis fits in with intersubjectivity, without which society would be impossible.

Current neuroscience work on mirror neurons gives strong empirical credence to a very old philosophical framework called pragmatism. This position was refined mainly at the University of Chicago after the First World War by the philosophers John Dewey and George Herbert Mead.[17] Mead called his approach 'social behaviorism' to avoid confusing it with the deterministic behaviorism of Watson.

A good way to define pragmatism is to contrast it with the two other competing epistemologies and how they linked the knower and the known.

Empiricism

The British Enlightenment empiricists saw this link between the knower and the known as given through the individual *bodily senses*. There was little social about this since all animals have senses. In fact, they saw the mind as passive and malleable blank tablets, 'written' on by the individual's experience with the environment. As such, this position is sometimes called 'copy theory' because the knower's mind is the deterministic one-way result of environmental factors. The knower and his linguistically enabled mind is absolutely passive. For the empiricists, the mind is like malleable cookie dough being molded by the cookie cutter of experience. Since the mind has to give way to the bodily senses as being all that is allowed in creating knowledge, we have a dualistic system with all its difficulties. Mind and body are inseparably contrasted and rendered into antithetical contrasts.

But as N. S. F. Northrop recognized so long ago, there are other problems.[18] As soon as we put words on the sensed impressions, we no longer have pure fact; we have *described* fact. This makes pure fact ineffable, that is, not describable linguistically to others. The taste of kale to one person is not the same to another, and some of us are color blind. There is no common world if we rely on pure sensation since it can be different from person to another person. We are stuck in our own ineluctable subjectivity, a condition that denies the reality of our social brains and makes society itself impossible. Northrop goes on to say that qualities of objects do not come with tags on them saying 'I come from an external object that exists beyond me, and I am purely subjective and illusionary in origin.'

Of course the idea of an illusory object is a theoretically inferred statement, not an immediately sensed one. Finally, since fact is dependent on immediate observation only, the pure empiricist must take care not to believe in his or her own permanent existence when not apprehending one's self. Empiricism and its appeal to the senses to copy the world 'as it is independently of us' faces its final demise through brain science and the biological notion that our senses are transducers. That is, our senses have to *change* incoming stimuli in a way that accommodates and thus changes them to what the brain can receive as meaningful. A transducer takes stimuli like light and sound and changes in colors, which are scientifically magnetic waves, into human greens and purples, as long as there is light to see. Surely, sound is as different from compression as a telephone number is to a subscriber. Sounds are really compressions of air reaching out from its source that our brain changes into something humanly meaningful. With no animal ears to 'transduce' these compressions into sounds, we have nature's compressions of air but no brain-given noise. The tree in the forest that falls with no ear to make this transduction is silent. For this we need both nature and nurture.

Our language misleads us at this point because it places this attribute into the object and the object alone. To make color also takes an animal brain. We say that grapefruit is yellow or pink. This places color in the fruit as a part of its nature when it is really the result of its ability to absorb light as well as our ability to accommodate it. This is called the 'stimulus error.' In this way, we have *relationism*, but it does not take us so far as a relativism in which 'any thing goes,' as Mead would say. This would mean, as the postmodernists used to contend, that 'there is no real error, just other stories.' The common ground between the two is what Mead called the 'objective reality of perspectives.' In sum, 'if the eye were not sun-like, it could not see the sun.'

Idealism

For the idealists, this causal direction was reversed, and mind was seen as a movie projector with the environment being the sole result of the *workings of the human mind.* As with the analytic philosophers, the link between the knower and the known was given through the correspondence between the reasoning mind and the same sweet reason that was inherent in the universe. Lakoff and Johnson hold that the analytic philosophers would reject everything we are saying about the mind's brain and embodiment. The authors go on to say that once the mind has become disembodied, the gap between the mind and the body that was closed by Roger Sperry's notion of mind as an emergent from the brain is insufferably opened and becomes unbridgeable. This flies in the face of neuroscience and an empirically responsible philosophy. Dualism and its many difficulties are brought into the picture when mind is seen in opposition to the body and its brain.

Social behaviorism

In sharp contrast to the epistemologies above, the Chicago pragmatists saw the link between the knower and the known as *behavior*. We know the world through how it responds to our *behavior* in it, and perception becomes a selective assessment of what action-possibilities an object affords for our intentions to act. The meaning of an object is not inherent in the object, but resides in how it responds to our various motor behaviors.

This also contrasts with a version of reinforcement theory wherein the stimulus comes first, then we perceive it, next we think (at least sometimes), and lastly, we act in a way dictated by our conscious thought.

Mead's four stages of the act

Mead's four stages of the act reversed this sequence and is more in line with mirror neuron research. First comes action – some impulse to behave. Second, what we perceive most clearly is what answers to or facilitates our interests and intentions to act. As Damasio wrote in *Descartes' Error*, 'Perception is more involved with action than we think. Perceiving is as much acting on the environment as it is about receiving signals from it.'[19] All perception is selective, and our behavioral intentions do the selecting. Books on the shelf that we have ignored for years come immediately and clearly into sight as we reach for them.

The next stage of the act is *manipulation* – *doing* something with the object or people, like taking the role of the other in social interaction – and, lastly, we have *consummation*. Here consummation, or the last stage of the act, reaches back teleologically to pull the different stages into being. This is teleological because the later stages are present in the first stages and control the others. The whole framework is voluntaristic because in the first stage, the image of consummation is symbolic and open to the actor's change.

This epistemology meets the criteria for a non-dualistic approach because one's own actions are required to obtain a worldly response, and yet there can be tension as pieces do not fit, and structures fail.

Transaction as a way to avoid dualisms

Transaction is a way of avoiding dualisms and its either/or thinking. It does not allow for 'hypothetisation' of thought detached from the body acting on the environment, and thus it seems consistent in this regard to the framework of Lakoff and Johnson's. Emirbayer brought the term back into sociology in 1997.[20] Though Mead did not use the term often, he assumed it and used it in his thinking. Its primary source is Dewey and Bentley's *Knowing and the Known*.[21] Transaction then refers to an epistemological framework that avoids looking at objects as having essentialist characteristics in and of themselves; these things cannot be understood independent of other things. As Ollman says, 'it chooses to view things as components of action that are never self-sufficient.'[22] Units involved in a transaction take on their identity from the fluctuating roles they play within the transaction. It provides us a view that sees the micro and the macro levels of analysis as different phases of common activity. Though assuming emergent structures, transactionalism avoids reifying structure into something *sui generis* independently from the everyday processes of the thinking and feeling of actual persons engaged in the process of communication. When, for example, we give emotions names and then view them as being separate unto themselves or as materially distinct systems in the brain, we are perfecting the art of reification. In sum, to overcome the mistakes of the Enlightenment, so much of which is dualistic in nature, we must see things in combination. As said above, in regard to the different sides of the dualisms – individual and society, knower and the known, subjective or objective – it is not a matter of dismissing one in favor of the other. It is a matter of seeing them both as one frame. What needs to be understood is how they are nonexistent without each other, but nonetheless tension will always appear.

Neuroscience and mirror neurons

It is high time now to define mirror neurons and show how findings about them support the pragmatic emphasis on behavior. A mirror neuron fires when we watch others perform a certain action. But it does more than that. It actively simulates other's actions unconsciously on the observer's motor cortex. In an important sense, we do behaviorally what we watch. These neurons are inherently social and directed toward the actions of others; they also actively combine motor activity and perception. Furthermore, and most importantly, mirror neurons fire when the action observed can be seen as *intentional on the part of the observed actor*, thus preparing us for the lightening quick responses that are made necessary by the speed of social interaction.

Mirror neurons and pragmatism

For researchers in mirror neurons, the relation between our motor intentions and the way we perceive the world is an 'affordance.' An affordance is something that the environment offers to an animal's behavior such as surfaces that provide support, objects to be manipulated, and substances that can be eaten. If I am not mistaken, the old Chicago pragmatists used the very same word. In fact, current research on mirror neurons confirms the pragmatic tradition.

Researchers Giacomo Rizzolatti and Corrado Sinigaglia argue that the visual perception of an object implies the immediate and automatic selection of those properties that facilitate our behavioral interaction with it.[23] I repeat: we perceive the world not so much as it is in some final sense, but in terms of how it answers to, or facilitates, our actions.

This means that what we sense and reason about is limited. The environment becomes constructed and objectified only in relation to the animal's limited motor capacities, senses, and brains. The German term *Umwelt* captures this organism/environment relation very well. *It refers to the world carved out for our attention by our own capacities, sensitivities, and motor repertoires.*

Let me give you an illustration of this term, motor 'repertoires.' The research on mirror neurons implies the existence and relevance of just this social behaviorist view. The mirror neurons of dancers responded most strongly to observations of familiar dance steps in line with the style of dancing the observers routinely used. More pointedly, when human subjects were shown videos of motor behaviors of different animals' communicating – like monkeys lip-smacking, dogs barking, and human's talking – mirror neurons became engaged only at the observation of humans talking, even though visually it was obvious that these all were communications with moving mouths. Since we have no spaces in our motor cortex for barking and lip-smacking, we cannot simulate them.

Mirror neurons and speech

For the pragmatists, the products of human thought and speech were clearly intangible universals. Words represent general terms subsuming all real particulars (another potential dualism). But the process is behavioral. Researchers have indeed found that as humans solve problems, they are really talking to themselves, making subliminal movements with those muscles associated with speech like the tongue, lips, and larynx. Deaf children privately thinking to themselves likewise make subliminal hand and body gestures that underlie their symbolic thought. The congenitally blind, who have never seen gestures, nonetheless make motor movements when they talk. Gestures, by the way, are not ancillary to talk but an inherent part of it, just as grammar is. We even gesture on the phone with no one to see us.

How do mirror neurons enter the picture at this point? Once again, we do not just passively hear talk – we actively *do* the talk – in this case, the talk of other people. Why? Because talk is behavioral, we simultaneously enact in our motor cortex the same speech movements others are making. Their talk literally becomes our talk with the provision that it is automatic and unconscious.

Interestingly enough, several social neuroscientists recently sent me some articles on mirror neurons that converge on the same findings. This allows for convergent validity and is a powerful tool to strengthen one's case. They also investigate the relationship of mirror neurons and conceptual *meanings* of spoken words. They call this 'embodied semantics' – semantics because the purpose of mirror neurons is to capture the meaningful intent of others. This position holds that concepts are represented in the brain by the same sensory motor circuitry on which the concept relies. The general *concept* of grasping would be represented by increased activity in the sensory motor areas devoted to grasping *actions*.[24] Neuroscientists Aziz-Zadeh and Damasio, and Sook-lie and Aziz-Zadeh also describe the last 20 years of research that gives evidence that language is biologically linked to action through mirror neurons in a way congruent with the priority given to action by the pragmatists.[25] They insist that 'Language is not separate from motor behavior, but rather inherently grounded in actions.' For example, kicking as a conceptual generality would be represented in the sensory motor space given over to the leg and foot. Even metaphors like 'We will kick off the new year with a party' will be represented in these foot and leg areas of the motor cortex. According to these researchers, the hypothesis of 'embodied semantics' has been confirmed by studies of patients with traumatized parts of the motor cortex, and fMRI studies that take actual movies of the activation of brain circuits. Other work involves studies of motor cortex activation and metaphors; these are more complicated in interpretation. Aziz-Zadeh and Damasio had subjects read phrases like 'bite the bullet,' 'grasp the meaning,' and 'kick the bucket.' As they read these phrases, the authors also showed them videos of hand, foot, and mouth actions. They then looked for mirror neuron activation for each of these three phrases, but none were found. Their interpretation of this finding was that once a metaphor becomes familiar, it no longer has to activate the sensory motor processing congruent with the metaphor. Recent studies however, while not testing the embodied semantic hypothesis, have nonetheless shown that novel and conventional metaphors are at least handled by distinct circuits of the brain. The right hemisphere handles the novel metaphors. The authors do not say what brain area is used for familiar metaphors but stress that they are activated quicker.

However this may be, the 'embodied semantic hypothesis' has been confirmed by those working in these different and independent perspectives strengthening their convergent validity.

Conclusion

Most people connect mirror neurons with empathy and for good reasons. There is strong evidence that we feel each other's pain in a real sense. There was some truth in our parent's statement that punishing us hurt them as it did us, but no one is fooled that it hurt them just as much!

Acknowledging *Homo sapien*'s ghastly penchant for doling out pain to his fellow creatures tells us, however, that much more also is going on. Culture often prevails over the brain, and though it emerges from humans, it is not necessarily humane.

I have attempted to show that an empirically responsible philosophy has to be an embodied philosophy informed by the nature of our brain and emergent minds. Ironically, a contemporary approach takes us back to the old social behaviorism of the early Chicago Pragmatists. It also leaves all 'copy' theories – analytic philosophy or empiricism – dead in the water.[26] Mirror neurons and their simulation of what others are doing on the motor cortex are both social and lodged in behavior that links us with the world and how it responds to our actions in pragmatic fashion. Neurosociology, the social nature of our brain, and mirror neurons are important considerations for evaluating our theories of knowledge, in this case seen as non-dualistically and transactionally as the knower and the known.

Please understand that I have only emphasized that part of mirror neuron research that relates to, and indeed supports, Chicago Pragmatism, although the mirror neuron concept is more acclaimed for creating empathy. Please also note that our knowledge of human mirror neurons is indirect because in usual cases we can only measure the increases of neuronal activity in large areas of the human brain. Even here it is indirect because in the vast majority of the cases, instead of measuring neural activity for particular single neurons, it taps into the amount of oxygen this activity takes up. This is what is actually measured. An extremely complicated system called BOLD measures the ratio of oxygenated to deoxygenated hemoglobin, which carries oxygen in the blood. This ratio is what we finally measure, but there is a positive note here – at least to me. One seldom measures an object directly in science. What we measure is some indicant of an object, and this indicant seldom subsumes the whole object. This is only one reason why absolute truth is always beyond our grasp – and that is good enough for me and at least some of my fallibilistic colleagues.

Notes

1. David D. Franks, and Thomas S. Smith, *Mind, Brain and Society: Toward a Neurosociology of Emotion* (Stamford, CT: JAI Press, 1999).
2. David D. Franks. *Neurosociology: The Nexus between Neuroscience and Social Psychology* (New York: Springer Press, 2010).
3. David D. Franks, and Jonathan Turner, *The Handbook of Neurosociology* (New York: Springer Press, 2013).
4. Marco Iacoboni, *Mirroring People: The New Science of How We Connnect with Others*. (New York: Farrar, Straus, 2008).
5. Giacomo Rizzolatti, and Corrado Sinigaglia, *Mirrors in the Brain: How Our Minds Share Actions and Emotions* (New York: Oxford University Press, 2008).
6. Robert Laughlin, *A Different Universe: Reinventing Physics from the Bottom Down* (New York: Basic Books, 2006).
7. See, for example, Michael Gazzaniga, *Who's in Charge? Free Will and the Science of the Brain* (New York: Harper Collins, 2012), p. 235.
8. Hannah Arendt, *The Human Condition* (Chicago: The University of Chicago Press, 1958/1998), p. 261.
9. Ronald De Sousa, *The Rationality of Emotion* (Cambridge, MA: MIT Press, 1989).
10. Antonio Damasio, *Descartes' Error: Emotion, Reason and the Human Brain* (New York: Avon Books, 1994).
11. George Lakoff and Mark Johnson, *Philosophy in the Flesh: The Embodied Mind and Its Challenge to Western Thought* (New York: Basic Books, 1999).

12. See Gazzaniga, *Who's in Charge?*, p. 235.
13. Leslie Brothers, *Friday's Footprint: How Society Shapes the Human Mind* (Oxford and New York: Oxford University Press, 1997).
14. Franks, *Neurosociology*, p. 56.
15. John Tredway, Stan Knapp, Louise Tredway, and Thomas Darwin, 'The Neurosociological Role of Emotions in Early Socialization, Reason and Ethics,' in *Mind, Brain and Society: Toward a Neurosociology of Emotion*, pp. 81–108.
16. Damasio, *Descartes' Error*.
17. Pragmatism was introduced by earlier American philosophers, Charles Sanders Peirce and William James. The Chicago school of Dewey and Mead initiated its social modification, described here.
18. N. S. F. Northrop, *The Logic of the Sciences and the Humanities* (New York: MacMillian, 1948).
19. Damasio, *Descartes' Error*, p. 225.
20. Mustafa Emirbayer, 'Manifesto for a Relational Sociology,' *The American Journal of Sociology*, 103(2) (1997): 281–317.
21. John Dewey, and Arthur Bentley, *Knowing and the Known* (Boston: The Beacon Press, 1949).
22. Bertell Ollman, *Alienation: Marx's Conception of Man in a Capitalist Society* (Cambridge, UK: Cambridge University Press, 1971).
23. Rizzolatti and Sinigaglia, *Mirrors in the Brain*.
24. See Lisa Aziz-Zadeh and Antonio Damasio, 'Embodied Semantics for Actions: Findings from Functional Brain Imaging,' *Journal of Physiology Paris* 102(1–3) (2008): 335–376; and Liew Sook-lie, and Lisa Aziz-Zadeh, 'The Neuroscience of Language and Action in Occupations: A Review of Findings from Brain and Behavioral Sciences,' *The Journal of Occupational Science Incorporated* 18(2) (2011): 1–18.
25. Ibid.
26. Lakoff and Johnson, *Philosophy in the Flesh*.

7
Dewey's Rejection of the Emotion/Expression Distinction
Joel Krueger

Introduction

In two short essays, Dewey develops what we might call a 'hybrid' model of emotions.[1] Drawing on both Darwin and James, Dewey rejects the supposition – still common in emotion research today – that emotions are comprised of states or processes physically located within the individual subject, and specifically within the subject's brain.[2] Instead, Dewey argues that emotions are made up of both internal (neural and physiological activity, phenomenal properties, cognitive judgments) *and* external processes (expressive behavior, ongoing 'transactions' with the surrounding environment). All of these aspects are composite parts of emotions. Accordingly, to give one aspect explanatory priority at the expense of the others is to artificially sever a dynamically interrelated, distributed process spanning brain, expressive body, and world – what Dewey terms the 'concrete whole' of emotional experience.

I specifically consider Dewey's rejection of the distinction between an emotion and its behavioral expression. I show that, for Dewey, the latter is a constituent part of the former; the expressive behavior is part of the ontology of some emotions. I argue that Dewey's hybrid model not only receives support from current research in cognitive and neuroscience but, additionally, that it highlights the central role that agency and the social world play in the development and experiential character of our emotional life.

Darwin, James, and the 'functional coordination' between emotion and expression

Dewey's hybrid model of emotions stems from the integration of what he thinks are the most valuable parts of theories of emotion developed by Darwin[3] and James.[4] Dewey accepts Darwin's evolutionary naturalism as a general principle. As he writes elsewhere, '[t]o see the organism *in* nature, the brain in the nervous system, the cortex in the brain is the answer to the problems which haunt philosophy.'[5] More specifically, Dewey appropriates Darwin's naturalistic portrayal of emotions as reflected in bodily habits, or still-developing habits, oriented toward some practical end. For Dewey – following Darwin – emotions are aspects of ongoing patterns of action through which an organism successfully negotiates its biosocial world. This is their adaptive utility.[6]

Dewey argues further that this *agentive* characterization of emotions, as we might term it, helps clarify their intentional character. Emotions involve an orientation or attitude toward some object or state of affairs: '[T]he full emotional experience...is always "about" or "toward" something; it is "at" or "on account of" something, and this prepositional reference is an integral phase of the single pulse of emotion.'[7] Their teleological mooring is what distinguishes emotions from free-floating affect. For Dewey, emotions are thus both directed toward, and responsive to, features of the environment. Accordingly, highlighting the agentive and intentional character of emotions, Dewey insists, helps account both for the rationality of the behaviors associated with them as well as their capacity to further individual and communal life (that is, their evolutionary significance).[8] As Dewey puts it – again following Darwin – emotional behaviors have been selected by virtue of their usefulness, not merely for expressing felt experiences but rather for their utility '*qua* acts – as serving life': that is, as acts that have proven useful in the larger struggle for survival.[9] Emotions are in this way part and parcel of our situated agency.

Despite this positive appropriation, Dewey nevertheless rejects parts of Darwin's theory. Specifically, Dewey rejects Darwin's serial characterization of 'the relation of emotion to organic peripheral action [that is, expression], in that it assumes the former as prior and the latter as secondary.'[10] According to Darwin, the emotion itself – its affective core, its felt aspect – exists antecedently to and independently from its behavioral expression. While bodily habits and specific behaviors *reflect* emotions, they are not, strictly speaking, *parts* of the emotions they reflect. Accordingly, emotions are functionally discrete processes; they are distinct from both stimulus (a charging bear) and response (facially expressing fear; judgment that the bear is fearful, and so forth).[11]

Dewey rejects this assumption. He argues that such an atomistic and, in particular, *internalist* conception of emotions is problematic for several reasons.[12] First, it generates an artificially simplified, serial account of a phenomenon that, in its lived fullness, rather exhibits an integrated, dynamic, and multi-dimensional reality.[13] I say more about this below.

Second, this reductive distortion mischaracterizes the causal relations linking stimulus, mental state, and response. Dewy observes that, in order for a stimulus (a charging bear) to be judged as fearful, say, the stimulus must already be colored with an affective quality; it must already be perceived *as* fearful.[14] On its own, the cold cognitive judgment that a particular stimulus is fearful is not sufficient to trigger the resultant cascade of emotion-related responses (physiological, behavioral, and expressive reactions associated with fear). In other words, if the perceptual stimulus is to be part of the emotion, it must be given with an affective valence in order to explain how it is that we respond to it the way that we do.

It is here that Dewey turns to James, looking to establish an 'organic connection,' as he puts it, between Darwin's evolutionary naturalism and James's somatic theory of emotion – and in particular, James's dual-emphasis on the neurophysiological and action-oriented dimensions of emotions.[15] Dewey is particularly drawn to James because James rejects the thesis that an emotion is causally prior to and thus distinct from its expression.[16] Instead, he famously argues that emotional experience arises within, and is in fact *constituted* by, the associated neurophysiological responses and expressive action. James writes:

> Our natural way of thinking about these coarser emotions is that the mental perception of some fact excites the mental affection called the emotion, and that this latter state of mind gives rise to the bodily expression. *My theory, on the contrary, is that the bodily changes follow directly the perception of the exciting fact and that our feeling of the same changes as they occur IS the emotion.* Common-sense says...we meet bear, are frightened and run...The hypothesis here to be defended says that this order of sequence is incorrect, that the one mental state is not immediately induced by the other, that the bodily manifestations must first be interposed between, and that the more rational statement is that we first feel sorry because we cry, angry because we strike...Without the bodily states following on the perception, the latter would be purely cognitive in form, pale, colorless, destitute of emotional warmth. We might then see the bear, and judge it best to run...but we should not actually feel afraid or angry.[17]

For James, then, the perception of, say, a fearful stimulus (a bear) first triggers various instinctive neurophysiological responses. These responses – and the actions they feed into – form the basis of both our judgments (running away, we get the idea of 'bear-as-thing-to-be-run-away-from') as well as our emotional phenomenology.[18] As Dewey glosses James, our 'beating heart, trembling and running legs, sinking in stomach, looseness of bowels, etc.' are part of '*a certain act of seeing*, which by habit, whether inherited or acquired, sets up other acts' such as turning and running away or judging that the bear is fearful.[19] The important point is that, from the start, perception and action are coupled processes. The process of my seeing the bear as fearful – and *experiencing* the bear I see as fearful, or *judging* it to be so – is 'constituted via the organic co-ordination of certain sensorimotor (or ideo-motor) activities, on the one side, and of certain vegetative-motor activities on the other.'[20] In other words, our *agency* underwrites both our emotional experience and associated evaluative judgments.

Dewey does not accept James's view wholeheartedly, however. In particular, he accuses James of failing to offer a more comprehensive account of how 'to connect the emotional seizure [the experience] with the other phases of the concrete emotion-experience.'[21] For Dewey, instinctive responses are only a part of emotional functioning; other phases that must be accounted for include higher cognitive functioning, communication, and interpretation.[22] Nevertheless, Dewey notes that James was mainly concerned with the phenomenology of emotional experience and likely did not intend his theory to be taken as 'dealing with emotion as a concrete whole.'[23]

In sum, Dewey accepts Darwinian naturalism and, in particular, Darwin's association of emotions with adaptively useful actions. This characterization emphasizes both their agentive character and social function. But Dewey rejects Darwin's implicit assumption that an emotion is distinct from the behavior expressing it. This is where Dewey turns to James, who rather argues that behavioral expressions are a constitutive part of the emotion itself. Experience and expression are 'functionally coordinated' within the larger dynamic of the emotion considered as a concrete whole.

The character of expression

For Dewey, then, an emotion is partially constituted by its bodily expression. Before considering empirical evidence that appears to support this idea, I first consider how we might understand 'expression' in this context. I also situate Dewey next to some similar ideas developed within the phenomenological tradition.

Peter Goldie argues that an expression of emotion is only genuine if it is not performed as a means to some further end.[24] There are three ways that an expression of emotion can fail to satisfy this condition. First, it can be done insincerely – such as when I smile and feign happiness upon shaking hands with someone I strongly dislike. In this situation, there is no authentic emotion motivating the expression. Rather, it is performed mechanically out of respect for the social norms governing that encounter. Second, an expression can be genuine (a frown expressing anger) but performed calculatingly, that is, to intentionally convey to others that one is experiencing *this* emotion – a communicative intention which becomes the true end of the expressive act. This calculative performance removes the spontaneity at the heart of authentic emotional expression and transforms the expression from an end to a means. Third, an expression cannot be done simply for pleasure: for example, kicking a table leg out of anger to feel better. Again, this would transform the expressive act from an end to a means – namely, the goal of slightly alleviating one's anger. So, for Goldie, a genuine expression of emotion must be *sincere*, *spontaneous*, and *self-contained* (that is, an end in itself).

Even with these conditions in place, the class of actions which are authentic expressions of emotion is fairly broad. It includes things like facial expressions, gestures, whole-body movements, spontaneous touches, gaze or breathing patterns, and prosody. Characterized thusly, genuine expressions of emotion seem to inhabit a middle space between mere *bodily changes* (responses of the autonomic nervous system, hormonal changes, muscular reactions, and so forth) and *reasoned actions* that flow from emotions (actions made rationally intelligible by appealing to some combination of beliefs and desires).[25] They have an experiential significance that renders them more meaningful than the former, yet they don't lend themselves to belief-desire rationalization quite like the latter.[26] However, as we will see – and as Dewey emphasizes – they seem to be a constitutive aspect of emotional experience and thus require explicit consideration.

A number of phenomenologists make a similar claim.[27] Consider the following well-known remark by Max Scheler:

[W]e certainly believe ourselves to be directly acquainted with another person's joy in his laughter, with his sorrow and pain in his tears, within his shame in his blushing, with his entreaty in his outstretched hands, with his love in his look of affection, with his rage in his gnashing of his teeth, with his threats in the clenching of his fist, and with the tenor of this thoughts in the sound of his words. If anyone tells me that this is not 'perception' [of the emotion itself], for it cannot be so, in view of the fact that a perception is simply a 'complex of physical sensations', and that there is certainly one sensation of another person's mind nor any stimulus from such a source, I would be him to turn aside from such questionable theories and address himself to the phenomenological facts.[28]

According to Scheler, we see the mental states of others within the dynamics of their expressive behavior. This is significant because it means that there is no need to posit an additional extra-perceptual cognitive mechanism (analogical inference, and so forth) purportedly responsible for our detection of others' mentality. Rather, since mental states are observable, they can be directly perceived. This sort of direct, non-inferential social perception, Scheler argues, is sufficient for accessing others minds directly and securing our knowledge of them.[29]

Maurice Merleau-Ponty defends a similar view. Although he does not say much about emotions explicitly – his discussions of emotions are generally part of his larger treatments of aesthetics, infant cognition, intersubjectivity, and bodily expressivity[30] – his dogged advocacy of a thoroughly embodied approach to cognition makes him highly relevant. Like Scheler, Merleau-Ponty insists that mental phenomena are often directly visible in another's expressive behavior and manner of bodily comportment. He writes that

> We must abandon the fundamental prejudice according to which the psyche is that which is accessible only to myself and cannot be seen from the outside. My 'psyche' is not a series of 'states of consciousness' that are rigorously closed in on themselves and inaccessible to anyone but me. My consciousness is turned primarily toward the world, turned toward things; it is above all a relation to the world. The other's consciousness as well is chiefly a certain way of comporting himself toward the world. Thus it is in his conduct, in the manner in which the other deals with the world, that I will be able to discover his consciousness.[31]

But not only is our body 'our general medium for having a world,' according to Merleau-Ponty; additionally, 'the body is essentially an expressive space.'[32] He writes elsewhere that

> ...I do not see anger or a threatening attitude as a psychic fact hidden behind the gesture, I read anger in it. The gesture *does not make me think* of anger, it is anger itself.[33]

I perceive the grief or anger of the other in his conduct, in the face or his hands, without recourse to any 'inner' experience of suffering or anger, and because grief and anger are variations of belonging to the world, undivided between the body and consciousness, and equally applicable to the other's conduct, visible in his phenomenal body, as in my own conduct as it is presented to me.[34]

By insisting that anger, for example, is not a psychic fact hidden behind the gesture, but that it is, rather, the gesture itself – and that emotions such as anger and grief are thus 'undivided between the body and consciousness' – Merleau-Ponty rejects the idea of a split between an inner emotion and its outer expression. Like Dewey and Scheler, Merleau-Ponty denies that emotion and expression are merely *causally* related (in the sense that the former is causally antecedent to the latter) but instead insists that their relation is one of *constitution*.[35] Seeing another's angry gesture is therefore to see an outward-facing feature of their anger itself.

To return to Dewey: one argument Dewey gives in support of this idea is that, from the standpoint of the subject undergoing the emotion, there is, strictly speaking, no such thing as an expression of emotion. Rather, there is simply the emotion *as lived through* – part of which includes an 'expressive' component: the emotional experience as articulated through a particular piece of overt behavior or pattern of 'serviceable associated habits.'[36] But Dewey argues further that '[t]o rate such movements as primarily expressive is to fall into the psychologist's fallacy: it is to confuse the standpoint of the observer and explainer with that of the fact observed.'[37] From the first-person perspective we simply live through our emotions. We only speak of an emotion's 'expression' when 'looking at it from the standpoint of an observer – whether a spectator or the person himself as scientifically reflecting upon his movements, or aesthetically enjoying them.'[38] But this distinction between an emotion and its expression – insofar as it is taken to refer to a basic fact about the ontology of emotions – conflates the first- and third-person perspective. It 'names the facts not as they are, but in their second intention' (that is, from an external vantage point).[39] This is because an emotion and its expression are originally 'one organic pulse...whose reality is the whole concrete co-ordination of eye-leg-heart, etc....within this one whole of action.'[40] Again, emotions for Dewey are temporally extended modes of behavior constituted by the functional coordination of inner and outer components. This is their concrete reality.

With this phenomenological argument against the emotion/expression distinction, Dewey is thus in good philosophical company. As we will see later on, Dewey's analysis also adds an important social dimension to emotional experience that does not appear to be explicitly present in phenomenological treatments of emotions. However, I now discuss some work in cognitive and neuroscience that appears to provide empirical support for Dewey's hybrid model of emotions.

Empirical support

Emotions without a face

The first line of evidence I want to consider involves Moebius Syndrome (MS). MS is a rare form of congenital facial paralysis. It is normally complete and bilateral, and results from maldevelopment of the sixth and seventh cranial nerves.[41] People with MS lack facial animation; they also lack ocular abduction and tend to move their entire head when tracking objects in the environment.[42] People with MS thus lack access to basic physical resources that most of us take for granted when expressing emotion: the ability to facially articulate the emotion, and in so doing, provide face-related social cues to others.

This lack of facial animation affects how people with MS experience emotions. They often report feeling that, generally speaking, their emotional life is somehow less robust than it ought to be; they also report that the phenomenology of particular emotions is diminished. They feel this diminishment is connected to their lack of facial expressiveness.

Consider the following quotes. One individual with MS, James, puts the point this way:

> I have a notion which has stayed with me over much of my life – that it is possible to live in your head, entirely in your head. I think I get trapped in my mind or my head. I sort of *think* happy or I *think* sad, not really saying or recognizing actually feeling happy or feeling sad...maybe I have to intellectualize mood...I am *thinking* rather than *feeling* it.[43]

James's narrative suggests that he has adopted a kind of mentalizing stance with respect to his emotion; it becomes a process he reflectively thinks through, rather than something he experiences. Accordingly, the qualitative component of his emotional experience, if not missing entirely, is diminished. Even the process of falling in love with his wife becomes, at least at the beginning, a predominantly cognitive enterprise: 'I was probably thinking [of being in love] initially. It was some years later when I really felt in love.'[44] This feeling of disconnectedness from his emotions – which carries over into a more pervasive feeling of disconnectedness from others – leads James to summarize his relationship to his own emotions this way: 'I've often thought of myself as a spectator rather than a participant.'[45]

Other individuals with MS report similarly diminished emotional experience. Speaking of her childhood, one woman reports that, 'I did not express emotion. I am not sure that I felt emotion, as a defined concept.'[46] Another woman describes the experience of playing the piano:

> By 13 I was quite competent and I found that my fingers unleashed emotion and expression in me, even though I did not know what they were. I would play one piece again and again in various ways; happy, sad, cheeky, all jumbled up inside...I might have been in one mood but another would come out through my fingers, there were channels of all sorts of different things inside me.[47]

This mode of expression allowed her to discover a novel way to articulate – and thus experience – previously inaccessible emotional experiences.

The idea of recruiting compensatory strategies to overcome a lack of facial animation – and, in so doing, recalibrate the phenomenology of emotional experience – appears in other MS narratives. For example, another woman reports that she learned to experience certain emotions only after traveling to Spain and, after careful scrutiny, teaching herself to mimic the gestures she encountered there. She writes:

> I do not think I had emotion when I was a child but now I have it. How did I get it? From Spain. I learnt Spanish in two months but – more – they are very graphic in their emotional expression. The body language I had learnt and used at university could be exaggerated in Spain, using the whole body to express one's feelings...because of this I learnt to feel within me...because of the cultural up regulation of feeling in gesture I learnt to feel. I am not sure how I mapped gesture and feeling onto my body, but I was starting to feel then. I could feel really ecstatic, happy, for the first time ever.[48]

Others with MS tell similar stories. Many adopt alternative compensatory strategies of embodied expression – prosody, gestures, and verbalization, along with energetic artistic activities such as painting, dancing, and (as we have seen) playing the piano – to scaffold their emotional experience and support its emergence in a new and more intense format.[49]

At this point, one might object that people with MS lack a reliable criterion for assessing their purported emotional diminishment (or, for that matter, its subsequent recalibration). In other words, since MS is congenital, how can they know if their emotional experience is genuinely diminished? They simply are not in a position to provide a reliable assessment when it comes to this part of their psychological economy. However, there are other sources of evidence that appear to support MS narratives.

Consider the case of Oliver, an individual who developed Bell's Palsy – a form of facial paralysis affecting the seventh cranial nerve – while at university (Cole 1999). Oliver is instructive because he systematically developed facial immobility whilst at university, lived with it for a period of six months, and then recovered. Like those with MS, Oliver experienced bilateral paralysis; over a period of several weeks, he lost the ability to move either side of his face except for some slight movement around the eyes and eyebrows. What is particularly illuminating about this case is that, because his condition was progressive and not congenital, Oliver was in a unique position to track the diminishment of his emotional life commensurate with his gradual loss of facial animation. He describes settling into a kind of emotional limbo with his loss of facial expressivity:

> I suppose I didn't feel constantly happy, but then I didn't feel constantly sad...I felt almost as if in a limbo between feelings – just non-emotional...it was within myself, an emotion limbo. I still felt happy to see or hear something I liked, but I didn't think that I felt it as much because I was not actually smiling. I started to write a diary...writing helped a lot. Such and such has happened and I feel this. Writing allowed me to express.[50]

For Oliver, it thus appears that the physical act of writing is another kind of compensatory strategy – a 'surrogate scaffolding,' as we might put it, functioning in place of an animated face – enabling Oliver to externally express his emotions and recalibrate at least part of their diminished phenomenology. Moreover, his experience seems to lend further support to the idea that the facial expression of emotion is in fact part of the emotion itself, a kind of external material scaffolding that plays a central role in supporting its emergence.

Further evidence

In considering the idea of a reciprocal link between emotional expression and experience, we need not confine ourselves to MS narratives or other forms of facial paralysis. A wealth of other studies indicates that the manipulation of expressive behavior produces a corresponding change in emotional phenomenology.

The largest and most consistent body of evidence concerns facial expressions. For example, multiple studies have found that when subjects are induced to adopt a particular posture or emotion-specific facial expression (grimacing, frowning, and so on), they report experiencing the corresponding emotion (disgust, anger, and so forth).[51] Paula Niedenthal has surveyed extensive research indicating both that (1) that adopting emotion-specific facial expressions and postures influences preferences and attitudes (for example, subjects judge comic strips funnier when smiling, less funny when frowning, and so on), and (2) inhibition of bodily expression leads to diminished emotional experience, as well as interference in processing emotional information.[52] With respect to (1), unconsciously mimicking a conversation partner's facial expressions appears to generate a similar experience as well as a mutual feeling of empathy and rapport.[53] However, with respect to (2), deficits in spontaneous facial mimicry – in autism[54] or some instances of severe depression or melancholia[55] – can lead to diminished affect more generally, as well as an impaired ability to process emotional expressions in others.[56]

A related strand of research has investigated the experiential effects of exaggerating or minimizing emotional expressions. In one study, subjects were asked to endure a series of painful electric shocks.[57] During some of the trials, subjects were asked to exaggerate their expressive reactions to the shocks; during other trials, they were asked to inhibit them. Skin conductance was measured during these different trials. In the exaggeration condition, skin conductance was higher, and subjects reported more intense feelings of pain. In the inhibition condition, skin conductance and pain reports were diminished. Other studies of a similar form replicated these results.[58] These findings suggest that exaggerating emotional expressions intensifies emotional phenomenology whereas inhibiting expressive behavior diminishes it.

The idea of a reciprocal link between facial expression and the experience of emotion receives further powerful support from studies of individuals who receive Botox injections, which inhibit facial expressiveness (that is, a kind of voluntary MS). Individuals who receive Botox injections report a decrease in the intensity of emotional experience[59] and are slower in processing emotional language referring to expressions (such as anger and frowning) requiring the paralyzed muscle.[60] Hennenlotter et al. found that patients who receive Botox injections in frowning muscles show reduced activity in the amygdala and brainstem during the imitation of angry facial expressions.[61] As a result, both the sensory input and visceromotor output controlled by these structures is inhibited, leading to a diminished emotional experience.[62]

But it is likely that emotional experience is not just tied to facial expressivity. The 'sensorimotor activities'[63] that Dewey argues are constitutive of certain emotional experiences extend beyond an emotion's facial signature into gestures, movement, and more general modes of bodily comportment. Additional evidence seems to support the idea that emotional experiences harbor a motor component. For example, individuals who have suffered severe spinal cord injuries report less intense feelings of high-arousal emotions such as fear, anger, or sexual arousal.[64] A similar effect can be found in cases that do not involve paralysis. A number of other studies have manipulated postures and observed changes in feeling of depression, anger, fear, sadness, confidence, and pride.[65]

Finally, Caruana et al. found that intracortical microstimulation (ICMS) of two different sectors of the macaque insula – a part of the brain involved in the representation of bodily sensations, as well as emotion and emotion recognition – resulted in distinct emotion-specific behaviors (disgust and affiliative states) and autonomic responses.[66] ICMS of the anterior sector of the insula evoked disgust-related behavior comprised of both a specific motor component (grimace) and more complex context-dependent behavior (refusal of food). ICMS applied to a posterior sector evoked affiliative behavior (lip-smacking). This functional data suggests the insula plays an important role in determining both the external and internal aspects of a given emotion (that is, both experience and behavior). The more general point is that emotional experience is shaped even at the neural level by a motor component.

In sum, the wide-ranging evidence surveyed above provides compelling support for Dewey's hybrid model of emotions. While the empirical details of the precise relation between emotion and expression have yet to be clarified, it does seem that the expression of an emotion is sufficient to bring about its experience. There exists, to return to Dewey's expression, a functional coordination between internal (neurophysiological and phenomenological) and external (expressive and behavioral) parts of the concrete whole of emotion. The outward-facing visible expression of emotion is thus part of the emotion itself. Publicly perceivable behavior does not merely carry information about 'private' emotions (that is, my facial expression conveys the anger I feel inside). Rather, the latter is, at least at times, a proper part of the former; to see the expression is to see part of the emotional mind in action.

Beyond expression: emotions, agency, and the social niche

There is a further implication of Dewey's hybrid model worth discussing. It stems from the Dewey's more general contention that agency is a core feature of emotional experience. Emotions aren't things that simply happen to us. They are often things that we *do*. We enact emotions.[67] And, as Dewey reminds us with his criticism of James – a criticism that might also apply to some phenomenologists – there is generally more to an emotion than an involuntary neurophysiological or even a gestural response. Since agency is always situated – that is, embedded in encompassing physical and social contexts – other people play a material role in helping us to enact our shared emotional experiences. However, by focusing exclusively on the neurophysiological and/or gestural aspects of emotional experience, there is a danger of overlooking the extent to which emotions are socially mediated over both short- and long-term time scales. But by following Dewey's lead and stressing the agentive and socially situated character of emotional performance, we can get a clearer picture of how social and emotional processes intertwine dynamically in real time.

Dewey emphasizes this point by first pointing out the *dialogical* character of infant emotions. From the start of life, others are intimately involved in the development and negotiation of our emotional experiences. Infants use laughter and positive emotions, for example, as affiliative and communicative displays intended to prompt further interaction. Dewey observes that, within free play episodes, '[r]hythmical activities, as peek-a-boo, call out a laugh at every culmination of the transition, in an infant. A child of from one and a half to two years uses the laugh as a sign of assent; it is his emphatic "I do" or "yes" to any suggested idea to which he agrees or which suddenly meets his expectations.'[68] But the caregiver's physical interventions (gestures, facial expressions, direction of gaze, body orientation, patterns of touch and vocalization, and so forth) play a critical role in mediating this process.[69] Both agents thus co-regulate this exchange. The experience and expression of happiness arises within this shared dynamic; the emotion has both a social origin and function. Much evidence from developmental psychology supports Dewey's observation.[70]

This dialogical process does not end in childhood. As Dewey observes, emotions remain socially mediated throughout our lives.[71] Consider the following scenario: I am angry. I suspect that my wife has been unfaithful. As I brood, my imagination swells with images of how I suspect this betrayal has unfolded. Each new image intensifies my anger. But things aren't quite that simple. For anger is rarely a freestanding state. Along with my anger, I actually experience an interrelated constellation of various other emotions: *jealousy* in the face of her betrayal; *shame* at my naïve trust; *humiliation* at the thought of others finding out; *sadness* at the dissolution of a long-term commitment; *disgust* at the thought of her being physically intimate with another, and so on. Within the throes of this episode, any of these emotions may at any moment take precedence over the others without thereby cancelling out their phenomenal presence. The particular phenomenology of my anger in this context is thus conditioned by the simultaneous upwelling of a flurry of other emotions. Later, however, after some reflection and cooling off, a weary sadness may assume phenomenological prominence without completely effacing the anger that had previously burned so intensely. And when discussing the situation even later with friends, my shame and humiliation may come to the fore, preserving the anger but modifying its felt texture by diminishing its intensity and introducing a more prominent shame-dimension.

An important lesson here is that emotions are often structurally complex both in terms of their ontology (that is, they are hybrid, as we have previous discussed, composed of features both internal and external to the skin) as well as their phenomenology. It is therefore misleading to speak of 'anger' or 'sadness' as though these terms pick out neatly circumscribed experiential states. The phenomenology of our emotional life is much messier than this way of speaking would suggest. This is because emotions are very often long-term processes 'lasting even for years or a lifetime and occupying several levels or dimensions of consciousness.'[72] Moreover, as the above example affirms – and as Dewey also observes[73] – emotions evolve as we negotiate various social contexts. They are interactively constituted in the sense that they are deeply interwoven with those of other people, along with the material and ideological structures of our social niche.[74]

These observations further affirm Dewey's emphasis on the *dynamic* and *transactional* character of emotional experience – the idea, once more, that emotions are both structurally complex (that is, interwoven with other emotions, and comprised of different dimensions like physiological arousal, cognitive judgments, intentionality, felt affect, and so forth) as well as essentially temporal (that is, they evolve and develop over time).[75] When sharing my anger over my wife's infidelity with friends, my anger solicits an angry response from them, which heightens my own anger, which in turn further animates theirs, and so on. Many emotions thus emerge quite literally *between* interactants, within this ongoing mutual adjustment of action, emotion, expression, and intention.[76]

Moreover, as interactively constituted, emotions are forms of engagement or 'variations of belonging to the world.'[77] They are part of our social agency. We use emotions to construct, modify, and negotiate various aspects of our relationships with other people and with the surrounding context.[78] This is their social function, their adaptive utility. Emotional expressions are therefore more than simply the external aspect of an intensely felt feeling or physiological reaction; rather, they motivate interaction and sustain our ongoing relationships with others. They enable us to negotiate our social niche.[79]

As with Dewey's contention that an expression is part of the emotion, there is empirical support for this view. Consider audience effects. Ten-pin bowling players, for example, smile significantly more after producing a positive event (such as bowling a strike or spare) when they turn to face their friends than when they are still facing the pins.[80] The physical presence of others provides a social niche in which a smile articulates a strong social motivation: an intention to share one's happiness and to relish the further development of this experience as mediated by the affiliative displays of others. A similar effect was observed in Spanish soccer fans who issue authentic ('Duchenne') smiles in response to goals only when facing one another.[81] Even Olympic athletes, whom one would presume could barely contain their joy at reaching the pinnacle of their field, smile during medal ceremonies almost exclusively when actually receiving their gold medal – that is, when interacting with officials and the public – as opposed to non-interactive contexts such as before the ceremony (by themselves in the tunnel, away from TV cameras) or while facing their country's flag during the playing of the national anthem.[82]

Studies of audience effects on emotional experience are in this way one line of research suggesting that facial displays and other bodily expressions of emotion are mediated by the extent to which individuals can fully interact in social situations.[83] It often takes the presence of others – as well as the appropriate context – to draw an emotion out of us and help us complete it. Of course, the emotion may be initially comprised of an instinctive physiological response and behavioral expression, which gives way to subsequent appraisals, judgments, and further associated behavior. These, too, are constituents of some emotional experiences. But the central point is that, very often, we sustain this initial impulse by following through with the emotion's affective and evaluative trajectory. We exert our agency and enact the experience by falling to our knees and bodily giving ourselves over to our sobbing, unleashing our rage in an extravagant display of clenched fists and contorted facial expressions, or willfully subduing our rising joy by taking a deep breath and adopting a relaxed posture. And crucially, we do this with others, within varying social contexts. Again, our emotional agency is always situated. Others thus enter into and play a central role in shaping the affective, evaluative, and agentive trajectory of our emotional performances. The external processes within Dewey's hybrid model of emotions are therefore not merely circumscribed by the gestures and movements of individual agents. Beyond this, the people and structures of our social niches are likewise part of the external processes responsible for the emergence of many of our emotional experiences. They, too, are part of our hybrid emotional mind.

To conclude, I have tried to sketch out Dewey's hybrid model of emotion and show not only that it receives robust support from current research in cognitive and neuroscience but also, additionally, that it highlights the central role that agency and the social world play in the development and character of our emotional life. This is not to deny, of course, that emotions are partially composed of intraindividual or nonsocial components or processes. Nor am I suggesting that an individual-centered approach to emotion research is never appropriate. Nevertheless, Dewey's hybrid model reminds us that emotions are dynamic, structurally complex, and multi-dimensional. Ultimately, then, for a clearer picture of the 'concrete whole' of our emotional life, we must resist an excessively brain-centered prejudice and be mindful of the larger embodied, social, and interactive contexts within which we are always situated and from which our emotions inevitably emerge and take shape.

Notes

1. Dewey, 'The Theory of Emotion I: Emotional Attitudes,' *Psychological Review* 1(6) (1894): 553–569; and 'The Theory of Emotion II: The Significance of Emotions,' *Psychological Review* 2 (1895): 13–32.

2. See, for example, Antonio Damasio, *The Feeling of What Happens: Body and Emotion in the Making of Consciousness* (New York: Harcourt Brace, 1999); Joseph E. LeDoux, *The Emotional Brain* (New York: Simon and Shuster, 1996); Martha Nussbaum, *Upheavals of Thought: The Intelligence of Emotions* (Cambridge: Cambridge University Press, 2001); Jaak Panksepp, *Affective Neuroscience: The Foundations of Human and Animal Emotions*, 1st ed. (New York: Oxford University Press, 2004); Jesse J. Prinz, *Gut Reactions: A Perceptual Theory of Emotions* (Oxford: Oxford University Press, 2004); James A. Russell, 'Emotion, Core Affect, and Psychological Construction,' *Cognition & Emotion* 23(7) (2009): 1259; and N. Schwarz, and G. L. Clore, 'How Do I Feel about It? Informative Functions of Affective States,' in *Affect, Cognition, and Social Behavior*, ed. K. Fiedler and J. P. Forgas, (Toronto: Hogrefe, 1988), pp. 44–62.
3. Charles Darwin, *The Expression of the Emotions in Man and Animals* (Chicago: University of Chicago Press, 1872).
4. William James, 'What Is an Emotion?' *Mind* 9 (1884): 188–205; and *The Principles of Psychology*, vols 1 and 2 (New York: Dover, 1890).
5. Dewey, 'Nature, Life, and Body-Mind' in *The Later Works of John Dewey*, vol. 1, ed. Jo Ann Boydston (Carbondale: Southern Illinois University Press, 2008), p. 224.
6. Dewey, 'The Theory of Emotion,' in *The Early Works of John Dewey*, vol. 4, ed. Jo Ann Boydston (Carbondale: Southern Illinois University Press, 2008), p. 155.
7. Dewey, 'The Theory of Emotion,' p. 173.
8. Suzanne Cunningham, 'Dewey on Emotions: Recent Experimental Evidence,' *Transactions of the Charles S. Peirce Society* 31(4) (1995): 865–874.
9. Dewey, 'The Theory of Emotion,' pp. 154, 160.
10. Ibid., p. 152.
11. Cf. Jim Garrison, 'Dewey's Theory of Emotions: The Unity of Thought and Emotion in Naturalistic Functional '"Co-Ordination" of Behavior,' *Transactions of the Charles S. Peirce Society* 39(3) (2003): 405–443, p. 406; Susan Hurley terms this the 'classical sandwich model' of cognition: the idea that the 'meat' of cognition (including, for our purposes, emotion) is distinct from the 'bread' of perception and action (*Consciousness in Action* (Cambridge, MA: Harvard University Press, 1998)).
12. Cunningham 'Dewey on Emotions: Recent Experimental Evidence.'
13. Dewey, 'The Theory of Emotion,' pp. 176–180.
14. Ibid., p. 175.
15. Ibid., p. 152.
16. Ibid., p. 152.
17. James, *The Principles of Psychology*, vol. 2, p. 449.
18. Dewey, 'The Theory of Emotion,' p. 178.
19. Ibid., p. 175.
20. Ibid., p. 180.
21. Ibid., p. 172.
22. Garrison 'Dewey's Theory of Emotions,' p. 409.
23. Dewey, 'The Theory of Emotion,' pp. 171–172.
24. Peter Goldie, 'Explaining Expressions of Emotion,' *Mind* 109(433) (2000): 25–38; and *The Emotions: A Philosophical Exploration* (Oxford: Oxford University Press, 2000).

25. Goldie, *The Emotions*.
26. Goldie (in *The Emotions*) further divides expressions of emotions into those which are actions (voluntary behavior like stroking the face of one's beloved) and those which are not (involuntary behavior like facial expressions). For reasons I discuss later, within the context of emotional experience, this distinction is helpful but relatively fuzzy.
27. See Joel Krueger, and Søren Overgaard, 'Seeing Subjectivity: Defending a Perceptual Account of Other Minds,' *ProtoSociology: Consciousness and Subjectivity* 47 (2012): 239–262.
28. Scheler, *The Nature of Sympathy*, trans. Peter Heath (London: Routledge and Kegan Paul, 1954), p. 260.
29. Shaun Gallagher, 'Direct Perception in the Intersubjective Context,' *Consciousness and Cognition* 17(2) (2008): 535–543; and Krueger, 'Seeing Mind in Action,' *Phenomenology and the Cognitive Sciences* 11(2) (2012): 149–173.
30. Suzanne L. Cataldi, 'Affect and Sensibility,' in *Merleau-Ponty: Key Concepts*, ed. Rosalyn Diprose and Jack Reynolds (Stocksfield: Acumen, 2008), pp. 163–173.
31. Merleau-Ponty, "The Child's Relations with Others", in *The Primacy of Perception and Other Essays on Phenomenological Psychology, the Philosophy of Art, History, and Politics*, ed. James Edie (Evanston: Northwestern University Press, 1964), pp. 116–117.
32. Merleau-Ponty, *Phenomenology of Perception*, trans. Colin Smith (New York: Routledge, 2002), p. 169.
33. Merleau-Ponty, *Phenomenology of Perception*, p. 214.
34. Ibid., p. 415.
35. To be clear, Merleau-Ponty – like Dewey – is not saying that one's anger is *identical* to one's gesture in the sense that it is wholly *reducible* to it. This would be a crude behaviorism; Merleau-Ponty's view is more subtle than this, and readily concedes that interiority is an essential part my own and others' experience (Merleau-Ponty, *Phenomenology of Perception*, pp. 415, 424). See Krueger and Overgaard, 'Seeing Subjectivity,' for further discussion.
36. Dewey, 'The Theory of Emotion,' p. 154.
37. Ibid., p. 154.
38. Ibid., p. 154.
39. Ibid., p. 154.
40. Ibid., p. 177.
41. Wolfgang Briegel, 'Neuropsychiatric Findings of Mobius Sequence: A Review,' *Clinical Genetics* 70(2) (2006): 91–97.
42. Other parts of the condition include small tongue (which can lead to difficulties with feeding and speaking), breathing difficulties, malformation of arms or legs (missing fingers, underdeveloped calf muscles and extremely high arched feet), associated movement difficulties (clumsiness, late development sitting and standing, difficulties in running, jumping, and hopping, and so forth.) (Jonathan Cole, and Henrietta Spalding, *The Invisible Smile: Living without Facial Expression* (Oxford: Oxford University Press, 2009), pp. 2–5).
43. Jonathan Cole, 'Empathy Needs a Face,' *Journal of Consciousness Studies* 5–7 (2001), p. 62.
44. Jonathan Cole, *About Face* (Cambridge, MA: MIT Press, 1998) p. 122.
45. Cole, *About Face*, p. 28.

46. Jonathan Cole, 'Impaired Embodiment and Intersubjectivity,' *Phenomenology and the Cognitive Sciences* 3 (2009), p. 354.
47. Cole, 'Impaired Embodiment and Intersubjectivity,' p. 354.
48. Ibid., p. 355.
49. Kathleen Bogart, and David Matsumoto, 'Living with Moebius Syndrome: Adjustment, Social Competence, and Satisfaction with Life,' *The Cleft Palate-Craniofacial Journal* 47(2) (2010): 134–142; cf. Krueger, 'Affordances and the Musically Extended Mind,' *Frontiers in Theoretical and Philosophical Psychology* 4 (2013): 1003; and Krueger and John Michael, 'Gestural Coupling and Social Cognition: Möbius Syndrome as a Case Study,' *Frontiers in Human Neuroscience* 6 (2012): 1–14.
50. Jonathan Cole, "On Being Faceless: Selfhood and Facial Embodiment", in *Models of the Self*, ed. S. Gallagher and J. Shear (Exeter: Imprint Academic, 1999), p. 310.
51. Sandra E. Duclos and James D. Laird, 'The Deliberate Control of Emotional Experience through Control of Expressions,' *Cognition & Emotion* 15(1) (2001): 27–56; Sandra E. Duclos, James D. Laird, Eric Schneider, Melissa Sexter, Lisa Stern, and Oliver Van Lighten, 'Emotion-Specific Effects of Facial Expressions and Postures on Emotional Experience,' *Journal of Personality and Social Psychology* 57(1) (1989): 100–108; B. Edelman, 'A Multiple-Factor of Body Weight Control,' *Journal of General Psychology* 110 (1984): 99–114; William Flack, James D. Laird, and Lorraine A. Cavallaro. 'Separate and Combined Effects of Facial Expressions and Bodily Postures on Emotional Feelings,' *European Journal of Social Psychology* 29(2–3) (1999): 203–217; Joan M. Kellerman, and James D. Laird, 'The Effect of Appearance on Self-Perceptions,' *Journal of Personality* 50(3) (1982): 296–351; for reviews, see James D. Laird, *Feelings: The Perception of Self* (Oxford: Oxford University Press, 2007); Paula M. Niedenthal, 'Embodying Emotion,' *Science* 316(5827) (2007): 1002–1005; Paula M. Niedenthal and Marcus Maringer, 'Embodied Emotion Considered,' *Emotion Review* 1(2) (2009): 122–128.
52. Niedenthal 'Embodying Emotion'; Paula M. Niedenthal, Lawrence W. Barsalou, Piotr Winkielman, Silvia Krauth-Gruber, and Francois Ric, 'Embodiment in Attitudes, Social Perception, and Emotion,' *Personality and Social Psychology Review* 9(3) (2005): 184–211.
53. Tanya Chartrand, and John. A. Bargh, 'The Chameleon Effect: The Perception-Behavior Link and Social Interaction,' *Journal of Personality and Social Psychology* 76(6) (1999): 893–910.
54. Erno J. Hermans, Guido van Wingen, Peter A. Bos, Peter Putman, and Jack van Honk, 'Reduced Spontaneous Facial Mimicry in Women with Autistic Traits,' *Biological Psychology* 80(3) (2008): 348–353
55. Thomas Fuchs, 'Corporealized and Disembodied Minds: A Phenomenological View of the Body in Melancholia and Schizophrenia,' *Philosophy, Psychiatry, & Psychology* 12(2) (2005): 95–107.
56. See also Tedra F. Clark, Piotr Winkielman, and Daniel N. McIntosh 'Autism and the Extraction of Emotion from Briefly Presented Facial Expressions: Stumbling at the First Step of Empathy,' *Emotion* 8(6) (2008): 803–809; Daniel N. McIntosh, Aimee Reichmann-Decker, Piotr Winkielman, and Julia L. Wilbarger, 'When the Social Mirror Breaks: Deficits in Automatic, but Not

Voluntary, Mimicry of Emotional Facial Expressions in Autism,' *Developmental Science* 9(3) (2006): 295–302.
57. John T. Lanzetta, Jeffrey Cartwright-Smith, and Robert E. Eleck, 'Effects of Nonverbal Dissimulation on Emotional Experience and Autonomic Arousal,' *Journal of Personality and Social Psychology* 33(3) (1976): 354.
58. James J. Gross, 'Emotion Regulation: Past, Present, Future,' *Cognition & Emotion* 13(5) (1999): 551–573; Robert E. Kleck, Robert C. Vaughan, Jeffrey Cartwright-Smith, Katherine Burns Vaughan, Carl Z. Colby, and John T. Lanzetta, 'Effects of Being Observed on Expressive, Subjective, and Physiological Responses to Painful Stimuli,' *Journal of Personality and Social Psychology* 34(6) (1976): 1211–1218; Robert E. Kraut, 'Social Presence, Facial Feedback, and Emotion,' *Journal of Personality and Social Psychology* 42(5) (1982): 853–863; and Miron Zuckerman, Rafael Klorman, Deborah T. Larrance, and Nancy H. Spiegel, 'Facial, Autonomic, and Subjective Components of Emotion: The Facial Feedback Hypothesis Versus the Externalizer–Internalizer Distinction,' *Journal of Personality and Social Psychology* 41(5) (1981): 929–944.
59. Joshua Ian Davis, Ann Senghas, Fredric Brandt, and Kevin N. Ochsner, 'The Effects of BOTOX Injections on Emotional Experience,' *Emotion* 10(3) (2010): 433–440.
60. D. A. Havas, A. M. Glenberg, K. A. Gutowski, M. J. Lucarelli, and R. J. Davidson, 'Cosmetic Use of Botulinum Toxin-A Affects Processing of Emotional Language,' *Psychological Science* 21(7) (2010): 895–900.
61. Andreas Hennenlotter, Christian Dresel, Florian Castrop, Andres O. Ceballos-Baumann, Afra M. Wohlschlager, and Bernhard Haslinger, 'The Link between Facial Feedback and Neural Activity within Central Circuitries of Emotion – New Insights from Botulinum Toxin-Induced Denervation of Frown Muscles,' *Cerebral Cortex* 19(3) (2009): 537–542.
62. Fausto Caruana, and Vittorio Gallese, 'Overcoming the Emotion Experience/Expression Dichotomy,' *Behavioral and Brain Sciences* 35(3) (2012): 145–146.
63. Dewey, 'The Theory of Emotion,' p. 180.
64. K. Chwalisz, E. Diener, and D. Gallagher, 'Autonomic Arousal Feedback and Emotional Experience: Evidence from the Spinal Cord Injured,' *Journal of Personality and Social Psychology* 54(5) (1988): 820–828; George W. Hohmann, 'Some Effects of Spinal Cord Lesions on Experienced Emotional Feelings,' *Psychophysiology* 3(2) (1966): 143–156; see also Laird, *Feelings*, pp. 74–76; H. Mack, N. Birbaumer, H. P Kaps, A. Badke, and J. Kaiser, 'Motion and Emotion: Emotion Processing in Quadriplegic Patients and Athletes' *Zeitschrift Für Medizinische Psychologie* 14(4) (2005): 159–166.
65. See L. Berkowitz, 'Is Something Missing? Some Observations Prompted the Cognitive-Neuroassociationist View of Anger and Emotional Aggression,' in *Aggressive Behavior: Current Perspectives*, ed. L. R. Huesmann (New York: Plenum, 1994), pp. 35–58; Duclos and Laird, 'The Deliberate Control of Emotional Experience through Control of Expressions'; Duclos et al., 'Emotion-Specific Effects of Facial Expressions and Postures on Emotional Experience'; Flack et al., 'Separate and Combined Effects of Facial Expressions and Bodily Postures on Emotional Feelings'; John H. Riskind, 'They Stoop to Conquer: Guiding and Self-Regulatory Functions of Physical Posture after Success and Failure,' *Journal of Personality and Social Psychology* 47(3) (1984): 479–493; Riskind, and Carolyn C. Gotay, 'Physical Posture: Could It Have

Regulatory or Feedback Effects on Motivation and Emotion?' *Motivation and Emotion* 6(3) (1982): 273–298; Sabine Stepper, and Fritz Strack, 'Proprioceptive Determinants of Emotional and Non-emotional Feelings,' *Journal of Personality and Social Psychology* 64(2) (1993): 211–220.
66. Fausto Caruana, Ahmad Jezzini, Beatrice Sbriscia-Fioretti, Giacomo Rizzolatti, and Vittorio Gallese, 'Emotional and Social Behaviors Elicited by Electrical Stimulation of the Insula in the Macaque Monkey,' *Current Biology* 21(3) (2011 8): 195–199.
67. Giovanna Colombetti and Evan Thompson, 'The Feeling Body: Toward an Enactive Approach to Emotion,' in *Developmental Perspectives on Embodiment and Consciousness*, ed. Willis F. Overton, Ulrich Müller, and Judith Newman (New York: Lawrence Erlbaum Associates, 2008), pp. 45–68.
68. Dewey, 'The Theory of Emotion,' p. 157.
69. Cf. Krueger, 'Ontogenesis of the Socially Extended Mind,' *Cognitive Systems Research* 25–26 (2013): 40–46.
70. See Vasudevi Reddy, *How Infants Know Minds* (Cambridge: Harvard University Press, 2008); Daniel Stern, 'Vitality Contours: The Temporal Contour of Feelings as a Basic Unit for Constructing the Infant's Social Experience,' in *Early Social Cognition: Understanding Others in the First Months of Life*, ed. Philippe Rochat (Mahwah, NJ: Earlbaum, 1999), pp. 67–80; and Colwyn Trevarthen, 'The Functions of Emotion in Early Infant Communication and Development,' in *New Perspectives in Early Communicative Development*, ed. Jacqueline Nadel and L. Camaioni (London: Routledge, 1993), pp. 279–308.
71. Dewey, 'The Theory of Emotion,' p. 158.
72. Robert Solomon, 'Emotions in Phenomenology and Existentialism,' in *A Companion to Phenomenology and Existentialism*, eds. H. Dreyfus and M. Wrathall (Oxford: Blackwell Publishing, 2006), p. 303.
73. Dewey, 'The Theory of Emotion,' pp. 155–156.
74. George Downing, 'Emotion Theory Reconsidered,' in *Heidegger, Coping, and Cognitive Science: Essays in Honor of Hubert L. Dreyfus*, vol. 2, ed. Mark Wrathall and Jeff Malpas (Cambridge: MIT Press, 2000), pp. 245–270; Brian Parkinson, Agneta H. Fischer, and Antony S.R. Manstead, *Emotions in Social Relations* (New York: Psychology Press, 2005).
75. This Deweyan perspective contrasts with Carroll Izard's characterization of emotions as 'brief...responses" ("Emotions, Human," *Encyclopedia Britannica*, 1974) and Joseph LeDoux's characterization of emotions as rapid neurological (amygdala) responses distinct from the cerebral activity that generally follows them (LeDoux, *The Emotional Brain*; cf. Damasio *The Feeling of What Happens*; Panksepp, *Affective Neurology* (New York: Oxford University Press, 1992).
76. Alan Fogel, and Andrea Garvey, 'Alive Communication,' *Infant Behavior and Development* 30(2) (2007): 251–257.
77. Merleau-Ponty, *Phenomenology of Perception*, p. 415.
78. Robert A. Hinde, 'Was "The Expression of the Emotions" a Misleading Phrase?' *Animal Behaviour* 33(3) (1985): 985–992; Kym Maclaren, 'Emotional Clichés and Authentic Passions: A Phenomenological Revision of a Cognitive Theory of Emotion,' *Phenomenology and the Cognitive Sciences* 10(1) (2011): 45–65.

79. Krueger, 'Emotions and the Social Niche,' in *Collective Emotions*, ed. Christian Von Scheve and Mikko Salmela (New York: Oxford University Press, forthcoming).
80. Robert E. Kraut, and Robert E. Johnston, 'Social and Emotional Messages of Smiling: An Ethological Approach,' *Journal of Personality and Social Psychology* 37(9) (1979): 1539–1553.
81. José-Miguel Fernández-Dols and María-Angeles Ruiz-Belda, 'Spontaneous Facial Behavior during Intense Emotional Episodes: Artistic Truth and Optical Truth,' in *The Psychology of Facial Expression*, ed. James A. Russell and José-Miguel Fernández-Dols (Cambridge: Cambridge University Press, 1997), pp. 255–294.
82. Fernández-Dols, and Ruiz-Belda, 'Are Smiles a Sign of Happiness? Gold Medal Winners at the Olympic Games,' *Journal of Personality and Social Psychology* 69(6) (1995): 1113–1119.
83. See Nicole Chovil, 'Social Determinants of Facial Displays,' *Journal of Nonverbal Behavior* 15(3) (1991): 141–154. This is not to deny that we never smile or feel happy, for example, when alone. But audience effects are also present in these solitary contexts, which are often shaped by an implicit sociality (Alan J. Fridlund, 'Sociality of Solitary Smiling: Potentiation by an Implicit Audience,' *Journal of Personality and Social Psychology* 60(2) (1991): 229–240). Even when alone, we interact with others via imagination or memory, anticipation or forecast – or we might even take ourselves as an interactant.

Part III
Creativity, Education, and Application

8
Finding Unapparent Connections: How Our Hominin Ancestors Evolved Creativity by Solving Practical Problems

Robert Arp

Introduction

John Dewey and other noted pragmatists of his time utilized evolutionary insights to inform the 'lived experience' of human psychology, which included responding to one's environment in novel ways. Dewey speaks of 'finding unapparent connections' to be an important human mental ability that is at times utilized for solving problems. In this chapter, I argue that the lived, practical experiences of our hominin ancestors occasioned the emergence of an ability to solve problems creatively, in a way unique among primate species. Specifically, in this chapter I first present the ideas and arguments put forward by evolutionary psychologists that our hominin ancestors evolved certain capacities to solve nonroutine, vision-related problems creatively. I then argue that *scenario visualization* – namely, a mental activity whereby visual images are selected, integrated, and then transformed and projected into visual scenarios for the purposes of solving problems in the environments in which one inhabits – emerged in our hominin past and accounts for certain kinds of vision-related creativity. The kinds of problems with which our hominin ancestors were confronted most likely were of the practical, spatial-relation and depth-relation types, so the capacity to scenario visualize would have been useful for their survival. Thus, scenario visualization has been and still continues to be relevant for *vision-related* forms of creative problem solving.

Prepared to act

There is an oft-quoted passage from Charles Peirce where he claims that Alexander Bain (1818–1903) is probably the 'grandfather of pragmatism' because of Bain's definition of belief as 'that upon which a man is prepared to act.'[1] Our early hominin ancestors of around 2 million years ago (mya) certainly were pragmatists in the colloquial sense of the word, since the beliefs they formed almost always had to do with assessing situations and solving problems in their environments in practical, matter-of-fact ways. Simple beliefs associated with the perception that, 'He's trying to take my food,' 'We need shelter from the cold,' 'That wooly mammoth is getting away,' or 'That saber-toothed cat is coming at me' are ones upon which many an early hominin were prepared to act, with the time between preparation and action being quite short in some cases. Much of the time, *not* being prepared to act meant death of individuals and/or even entire species.

But had they the opportunity to speculate on the very nature of their beliefs – when they weren't feasting, fighting, fleeing, foraging, or engaged in other basic life-preserving activities – our early hominin ancestors likely would have subscribed to classical and other contemporary forms of *philosophical* pragmatism provided, of course, Peirce is correct about the interconnection between pragmatism and Bain's definition. Peirce, along with William James, George Herbert Mead, John Dewey, and other noted classical pragmatists around the turn of the 20th century actually utilized evolutionary insights to inform the development of human psychology in terms of belief formation and preparedness to act. In this chapter, utilizing evidence from biology, neuroscience, and evolutionary psychology, I argue that the lived, practical experiences of our hominin ancestors occasioned the emergence of an ability to form beliefs and act to solve problems creatively, in a way unique among primate species. The upshot, then, is that many of the insights of classical pragmatism (as well as many insights of contemporary pragmatists) are indeed buttressed by modern cognitive science, while at the same time these insights can continue to inform cognitive science in the neuropragmatic fashion outlined by the editors of this volume.

One such insight that is important here comes from John Dewey and his ideas surrounding what is called *lived experience*. In *Experience and Nature* (1925) Dewey notes that, given the fact that humans are organisms interacting with multiple environments, we constantly are experiencing resistance from these environments, which causes stimulation and pain that leads to either death or growth, reminiscent of Friedrich Nietzsche's 'that which does not kill us makes us stronger.' Our lived experience of the environment provides us with times when we are faced with some resistance, some problem to overcome of which we become directly cognizant. We can then use our imagination to solve the problem and make our situation a better one. 'Undergoing' such experiences, Dewey tells us, leads to 'the provocation and invitation to thought – seeking and finding unapparent connections.'[2] And as I will argue below, the finding of these unapparent connections in our imagination is akin to solving a problem creatively, in a unique way not solved by anyone in the past. And further, it is precisely creative problem solving that enabled our early hominin ancestors to outwit and outcompete other hominin species (and other species) so as to survive and endure as modern humans today.

Resourceful animals

We humans are resourceful animals. We can engage in *routine problem solving*, whereby someone recognizes many possible solutions to a problem given that the problem was solved through one of those solutions in the past. People constantly perform routine problem solving activities that are concrete and basic to their survival, equipping them with a variety of ways to 'skin' the proverbial cat, as well as enabling them to adapt to situations, and reuse information in similar environments. However, we also can engage in activities that are more abstract and creative – finding those 'unapparent connections' Dewey talks about – such as invent new tools based upon mental blueprints, synthesize concepts that, at first glance, seemed wholly disparate or unrelated, and devise novel solutions to problems. If a person decides to pursue a *wholly new way* to solve a problem, one that has not been tried before, then that would be an instance of *nonroutine creative problem solving* (NCPS). Because a person does not possess a way to solve the problem already, NCPS involves finding a solution to a problem that has not been solved previously. Not only do humans invent products, they manufacture space shuttles, successfully negotiate environments, hypothesize, thrive, flourish, and dominate the planet by coming up with wholly novel solutions to problems. In short, they solve all kinds of nonroutine problems in creative ways. How is this NCPS possible?

In field observations and in controlled laboratory experiments, we witness and document numerous animals engaging in fairly sophisticated forms of problem solving; however, even the most advanced animals – such as chimps – achieve the problem-solving capacities of a normal human 3- or 4-year old, and no animal to date has been able to solve the kinds of problems that even our earliest hominin ancestors apparently were able to solve, like the problem of how to kill a mammoth without getting yourself killed, which was solved simply by placing a flake on the end of a stick to produce a projectile such as the spear.[3]

Archeologist Steven Mithen has offered a plausible evolutionary account of how the human mind evolved an ability to use mental images creatively so as to generate novel pieces of artwork, invent tools, and solve *nonroutine* problems.[4] Several evolutionary psychologists believe that these complex cognitive abilities are the result of specified Swiss-Army-knife-like mental modules (There are numerous versions of this idea.) having evolved in our early hominin Pleistocene past (approximately 2.5 mya to 12,000 years ago) to deal with the various and sundry problems a human might have experienced. Mithen shows the deficiency in this position and argues that creativity is possible because the mind has evolved what he calls *cognitive fluidity*, an ability to exchange information flexibly between mental modules – or, to use a term from Arthur Koestler, an ability to *bissociate*. Cognitive fluidity and bissociation are akin to Dewey's notion of the human imagination's ability to find unapparent connections. In fact, according to Mithen, cognitive fluidity *is* conscious reasoning, our uniquely human mental ability. This is a plausible view that has been well received in the literature concerning the evolution of consciousness, imagination, and creativity.[5]

Mere cognitive fluidity and bissociation cannot be the full story, however. My claim in this chapter is that what I call *scenario visualization* emerged as a mental property to act as a kind of metacognitive process that selects and integrates relevant visual information from psychological modules, in order to perform vision-related, NCPS tasks in environments. However, if this kind of mental activity were *merely* free flow of information – as suggested by Mithen – there would be no mental coherency; the information would be chaotic and directionless, and not really *informative* at all. Data need to be segregated and integrated so that they can become informative for the cognizer; in fact, selecting and integrating visual information from mental modules is the function of scenario visualization.

I have argued for my scenario visualization view in the past, and not only has it been applauded as 'innovative and interesting,' and even 'ambitious,' it also has been utilized by numerous philosophical psychologists, cognitive scientists, AI researchers, and others in their own work.[6]

Thus, the view likely has *at least* initial plausibility. Still, I have critics, and I welcome the continued dialogue concerning the evolution of the human mind. Although I desire to explain the specific ways various researchers have utilized my scenario visualization view, as well as offer numerous responses to my critics, given space limitations here – coupled with the nature of this book – I will stick to the basic plan of explaining, and arguing for, scenario visualization as a plausible hypothesis associated with the evolution of our mental architecture.

Evolutionary psychology and the Swiss Army knife modular mind

According to many evolutionary psychologists, the mind is like a Swiss Army knife loaded with specific mental *tools* that evolved in the Pleistocene epoch (which began some 2.5 mya and lasted until about 12,000 years ago) to solve specific problems of survival, such as face recognition, mental maps, intuitive mechanics, intuitive biology, kinship, language acquisition, mate selection, and cheating detection. The list of mental tools could be longer or shorter, and there are many variations of the Swiss Army knife model, with the human mind having evolved one larger all-purpose tool to complement the more-specified tools, or several dual-purpose tools coexisting with several more-specified tools, or any combination thereof.

Evolutionary psychologists speak of these mental modules as domains of specificity. What this means is that any given module handles only one kind of adaptive problem to the exclusion of others. Modules are encapsulated in this sense and do not share information with one another. For example, one's cheater-detection module evolved under a certain set of circumstances and had no direct connection to one's fear-of-snakes module, which evolved under a different set of circumstances. This kind of encapsulation works best for environments where the responses need to be quick and routine; such developments enabled these organisms to respond efficiently and effectively in their regular or accustomed environments.[7]

A problem for the Swiss Army knife modular mind

There seems to be a fundamental flaw, however, in the evolutionary psychologist's reasoning. If mental modules are encapsulated, and are designed to perform certain *routine* functions, how can this modularity account for *novel* circumstances? When routine perceptual and knowledge structures fail, or when atypical environments present themselves, it is *then* that we need to be innovative in dealing with this novelty. Imagine the Pleistocene epoch. The climate shift in Africa from jungle life to desert savanna life forced our early hominins to come out of the trees and survive in totally new environments. Given a fortuitous genetic code, some hominins re-adapted to the new African landscape, some migrated elsewhere to places like Europe and Asia; and most died out. This environmental shift had a dramatic effect on modularity, since now the specific content of the information from the environment in a particular module was no longer relevant. *The information that was formerly suited for jungle life could no longer be relied upon in the new environment of the savanna.* Appeal to modularity alone would have led to certain death and extinction for our hominin ancestors.

The successful progression from the typical jungle life to the atypical and novel savanna life of our early hominin ancestors would have required some other kind of mental capacity to emerge that could creatively handle the new environment. But how is it that we can be creative?

Mithen and cognitive fluidity

Steven Mithen advanced the evolutionary psychologists' modular mind by introducing *cognitive fluidity*, which enables one to respond creatively to novel environments. Mithen sees the evolving hominin mind as going through a three-step process beginning prior to 6 mya when the primate mind was dominated by what he calls a *general intelligence*. This general intelligence was similar to chimpanzee mindedness in that it consisted of an all-purpose, trial-and-error learning mechanism that was devoted to multiple tasks where all behaviors were imitated, associative learning was slow, and there were frequent errors made.

The second step coincides with the evolution of the *Australopithecine* line (4 to 2 mya), and continues all the way through the *Homo* lineage (beginning 2.4 mya) to *H. neandertalensis* (400,000 to 30,000 ya). In this second step, multiple *specialized intelligences*, or modules, emerge alongside general intelligence. Associative learning within these modules was faster, so more complex activities could be performed. Compiling data from fossilized skulls, tools, foods, and habitats, Mithen concludes that *H. habilis* (2.4 mya) probably had a general intelligence as well as modules devoted to social intelligence (because they lived in groups), natural history intelligence (because they lived off the land), and technical intelligence (because they made tools). *Neandertals* and *H. heidelbergensis*

(800,000 to 300,000 ya) would have had all of these modules, including a primitive language module, because their skulls exhibit bigger frontal and temporal areas – areas that in the modern human brain are engaged in language functioning. According to Mithen, the *neandertals* and *H. heidelbergensis* had the Swiss Army knife mind that the standard evolutionary psychology account describes.

Now, a problem arises of which Mithen, too, is aware: it cannot be the case that the emergence of distinct mental modules that evolutionary psychologists today postulate as accounting for learning, negotiating, and problem solving took place *during the Pleistocene*. The potential variety of problems encountered in generations subsequent to the Pleistocene was too vast for a limited Swiss Army knife mental repertoire; there were too many hypothetical situations for which *nonroutine creative problem solving* would have been needed in order to survive and dominate the earth. There are potentially an *infinite number* of problems confronting animals constantly as they negotiate environments. That we negotiate environments so well shows that we have some capacity to handle the various and sundry *potential nonroutine* problems that arise in our environments.

Here is where the third step in Mithen's evolution of the mind, known as *cognitive fluidity*, comes into play. In this final step – which coincides with the emergence of modern humans – the various mental modules are working together with a flow of knowledge and ideas between them. The modules can now influence one another, resulting in an almost limitless capacity for imagination, learning, and problem solving. The working together of the various mental modules as a result of this cognitive fluidity *is* consciousness for Mithen and represents the most advanced form of mental activity.[8]

Cognitive fluidity and creativity

Mithen notes that his model of cognitive fluidity accounts for human creativity in terms of problem solving, art, ingenuity, and technology. His idea has initial plausibility, since it is arguable that humans would not exist today if they had not evolved consciousness to deal with novelty. No wonder, then, Francis Crick maintains that 'without consciousness, you can deal only with familiar, rather routine situations or respond to very limited information in new situations.'[9] Also, as John Searle observes: 'one of the evolutionary advantages conferred on us by consciousness is the much greater flexibility, sensitivity, and creativity we derive from being conscious.'[10]

Mithen's idea resonates with what researchers refer to as *bissociative creativity* and creative problem solving. It also resonates with Dewey's idea of the human ability to find unapparent connections in our thinking and imagining. Scientists have documented chimps looking pretty creative in their problem solving by trying a couple of different ways to get at fruit in a tree – like jumping at it from different angles or jumping at it off tree limbs – before finally using a stick to knock it down. In fact, several observations have been made of various kinds of animals engaged in imitative behaviors that look like creative problem solving.[11]

However, the number of possible solutions is limited in these examples of routine problem solving because the mental repertoire of these animals is environmentally fixed, and their tool usage (if they have this capacity) is limited. In fact, all attempts to get chimpanzees and other primates to imitate the basic knapping method utilized by *Homo habilis* (2.33 to 1.4 mya), for example – where essentially a stone tool is used to knap (strike and chip) to make another stone tool – have failed.[12]

Unlike routine problem solving, which deals with associative connections within familiar perspectives, nonroutine creative problem solving entails an innovative ability to make connections between *wholly unrelated* perspectives or ideas. A human seems to be the only kind of being who can solve nonroutine problems *on her or his own, without imitation or help*. Arthur Koestler referred to this quality of the creative mind as a *bissociation of matrices*. When a human bissociates, that person puts together ideas, memories, representations, stimuli, and the like in wholly new and unfamiliar ways *for that person*. Humans *bissociate*, and are able to ignore normal associations, trying out *novel* ideas and approaches in solving problems. Bissociation also has been pointed to as an aid in accounting for the ability to laugh, hypothesis-formation, art, technological advances, and the proverbial 'ah-hah,' creative insight eureka moments humans experience when they come up with a new idea, insight, or tool.

So, when we ask how it is that humans can be creative, part of what we are asking is how they bissociate, viz., *juxtapose formerly unrelated ideas in wholly new and unfamiliar ways for that person*. To put it colloquially, humans can take some visual perception, concept, or idea found 'way over here in the left field' of the mind and make some coherent connection with some other wholly disparate and unrelated visual perception, concept, or idea found 'way over here in the right field' of the mind. Again, in Dewey's terms, we are adept at finding unapparent connections between and among our thoughts. And humans seem to be the only species that can engage in this kind of mental activity.

Scenario visualization: advancing Mithen's view

Mithen's account of cognitive fluidity allows for the free movement of information between modules (Koestler's bissociation[13] or Dewey's finding unapparent connections). I believe this is important as a *precondition for* mental activities, such as imagination, that require the simultaneous utilization of several modules. So, for example, Mithen would think that totemic anthropomorphism associated with animals in, say, a totem pole made up of part-human and part-animal figures, derives from the free flow of information between a natural history module dealing specifically with animals and their characteristics, and a social module dealing specifically with people and their characteristics. A totem carved out of wood is the *material* result of the free flow of information between the natural history and social modules that occurred in the mind of the artist.

Mithen's model is unsatisfactory, however, because he makes consciousness out to be a passive phenomenon. On his account, consciousness is just a flexible fluidity, and this does not seem to me to be the full account of consciousness. When we are engaged in conscious activity, we are *doing* something. I am suggesting that mental activities associated with the selecting and integrating of visual information from mental modules for the purposes of negotiating environments are essential to creative problem solving, *and* that Mithen's account of cognitive fluidity acts as a precondition for the possibility of the information contained in these modules to intermix. So on one hand, Mithen is correct about the possibility of information between and among mental modules as intermixing, and he is correct that cognitive fluidity probably is a better description of our mental architecture, given the early hominin ability to survive in the ever-changing Pleistocene environments. On the other hand, I am transforming, and adding to, Mithen's account by arguing that possible intermixing of modular information is not the full story concerning vision-related, creative problem solving.

I am arguing for a view I call *scenario visualization*, which is

> *a mental process that entails selecting pieces of visual information from a wide range of possibilities, forming a coherent and organized visual cognition, and then projecting that visual cognition into some suitable imagined scenario, for the purpose of solving some problem posed by the environment which one inhabits.*

In my example of totemism (above), the images utilized had to be *selected from* other visual images as relevant. In the totem, visual information from both the social and the natural history modules is *synthesized*, allowing for something sublimated or innovative *to emerge anew* as a result of the process.

I think that scenario visualization comes to light most clearly when humans engage in vision-related forms of problem solving. I am not suggesting that people *always* visualize or *never* use semantic forms of reasoning, or other forms of reasoning, when solving nonroutine problems. Whether one utilizes scenario visualization most likely will depend upon the type of problem with which one is confronted. There are some problems – for example, certain mathematical problems – that can be solved without the use of scenario visualization. Other problems, like spatial relation or depth perception problems, may require scenario visualization. The kinds of problems with which our hominin ancestors were confronted most likely were of the spatial-relation and depth-relation types related to basic survival – such as judging the distance between an object and oneself, determining the size of an approaching object, matching an object to any number of associated memories, anticipating the need for a particular kind of tool to accomplish a task – and so the capacity to scenario visualize would have been useful for their survival. Thus, scenario visualization has been, and continues to be, relevant for *vision-related* forms of creative problem solving.

Scenario visualization and toolmaking

It is generally agreed upon by biologists, anthropologists, archeologists, and other researchers that a variety of factors contributed to the evolution of the modern human brain, including bipedalism, diversified habitats, social systems, protein from large animals, higher amounts of starch, delayed consumption of food, food sharing, language, and toolmaking.[14] It is not possible to get a complete picture of the evolution of the brain without looking at all of these factors, since brain development is involved in a complex coevolution with physiology, environment, and social circumstances. The emergence of language in our species clearly occupies a central place with respect to our ability to flourish and dominate the earth. However, I wish to focus on toolmaking as essential in the evolution of the brain and visual system, and I do this for four reasons:

1) First, toolmaking is the mark of intelligence that distinguishes the *Australopithecine* genus from the *Homo* genus in our evolutionary past.

Homo habilis was the first toolmaker, as meaning the Latin name, 'handyman,' denotes.
2) Second, tools offer us indirect but compelling evidence that psychological states emerged from brain states. In trying to simulate ancient toolmaking techniques, archeologists have discovered that certain tools can only be made according to *mental* templates.[15]
3) Third, as I mentioned above, our hominin ancestors weren't solving math problems; they were concerned with recognizing and discerning prey, predator, friend, and/or foe (and other basic survival activities), so the capacity to scenario visualize with respect to toolmaking would have been useful for their survival.
4) Finally, as I attempt to show, the evolution of toolmaking parallels the evolution of visual processing in terms of scenario visualization.

The breakthrough in tool technology that is central to my scenario visualization theory was the Mousterian industry that arrived on the scene with the *H. neandertalensis* lineage, near the end of the *H. heidelbergensis* lineage, around 300,000 ya. Mousterian techniques involved a more complex three-stage process of constructing (a) the basic core stone, (b) the rough blank, and (c) the refined finalized tool. Such a process enabled various kinds of tools to be created, since the rough blank could follow a pattern that ultimately could become cutting tools, serrated tools, flake blades, scrapers, or lances. Further, these tools had wider application as they were being used with other material components to form handles and spears, and were being used to make other tools, such as wood and bone artifacts. Consistent with the increase in complexity of toolmaking, the brain of *H. heidelbergensis* and *H. neandertalensis* increased to 1,200 cm and 1,500 cm in volume, respectively, up 300–600 cm from *H. erectus*.[16]

By 40,000 ya, some 60,000 years after anatomically modern *H. sapiens* evolved, we find instances of human art in the forms of beads, tooth necklaces, cave paintings, stone carvings, and figurines. This period in tool manufacture is known as the Upper Paleolithic, and it ranges from 40,000 ya to the advent of agriculture around 12,000 ya. Sewing needles and fish hooks made of bone and antlers first appeared, along with flaked stones for arrows and spears, burins (chisel-like stones for working bone and ivory), multi-barbed harpoon points, and spear throwers made of wood, bone, and antler.[17]

I suggest that scenario visualization emerged as a natural consequence of the development of a complex nervous system in association with environmental pressures that occasioned its evolution. In attempts to recreate early hominin tools from the later Mousterian and Upper Paleolithic industries, archeologists have shown that the construction of such tools would require several mental visualizations, as well as numerous revisions of the material, so as to attain optimal performance of such tools.[18] Such visualizations likely included the abilities to, at least, identify horizontal or vertical lines, select an image from several possible choices, distinguish a target figure embedded in a complex background, construct an image of a future scenario, project an image onto that future scenario, as well as recall from memory the particular goal of the project. If an advanced form of toolmaking acts as a mark of the most advanced mind, then given complex and changing Pleistocene environments, as well as the scenario visualization that is necessary to produce tools so as to survive these environments, what I am suggesting is that visual processing most likely was the primary way in which this advanced mind emerged on the evolutionary scene.

Evolution of the javelin

In what follows, I trace the development of the multi-purposed javelin from its meager beginnings as a stick, through its modifications into the spear, and finally, its specialization into a javelin equipped with a launcher. We need an example that illustrates the emergence of scenario visualization in our evolutionary past, and the development of this tool gives us concrete evidence. The following story is meant to be presented as a plausible account of how scenario visualization would have emerged in our early hominin past and, like most evolutionary stories, it is not meant to be an account for which we have *decisive* evidence.

Step 1: the stick

We can take present-day chimpanzee activities to be representative of early hominin life, and we can see that chimps in their native jungle environments do indeed use tools. The chimps use rocks, leaves, and sticks to crack open nuts, carry items, fish for termites, and hit in self-defense or in attack. This is probably what our early hominins did while in the jungles, savannas, and grasslands of Africa.[19]

As previously mentioned, chimps engage in trial-and-error and imitative learning. Baby chimps try to imitate the actions of older chimps, including the tool usage. Researchers have tried to get chimps to use tools to make other tools with cobbles and stones (the way early *H. habilis* is likely to have done) by flaking and edging, but they cannot do it. So it seems that chimps can form and recall visual images from memory when using tools. But they clearly do not have the capacity to produce tools like those found in the Upper Paleolithic industry; their tool usage merely is imitative, and wholly lacking in innovation.[20]

When the climate changed and early hominins moved from the jungles to forage food on African savannas, they constructed javelins they could throw from a distance in order to kill prey. One could continue to hit prey with a stick until it dies, as was done in jungle environments. This may work for some prey, but what about the ones that are much bigger than you? Imagine being stuck on the savanna with a stick as your only tool of defense against wool mammoths and saber-toothed tigers. Stated simply, you would need to become more creative in your toolmaking just to survive.

The progression from stick to javelin went through its own evolution that is indicative of the advance from visual processing to scenario visualization. The kind of toolmaking that our early *Homo* ancestors engaged in was likely to be little more than trial-and-error or imitative learning that was passed on from generation to generation. Flakes were constructed. So too, sticks were constructed. Apparently, however, it never occurred to members of these species to place one of their flaked stones on the end of a stick.[21]

Step 2: the spear

By the end of the Mousterian industry, archaic *H. heidelbergensis* and *H. neandertalensis* had adopted a three-step stone-forming process, which allowed for the construction of a variety of tools. Also, stone flakes were placed on the ends of sticks as spears. The most basic step in constructing a stone tool has to do with simply striking a flake from a cobble.[22]

When we consider that our early hominin ancestors not only had to select certain materials that were appropriate to solve some problem in a particular environment, but that they also utilized a diverse set of stone working techniques involving a number of steps that resulted in a variety of tool types, it becomes apparent that a fairly advanced form of mental activity had to occur. Striking a sequence of flakes (knapping) in such a way that each one aids in the removal of others demands much more control of the brain, as well as a hand equipped with a variety of grips. The various steps in the process must be evaluated, and it may be the case that previous steps are seen in light of future steps.[23]

It is safe to say that the variety of tools constructed is evidence that these hominins were visualizing future scenarios in which these tools could be used; otherwise, *what would be the point of constructing a variety of tools in the first place?* Chimps use the same medium of sticks or rocks to hit, throw, or smash. However, the construction of a variety of tools indicates that the tools have a variety of purposes. What is a purpose in this context, other than the formation of a visual image, the projection of that visual image onto some future scenario, and the intent to act on said visualization? The variety of tools is the material result of purposive scenario visualization.

Step 3: the javelin

Around 40,000 ya, 60,000 years after the emergence of modern humans, we find evidence of a variety of types of javelins, spears, and javelin launchers. Archeologists have shown that the construction of a javelin would require several mental visualizations, as well as numerous revisions of the material, so as to attain optimal performance of such a tool. Such visualizations likely included the abilities to: (a) identify horizontal or vertical lines; (b) select an image from several possible choices; (c) distinguish a target figure embedded in a complex background; (d) construct an image of a future scenario; (e) project an image onto that future scenario; and (f) recall from memory the particular goal of the project in the first place.[24]

Different types of javelins with different shaped heads and shafts were constructed, depending upon the kind of kill or defense anticipated. If our early hominin ancestors tried simply to walk up to and hit a large animal, they likely would have been killed. In fact, this is probably what happened on more than one occasion to the early hominin working out of the environmental framework of the jungle in this totally new environmental framework of the savanna. Eventually our ancestors, such as *H. neandertalensis*, developed the spear; however, the evidence suggests that they could only develop spears, and not javelins. *H. sapiens sapiens* developed javelins, equipped with launchers, that could be used in creative ways to not only throw from a distance, but also to spear at close range, hack, and cut.[25]

Our hominin ancestors were living in social groups, watching and learning from each other. I am not suggesting that scenario visualization occurs in some solipsistic vacuum. Just as with other primates, our ancestors would have learned a lot from trial-and-error, and other forms of memetic expression, in their social groups. At the same time, we can think of the proverbial 'mad scientist' who might lock him or herself away to work on some problem into which he or she has some insight. There are always innovators present in every social group. My suggestion is that, by 40,000 ya, the brains of our hominin ancestors were fortunate enough, through genetic variability, to have the right connections in their neural hardware so as to allow for the possibility of scenario visualization. With these *neural* connections already in place, all that was needed was some environmental cue to prompt the *psychological* connections, inferences, and insights to be made. All it takes is some psychologically creative 'good trick' (to use the words of Daniel Dennett) – implemented, possibly, by even one hominin – to get the creative juices flowing, so to speak, and prompt scenario visualization in our hominin ancestry. I would imagine that there would have been a complex interplay of trial-and-error, creative learning, and implementation occurring in our hominin lineage with respect to negotiating environments, just as there is today.

Through the fortunes of genetic variability and natural selection, the brains of our hominin ancestors would have needed all the right neural connections in place to allow for scenario visualization. The hominins were living in social groups, learning from each other, and implementing behaviors through trail-and-error. This good trick is just that: a *useful device* for handling certain vision-related problems encountered by our ancestors, and the ones who could utilize it survived so as to pass their genes and memes (trial-and-error kinds as well as more innovative kinds) on to the next generation. Those of us in our species living today still retain this capacity.

The harpoon

Below is a diagram that has to do with the construction of a harpoon. This schematization is supposed to represent the slower, intelligent processes associated with one of our early hominin ancestor's ability to scenario visualize.

The diagram (Figure 8.1) is based upon information gathered regarding the Angmagsalik hunters of Greenland and their construction of harpoons utilized to hunt seals. Their harpoons are fairly complex, having a spearhead equipped with a line attached to a flotation device, as well as several other parts designed to make the harpoon sturdy, accurate, and easy to throw. These hunters are an interesting case because it is likely that their harpoon technology has not changed much in thousands of years; thus, their technology can be studied to get a sense of what early hominin toolmaking may have been like.

In the schematization, I ask you to imagine that the problem to be solved has to do with throwing a projectile at a seal from a distance, for the purposes of killing it, skinning it, and using its body parts for food and warmth during the approaching winter months. I also ask you to imagine that this is the *very first instance* of some hominin coming up with the idea of the harpoon.[26] At first, this particular hominin has no prior knowledge of the harpoon, but through the process of scenario visualization, she or he eventually 'puts two and two together' and devises the mental blueprint for the harpoon. In other words, this is supposed to be a schematization of bissociative, nonroutine creative problem solving at work in the early hominin mind.

In the first step, the hunter has separate visual images associated with seal characteristics, the properties of objects in water, the manufacture of the bi-faced hand-axe, and projectiles moving through the air. Consistent

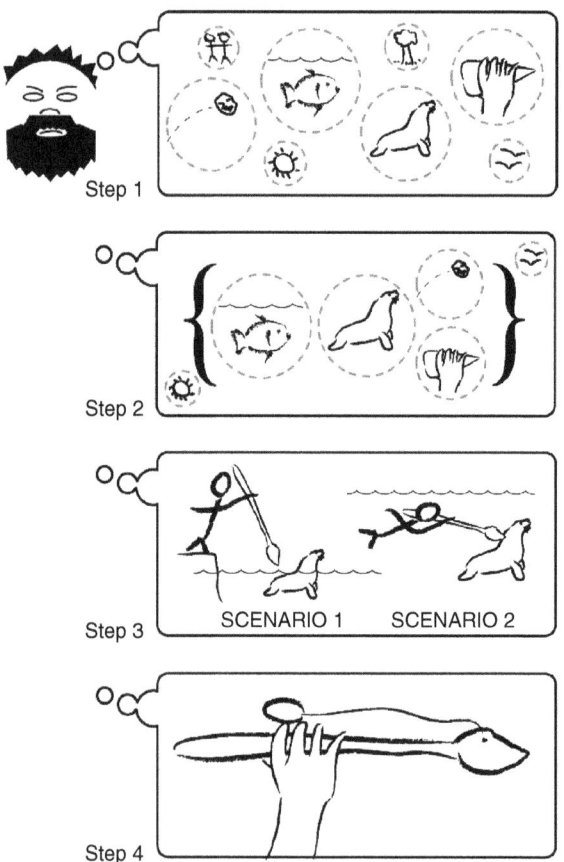

Figure 8.1 The construction of a harpoon

with Mithen's idea of cognitive fluidity, the visual information among these mental spheres has the potential to intermix and is represented by the dotted-line bubbles. Further, consistent with the data presented by developmental and evolutionary psychologists, there are several mental modules (dotted-line bubbles) that make up a person's mind. In the second step, scenario visualization is beginning as the animal, biological, technological, and intuitive physics modules are bracketed off or segregated from the other mental modules. In the third step, the process of visualization is continuing because the hominin is manipulating, inverting, and transforming the images as they are projected into a future imagined scenario. In the fourth step, these modules are actively integrated so that a wholly new image is formed that can become implemented in the actual production of the harpoon.

Conclusion

While in graduate school, I read Dewey's *The Quest for Certainty* (1929), and it was one of the most important books for my philosophical formation. I still have such positive feelings when I think of the book. Pragmatism always featured in my thinking, even when I began specializing in philosophy of mind and biology. It was refreshing and enlightening to go back to Dewey and other pragmatists recently to see what pearls of wisdom can be found in their works. As the reader can see from this chapter, Dewey's idea of 'finding unapparent connections' from *Experience and Nature* is quite the treasure and meshes nicely with my idea of scenario visualization. Further, it was a series of practical problems encountered by our early hominin ancestors over several millennia that occasioned creative problem solving, so that the pragmatic maxim of the interdependency of thought and action truly can be witnessed in the very evolution of humanity. I hope that I have given a plausible account concerning a certain aspect of our mental architecture. Finally, in their chapter on the definition and characteristics of neuropragmatism, the editors lay out Twelve Theses of Neuropragmatism, and the reader will notice that what has been communicated in my chapter buttresses and reinforces the following theses:

4. Cognitive systems are dynamically adaptive to organism–environment interactions, to deal with shifting conditions of situations as practical goals are pursued.
5. Under pressures from dealing with the environment, the brain modifies its neural connections to improve practical performance. The

measure of this neural learning is improved habitual efficiency at specific routine tasks.
6. Complex cognitive processes are the brain's work of effectively coordinating behavior for reliably achieving variable goals in a changing environment.

Notes

1. Charles Sanders Peirce, 'Pragmatism: Historical Affinities and Genesis,' in *The Collected Papers of Charles Sanders Peirce*, vol. 5, ed. Charles Hartshorne and Paul Weiss (Harvard: Harvard University Press, 1934), para. 12, p. 7.
2. John Dewey, *Experience and Nature*, in *The Collected Works of John Dewey: The Later Works, 1925–1953*, ed. Jo Ann Boydston (Carbondale, IL: Southern Illinois University Press, 1925/1981), p. 246.
3. See the numerous papers in Elizabeth Lonsdorf, Stephen Ross, and Tetsuro Matsuzawa, eds., *The Mind of the Chimpanzee: Ecological and Experimental Perspectives* (Chicago: University of Chicago Press, 2010); also John Pearce, *Animal Learning and Cognition: An Introduction* (New York: Psychology Press, 2008).
4. See Steven Mithen, *The Prehistory of the Mind: The Cognitive Origins of Art, Religion and Science* (London: Thames and Hudson, 1996); also *The Singing Neanderthals: The Origins of Music, Language, Mind and Body* (London: Weidenfeld and Nicolson, 2005).
5. See the bibliographies of my book and most recent paper: *Scenario Visualization: An Evolutionary Account of Creative Problem Solving* (Cambridge, MA: MIT Press, 2008); 'The Evolution of Scenario Visualization and the Early Hominin Mind,' in *Origins of Mind*, ed. Liz Swan (London: Springer, 2013), pp. 143–159.
6. Perform a Google Scholar search using my name and 'scenario visualization' to find the numerous researchers referring to, or utilizing, my idea.
7. Jaime Confer, Judith Easton, Diana Fleischman, Cari Goetz, David Lewis, Carin Perilloux, and David Buss, 'Evolutionary Psychology: Controversies, Questions, Prospects, and Limitations,' *American Psychologist* 65 (2010): 110–126; Simon Hampton, *Essential Evolutionary Psychology* (Thousand Oaks, CA: SAGE Publishers, 2010).
8. Mithen, *The Prehistory of the Mind*, and *The Singing Neanderthals*.
9. Francis Crick, *The Astonishing Hypothesis* (New York: Simon & Schuster, 1994), p. 20.
10. John Searle, *The Rediscovery of the Mind* (Cambridge, MA: MIT Press, 1992), p. 109.
11. Lonsdorf, et al., *The Mind of the Chimpanzee*; Pearce, *Animal Learning and Cognition*; also Jacob Norris and Mauricio Papini, 'Comparative Psychology,' in *The Corsini Encyclopedia of Psychology*, ed. Irving Weiner and Edward Craighead (Malden, MA: Wiley-Blackwell, 2010), pp. 507–520.
12. Lonsdorf, et al., *The Mind of the Chimpanzee*; also Sophie De Beaune, Frederick Coolidge, and Thomas Wynn, *Cognitive Archeology and Human Evolution* (Cambridge, UK: Cambridge University Press, 2009).

13. Leslie Merchant and William McGrew, 'Percussive Technology: Chimpanzee Baobab Smashing and the Evolutionary Modeling of Hominid Knapping,' in *Stone Knapping: The Necessary Conditions of a Uniquely Hominid Behaviour*, ed. Valentine Roux and Blandine Bril (Cambridge, UK: McDonald Institute Monograph Series, 2005), pp. 339–348; Iain Davidson and William McGrew, 'Stone Cultures and the Uniqueness of Human Culture,' *Journal of the Royal Anthropological Institute* 11 (2005) pp. 793–817; Stephen Ambrose, 'Paleolithic Technology and Human Evolution,' *Science* 291 (2001), 1748–1753.
14. Chris Stringer, *Lone Survivors: How We Came to Be the Only Humans on Earth* (New York, Henry Holt, 2012); Ian Tattersall, *Masters of the Planet: The Search for Our Human Origins* (New York: Palgrave Macmillan, 2013).
15. Leslie Merchant and William McGrew, 'Percussive Technology: Chimpanzee Baobab Smashing and the Evolutionary Modeling of Hominid Knapping,' in *Stone Knapping: The Necessary Conditions of a Uniquely Hominid Behaviour*, ed. Valentine Roux and Blandine Bril (Cambridge, UK: McDonald Institute Monograph Series, 2005), pp. 339–348; Iain Davidson and William McGrew, 'Stone Cultures and the Uniqueness of Human Culture,' *Journal of the Royal Anthropological Institute* 11 (2005) pp. 793–817; Stephen Ambrose, 'Paleolithic Technology and Human Evolution,' *Science* 291 (2001), 1748–1753.
16. Ambrose, 'Paleolithic Technology and Human Evolution'; also John McNabb and Nicholas Ashton, 'Thoughtful Flakers,' *Cambridge Archeological Journal* 5 (1995): 289–301.
17. Mithen, *The Prehistory of the Mind*; also William Calvin, *A Brief History of the Mind: From Apes to Intellect and Beyond* (Oxford: Oxford University Press, 2004); Jacques Pelegrin, 'A Framework for Analyzing Stone Tool Manufacture and a Tentative Application to Some Early Stone Industries,' in *The Use of Tools by Human and Non-Human Primates*, ed. Arlette Berthelet and Jean Chavaillon (Oxford: Clarendon Press, 1993), pp. 302–314.
18. Mithen, *The Prehistory of the Mind*; Ambrose, 'Paleolithic Technology and Human Evolution,' Davidson and McGrew, 'Stone Cultures and the Uniqueness of Human Culture'; also Glynn Isaac, 'Foundation Stones: Early Artifacts as Indicators of Activities and Abilities,' in *Stone Age Prehistory*, ed. G. N. Bailey and P. Callow (Cambridge, UK: Cambridge University Press, 1986), pp. 221–241.
19. See the papers in B. M. DeWaal, ed., *Tree of Origin: What Primate Behavior Can Tell Us about Human Social Evolution* (Cambridge, MA: Harvard University Press, 2001); also the papers in Elizabeth Lonsdorf, Stephen Ross & Tetsuro Matsuzawa, eds., *The Mind of the Chimpanzee: Ecological and Experimental Perspectives* (Chicago: University of Chicago Press, 2010).
20. Merchant and McGrew, 'Percussive Technology'; also the papers in Roux and Bril, *Stone Knapping: The Necessary Conditions of a Uniquely Hominid Behaviour*.
21. Pelegrin, 'A Framework for Analyzing Stone Tool Manufacture and a Tentative Application to Some Early Stone Industries'; also Steven Churchill and Jill Rhodes, 'The Evolution of the Human Capacity for "Killing at a Distance": The Human Fossil Evidence for the Evolution of Projectile Weaponry,' *Vertebrate Paleobiology and Paleoanthropology: Special Issue on the Evolution of Hominin Diets* (2009), pp. 201–210.

22. Merchant and McGrew, 'Percussive Technology'; also the papers in Roux and Bril, *Stone Knapping: The Necessary Conditions of a Uniquely Hominid Behaviour*.
23. McNabb and Ashton, 'Thoughtful Flakers'; Isaac, 'Foundation Stones'; Mithen, *The Prehistory of the Mind*.
24. Churchill and Rhodes, 'The Evolution of the Human Capacity for "Killing at a Distance"'; Merchant and McGrew, 'Percussive Technology'; also the papers in Roux and Bril, *Stone Knapping: The Necessary Conditions of a Uniquely Hominid Behaviour*.
25. Ibid.
26. Consider the first instance of a hominin as described in Churchill and Rhodes, 'The Evolution of the Human Capacity for "Killing at a Distance."'

9
Neuropragmatism and Apprenticeship: A Model for Education

Bill Bywater and Zachary Piso

Neuropragmatism takes seriously Dewey's admonition that to solve the problems of philosophy we have to put the cortex in the brain, the brain in the nervous system, the nervous system in the organism, and the organism in nature.[1] Further, we have to treat these items not as marbles in a box but as events in a historical process. This chapter is not going to solve all the problems of education. It does, however, present an approach to education which incorporates neuroscience and anthropology and presents a challenge to less holistic approaches by following Dewey's advice to contextualize. We proceed by arguing that neuroscience does have a role to play in educational theory when we set aside scientism and follow pragmatism's reconstruction of the role of science in society. We then offer an illustration of how neuroscience can be useful in thinking about education by challenging the approach to education implicit in Daniel Kahneman's recent book *Thinking, Fast and Slow*.[2] From there we move to a more holistic approach which employs anthropology and neuroscience to ground an apprenticeship model of education. We conclude with a discussion of some general implications of our apprenticeship model.

Neuropragmatists as liaisons and apprentices

A preliminary issue regarding neuroscience and education is how the two perspectives can intersect. There is robust conversation about the relation of neuroscience to pedagogy. This literature is teeming with discussions of the promise and practice of collaboration, though many researchers maintain that fundamental differences between neuroscience and educational theory are challenging.[3,4,5] We argue that this ongoing conversation can be meaningfully informed by the pragmatism of John Dewey, particularly his idea of philosophers as liaisons, 'a messenger of communication' between intellectual endeavors that appear to be at odds.[6] Dewey's prescience is reinforced by contemporary calls for liaisons to facilitate interaction between neuroscientists and educators.[7,8] For Dewey, philosopher liaisons do not discover new facts, nor do they have any special insight into what we should value.[9] 'Philosophy is inherently criticism,' he says, 'having as its distinctive position among various modes of criticism in its generality; a criticism of criticisms, as it were.'[10] Philosopher liaisons must assiduously avoid becoming 'a diplomatic agent of some special and partial interest.'[11] Liaisons must be alert for what Dewey calls 'selective emphasis'[12] in their criticism and in the work of others. Selective emphasis is always present, for example, in problem solving when we focus on certain aspects of the situation because we believe that working with or on them will yield a solution. In physics, mass is likely to be emphasized over color. In the composition of artwork, color may be emphasized over mass.

For Dewey, 'selective emphasis, choice, is inevitable whenever reflection occurs. This is not an evil. Deception comes only when the presence and operation of choice is concealed, disguised, denied.' Selection becomes concealed when 'whatever is capable of certainty is assumed to constitute ultimate Being'; when selected objects are turned into simple elements that have 'no potentialities in reserve'; when whatever is held invariant becomes fixed and final, eternal; and when what we value is made into a preexisting presence rather than 'something to be accomplished, to be brought about by the actions in which choice is manifested and made genuine.'

Each of these errors conceals choice by reifying some element of a complex situation. Dewey's solution to reification is to employ what he calls 'the empirical method.' Modeled on the method of science, the empirical method, applied in all contexts of inquiry, requires that the results of any inquiry, no matter how abstract, ultimately reconnect with 'the context of actual experience, there to receive their check, inherit their full content of meaning, and give illumination and guidance in the immediate perplexities which originally occasioned reflection.' Responding to a perplexity requires the use of imagination. Dewey observes that, 'Only imaginative vision elicits the possibilities that are interwoven within the texture of the actual.'[13] There are two aspects to imaginative vision. They are seeing-*with*-the-possible and seeing-*into*-the-actual: reaching-for-what-is-not-yet and experiencing what-is-available-at-the-moment. Thomas Alexander writes, '[i]ntelligence for Dewey is nothing less than the effort to see the actual in light of the possible and thereby to be responsive toward liberating ideals of conduct, which, in turn, give a fulfilling continuity, meaning and coherency to action,' and 'as projected completion of action, imagination seeks to understand the actual in light of the possible in a dramatic or experimental way.'[14]

The two aspects of Dewey's dynamic imagination focus on the possible and the actual. Neither is viable without the other. We cannot usefully project what is possible unless we have a good grasp of the actual; and the actual has limited meaning or usefulness unless we understand the potentiality within it. Alexander captures a sense of this interplay when he says, 'Imagination is neither merely an extension of the passive capacity of sensation, subsumable under pre-established rational categorical structures, nor is it a purely intuitive source of novelty. It is a mode of action and as such seeks to organize experience so that it anticipates the world in a manner that is meaningful and satisfying.'[15] Imagination organizes experience as a narrative. This is the basis of a dramatic rehearsal which uses imagination to test various narratives. Dewey says, 'Deliberation is actually an imaginative rehearsal of various courses of conduct. We give way, *in our mind*, to some impulse; we try, *in our mind*, some plan. Following its career through various steps, we find ourselves in imagination in the presence of the consequences that would follow...Deliberation is dramatic and active, not mathematical and impersonal...'[16]

As Steven Fesmire points out, our narrative deliberations will be more fruitful as they account for the reality and importance of human transaction. Imagination has to be 'socially responsive' in that it seeks out 'mutual growth' rather than focusing too exclusively upon oneself.[17] According to Fesmire, the narrative developed in imagination will 'discover how others' life-dramas can develop coordinately with one's own [by paying] attention to the constraints imposed and possibilities made available by other dramas' with which one is entangled. Without what Fesmire calls 'deep perception,' '[m]oral imagination may collapse into pseudo-empathy of the Golden Rule variety in which others' values and intentions are reduced to one's own.'[18] Fesmire's comment about coordinating life-dramas serves as a reminder that when we see into the actual, we also connect with the past. Life's dramas contain elements from our pasts which we are attempting to preserve, to abandon, or to change. All of our lives are informed by our pasts as we come to embody the habits of our close companions and our communities. Our dramas are played out in relation to these habits, many of which block the coordination of our dramas with one another. Yet coordination is the basis for community and, as James Campbell reminds us, community is fundamental to growth.[19] Whether our dramas involve preserving, abandoning, or changing (most likely varying combinations of all three), we are going to be employing our imaginations as we plot these events. As we design our ends-in-view, as we assess what resources the actual presents for their realization, we will be seeing-with-the-possible and into-the-actual. In order to accomplish this goal, we need to undertake apprenticeships.

We will say a great deal more about apprenticeship later in this chapter. Here we observe that neuropragmatists now have the data to develop compelling arguments that our imaginative capacities are closely connected with 'performing and observing intentional action.' Indeed, Jay Schulkin indicates that intentional action and imagination are intimately connected.[20] Human community requires our ability to 'interpret the behavior of those around us...to recognize the same as present in others, not only in ourselves.' Schulkin goes on to point out

> that imagining the intentional action of another person and actually performing one's own intentional action both engage many of the same underlying information-processing systems in the brain...In general imagined activity, simulated activity generates neuronal populations that would be in use during the actions themselves. This has become a general rule across a number of both simulated or imagined and real contexts.

When we think about the future we use some of the same neuronal connections that we have employed in the past when taking action *and* when understanding the action of others. According to Schulkin, mirror neurons are often activated in this process along with other systems. Environments enriched in various ways can be the source for material that challenges us from our cells and atoms to our most established habits of action. As we change, others are impacted as well. Our behavior can change their cells, atoms, and established habits of action. Within this discussion, Schulkin invokes the idea of apprenticeship. He introduces apprenticeship in an educational context in which the student is an apprentice learning from teachers by 'capturing a sense of their agency and the trajectory of their journeys in life.'[21] Schulkin's own argument, however, indicates that we are all apprentices to one another. The neurons and habits of teachers as well as of students are impacted by the agency and the journeys in the lives of both. Dewey observed in

Democracy and Education that teaching becomes mechanical unless both teachers and students are responding to one another so that the teacher is getting 'new points of view' and 'intellectual companionship.'[22]

The activity of philosopher liaisons puts apprenticeship into a broader context. As 'a messenger of communication,' liaisons must apprentice themselves to the activities of the parties between whom they desire to build bridges. They look into the actual work of each party with an eye for possible intersections between them. While they employ imagination, they must guard against any reification of selective emphasis that would block imagination's grasp of what is possible. Now that we have described the activity of philosopher liaisons, we plunge back into the details of the intersection of neuroscience and education. In what follows, we will be acting as liaisons, seeing-*with*-the-possible and seeing-*into*-the-actual, with the goal of changing atoms, neurons, thoughts, and imaginations about how neuroscience, education, and other disciplines can interpenetrate.

The intersection of neuroscience and education

Skepticism regarding the intersection of neuroscience and education is illustrated in the work of Boba Samuels and Daniel Willingham.[23, 24] Samuels expresses the 'hope for productive collaboration' but believes it is 'constrained, however, by the fields' histories, by societal limitations and philosophical values, and by disciplinary epistemic demands.' Divisiveness between philosophical values and disciplinary epistemic demands is exemplified in Daniel Willingham's 'Three Problems in the Marriage of Neuroscience and Education.' Willingham prefaces his objection with a clear delineation between the natural sciences, which are 'descriptive' and 'aim to discover principles that describe neural structure and function and in so doing to bring order and comprehensibility to data', and the artificial sciences, which are 'normative' and whose 'aim is not description of the natural world as it exists, but the creation of an artifact, designed to serve a specified goal, within a particular environment'.[25] This distinction affords three incommensurabilities, for Willingham, between neuroscience and educational theory that undermine their synthesis. First, there exists a 'goals problem' because the goals of education are outside of the scope of neuroscience, and 'it is clear that neuroscience will *never* provide a prescriptive solution.'[26] Willingham goes on to describe, second, the 'vertical problem,' that simplified laboratory experiments cannot be scaled up to the complexity of classroom use, and, finally, the 'horizontal problem,' which maintains that the micro-scope of neuroscience cannot be translated to the macro-scope of behavioral theory.

Fundamental to Willingham's argument is a reification of science. By claiming that natural science is descriptive, and all other inquiry is normative, Willingham has isolated science. He makes it impossible for science to impact any normative inquiry because science is descriptive and, so, cannot provide normative answers. He makes it impossible for normative theory to impact science because science is only descriptive. Willingham's 'science' becomes an inflexible monolith. It has lost touch with the context of actual experience, and the immediate perplexities which originally occasioned it, as Dewey would say. Mark Tschaepe has observed, 'Dewey provides an understanding of scientific activity that is not partitioned from other activities, thus avoiding the reification of science as an entity or enterprise.'[27] We should focus, Dewey argues, on how science works as a social practice. Its reification aside, science, *qua* scientific activity, has achieved over the centuries an increasingly careful and self-correcting method for solving problems.[28] Some of the problems are generated by science for science, but many of the problems are generated by us and by the social arrangements we have created. From this perspective, the reason why we would want to 'bring order and comprehensibility to data,' as Willingham says, is to solve the problems we confront in daily life. The order and comprehensibility of the data will be a function of what helps us with those problems. Scientific activity creates an artifact 'designed to serve a specified goal, within a particular environment,' to follow Willingham again, by making discoveries about the 'natural world.' What Willingham sees as a clear delineation between descriptive and normative, and what Samuels sees as 'disciplinary epistemic demands' that constrain cooperation, are phantoms of reification.

Now, when we no longer hold science aloof from human problem solving – when science is no longer 'abstracted into something beyond activities that are scientific or a particular mode of thought'[29] – Willingham's three incommensurabilities disappear. Instead of waiting for neuroscience to pronounce amazing truths to the rest of, educators can join with neuroscientists in reciprocal and non-hierarchical relationships to design studies in which findings 'are tested in an iterative fashion in classroom and lab; and [in which] neuroscience is just one source of evidence, in combination with behavioral data and supporting theory.'[30] Treating reciprocal and non-hierarchical scientific activity as the most useful way to achieve beliefs fruitful for meeting the challenges posed by our desires to survive and flourish, neuropragmatist liaisons can work to understand the activities of the neuroscientist, the educator, the sociologist, and the anthropologist as modes of inquiry which are commensurable with one another.

Willingham's 'goals problem' is removed because we are no longer looking to neuroscience for prescriptive solutions. Instead, such solutions will arise from the collaborative efforts of researchers in various fields abetted by the work of neuropragmatic liaisons. The goal of scientific activity need not be simply or merely descriptive. Willingham's 'vertical problem' is swept away by collaborative scientific activity because the collaborators will design experiments appropriate to the complexity of the research undertaken. Willingham's simplified laboratory experiments may play a role in the research, but it need not be a question of then scaling them up to the complexity of the classroom, as he worries. The data derived from such work could feed into the design of other experiments of larger scale. Willingham's 'horizontal problem' is a nonstarter. Later, we will discuss the hormone cortisol, the activity of which can be seen at the micro-level to have a significant impact at the behavioral level. Cortisol is far from the only substance whose micro-actions can be linked to behavioral events. Oxytocin and vasopressin are also examples.

Martin and Groff as well as Coch and Ansari hold that this collaborative effort requires reciprocal familiarity with the content and methodology of the disciplines involved in the research which necessarily leaves behind the traditional distinction between 'basic' and 'applied.'[31, 32] Interdisciplinary partners must recognize one another as collaborators, not as competitors. In fact, most studies of the neurophysiology of education have explored pedagogical instruction that is already documented as successful, and then employed neuroimaging to better understand the physiological changes that accompany the behavioral changes.[33, 34] Samuels, Christodoulou and Gaab, and Varma, McCandliss, and Schwartz, even argue that the interdisciplinary partners should work together throughout the full process of inquiry as opposed to working in parallel and sharing results.[35, 36, 37] Doing research this way is very desirable as coordination and communication are enhanced and experimental procedures can be fine-tuned on the spot. However, as the liaison function suggests, there will be many instances in which research done quite independently will later be seen, perhaps by parties far distant from the original work, to be connectible for important purposes. Inquiry begins with the experience of a problem to be addressed. Whether this experience is a teacher educating in a classroom, a school board member developing a budget, a high school principal formulating a disciplinary policy, or a neuroscientist studying the effects of cortisol in a specialized laboratory, we want each actor to be able to reach out to the fullest possible fund of resources. Because collaboration often occurs after the fact, so-to-speak, involving disparate actors removed in space and time, scientists, teachers, principals and board members can seldom foresee the significance of, or coordinate their actions in such a way that their works resonate with and form themselves around one another. Ideally, the collaborative process would begin prior to settling on an experimental design to ensure that inquiry considers the problem as articulated by different actors. Collaboration will be facilitated by interdisciplinary liaisons who will watch for the emergence of problematic situations and alert problem solvers to resources from across the spectrum of disciplines. Dewey sees liaisons as primarily engaged in criticism. In our context, this means exploring with the teachers, the board members, the principals, and the scientists how fully they have considered what they are about to undertake and whether they have engaged the resources available to them.

Neuroscientists have issued warnings that it is premature for educators to draw on neuroscience. A significant faction of neuroscientists are committed to disciplinary aims that are not aligned with the concerns of educational theorists, and there is little reason to expect that more time will lead to safe and accurate expropriations by educators. For instance, researchers have described the primary program of neuroscience as the localization of brain functions, irrespective of whether this would offer any guidance to educators.[38] It is not enough that neuroscience develops out of its supposed present prematurity toward a future where it might safely appeal to a complete and objective account of the neural mechanisms of learning. Rather, maturity includes placing neuroscience into its social context, by locating it in relation to the problems of society as does, for example, Marco Iacoboni when he discusses the significance of the study of mirror neurons for understanding autism. It is always desirable for researchers to be concerned with the products of investigation, the methods of experimentation and the social structure into which such products are received.[39] Yet, as we have argued, this will not be and cannot always be the case. Research which has no obvious connection in its *contemporary* social structure may later be found by a liaison to be very valuable.

Many educators see a natural connection between neuroscience and education given the brain's crucial role in the learning process.[40, 41] Without the activity of liaisons operating as honest brokers between neuroscientists and educators, educational practice and policy are rarely based on primary source evidence and commonly rely on secondary summaries and interpretations in the popular press.[42, 43] Misappropriations of neuroscience, such as oversimplifications of hemispheric specialization[44] or exaggerations of critical learning periods,[45] have dominated the exchange between the two disciplines.[46, 47] All too frequently, neuroscientific studies are received as a panacea for problems in the classroom, and neuroscience is misunderstood as resting at the top of an epistemological hierarchy against which traditional educational theory is discounted.[48]

This problem is not specific to educators; in general, individuals without expert knowledge of neuroscience will find neuroscientific explanations more satisfying than psychological explanations, even when the neuroscience is logically irrelevant to the argument at hand.[49] Thus, educators and the general public, parents of school children especially, become prey for third party for-profit groups that commercialize brain-based educational programs which are often grossly misleading about the potential of curricula informed by neuroscience. At a recent conference at the Center for Neuroscience in Education at the University of Cambridge, teachers reported that they had received more than 70 group emails per year advertising courses on brain-based learning.[50] The claims of these brain-based learning programs range from innocent misinterpretation to wanton fabrication bordering on the absurd. Some programs propagate 'neuromyths,' such as the dividing of a student body based on each students proclivity for visual, auditory, or kinesthetic learning style, others go so far as to inform educators that students have a simple 'brain button' that they must push prior to engagement in the classroom. Third party commercialization of these pedagogical tools often fails to either relay the applicable findings from neuroscience or offer actionable strategies based on educational theory. One common product, Brain Gym, ensures that 'true' education will only occur if and only if 'in technical terms, information is received by the brainstem as an "impress," but may be inaccessible to the front brain as an "express." This...locks the student into failure syndrome. Whole-brain learning draws out the potential locked in the body and enables students to access those areas of the brain previously unavailable to them. Improvements in learning...are often immediate.'[51]

Dewey was concerned that the activities of liaisons might be perverted by special interests. Impartiality can be ruined by economic interest. It can also be ruined if a researcher has an already existing commitment to some theory or set of facts that subsequently lead to a misunderstanding or a misrepresentation of the learning process or of the way we reason.

As in the examples of the neuromyths cited above, this work can be very convincing when it appears to support conclusions, couched in language taken from brain science – conclusions that challenge or support widely held beliefs. We turn now to such a case. Our analysis will illustrate a facet of the critical activity of neuropragmatist liaisons and will lead us subsequently into our proposals about apprenticeship.

Case study: birth of a neuromyth

Using the data he has spent decades gathering, Daniel Kahneman describes a large number of forms of human error which involve jumping to conclusions. Some of them are visual illusions like the Müller-Lyer, but the majority are what he calls 'cognitive illusions.'[52] These illusions are caused by priming events, including words and gestures, of which we are unaware, by confusing familiarity with truth, by mood, by applying causal reasoning when statistical reasoning is appropriate, by neglecting ambiguity, by the halo effect, and by assuming that what you see is all there is.[53] These are just seven examples from Kahneman's extensive discussion. He argues that even though we will always *see* one line as longer that the other in the Müller-Lyer illusion, we can *know* that the lines are of equal length and, thus, not be deceived into taking any action on the basis of what we see. But can we correct cognitive illusions as well? Kahneman is not sure that we can. He claims that the vigilance necessary to monitor all the results of the unconscious processes which produce cognitive illusions would overwhelm us. Routine decisions would become such drawn out affairs that our activities would come to a standstill. The best we can hope for is to recognize which mistakes are likely in situations where we are making the most important decisions. Otherwise, we must expect mistakes and make the best we can of this situation.[54]

For Kahneman, cognitive illusions are produced by the circumstance that brain processes work so quickly that we cannot be aware of them. David Franks observes that we can consciously process about 50 bits of information per second, while, for example, our visual system alone processes 11 million bits in a second.[55] Franks, appealing to Gazzaniga's *The Mind's Past*, argues that we can now speak of a 'new unconscious' that is composed of the myriad processes that establish the unity of objects and our perception of the intentions of others, for example.[56, 57] For Kahneman, it is some of these unconscious processes that create cognitive illusions. Kahneman calls unconscious brain processes System 1. They are a powerful influence on us, yielding, as we have seen, many errors. In contrast, our ability to consciously control our inclination to jump to conclusions is quite weak. The processes of conscious control, which Kahneman calls System 2, are responsible for self-correction. System 2 controls the thoughts and behaviors which are 'suggested,' as Kahneman writes, by System 1.[58] This requires attention and effort on the part of System 2, as it reasons about the deliverances of System 1; Kahneman suspects, 'frequent switching of tasks and speeded-up mental work are not intrinsically pleasurable, and people avoid them when possible.'[59] Soon after he concludes that many people – as many as 80% of college undergraduates – place 'too much faith in their intuitions' and 'find cognitive effort at least mildly unpleasant.'[60] Later he admonishes, 'Remember that System 2 is lazy and that mental effort is aversive.'[61]

The relatively slow pace of conscious deliberation supports Kahneman's position, as does the presence of the new unconscious. Further support even comes from neuropragmatism's use of Dewey's early work in 'The Reflex Arc Concept in Psychology' of 1896 and 'The Influence of Darwinism on Philosophy' of 1910.[62, 63] In the history of human evolution, our survival has depended on our ability to make quick discriminations and quick decisions about action. *We are designed for action, not contemplation.* We must act even to gather the input which our brains process to yield further action: we move our eyes, turn our heads, reach out with hands and arms, and travel with feet and legs. The nervous system is in the body, and the body is in the world. As Dewey points out, we are always located, and we are always exploring that location for its affordances and its dangers. Within the system of tensions that Dewey opposes to the reflex arc, we are never passive receivers of information: we are always in transaction, always entangled with the world. Yet Kahneman's work raises a vitally important issue. His data suggests that many of us are poor reasoners and are disinclined to think carefully about problems we confront. His conclusion is that this data is an accurate picture of us and of our potential. But how can this be if we have survived in a precarious environment where making too many rash and wrong decisions would have severely reduced the likelihood of our survival? To survive, we must act both quickly and accurately.

Recall that Kahneman moves rapidly from a personal suspicion, to samples of college undergraduates, and then to a bald assertion of laziness on the part of System 2. From the perspective of neuropragmatism, the question becomes whether Kahneman's rather grim portrait of us is one we must accept and adapt to even though it seems counterintuitive to a position which takes seriously the impact of Darwin. There is reason to question Kahneman's conclusion on the basis of the sample from which he generalizes. Henrich and colleagues raise wide-ranging questions about the legitimacy of any generalization about human psychology or behavior based solely on samples of undergraduate students from 'Western, Educated, Industrialized, Rich, and Democratic (WEIRD) societies.'[64] Playing on the acronym, Henrich and colleagues illustrate the many ways in which samples like Kahneman's are outliers rather than typical of all human populations. They even present evidence that the Müller-Lyer figure presents no illusion for certain groups of humans, leading to the suggestion that WEIRD environments establish visual habits which foster the illusion.[65] They also point out that WEIRD individuals see themselves as more independent, possess stronger self-serving biases, value choice more, have less motivation to conform, and think more analytically in the sense of attending more to objects than fields, explaining behavior in decontextualized terms, and relying more on rules than similarity in classifying objects.[66] WEIRD environments not only develop visual habits. They develop habits of thought which distinguish Westerners from many other groups of humans. We shall soon see more fully that there are reasons for distinguishing System 1 from System 2, but we also now see that these systems do not operate independently of environmental impact. Tibor Solymosi and John Shook have recently explored these connections, which they refer to as System 3, expanding on Kahneman's distinctions.[67]

Contextualizing Kahneman's findings leads us to our central problem. What has happened in the lives of college undergraduates in the United States that has made them such lazy reasoners? Even if there are cultural forces which promote this behavior, should not education correct this situation? The argument we pursue in the rest of this chapter is that there is a strong anthropological and neuroscientific evidence for the argument that to create careful and creative thinkers who enjoy solving the problems which confront them, our educational system must encourage students and teachers to undertake a series of apprenticeships. While Kahneman focuses on individual decision makers taking tests and being subjects in psychological experiments, he is removing individuals from the larger context of our engagement with one another and with the world. His results reflect an educational system which does the same by focusing on individual leaners and their private mastery of a subject matter. The educational arrangement that leads Kahneman to such a dismal conclusion is that we learn best when we are isolated from one another and, indeed, when we are in competition with one another.

By positing a discontinuity between the cognitive processes of System 1 and the cognitive processes of System 2, Kahneman rejects the unity of human beings as organisms that can know, feel, and act intelligently and as organisms which interact in communities. This discontinuity stands in stark contrast with Dewey's understanding of organisms which is incorporated into neuropragmatism. Whereas Kahneman's depiction of System 1 and System 2 treats different cognitive activities 'like marbles in a box,' neuropragmatism sees these systems as 'events are in a history, as a moving, growing never finished process.'[68] Our reflections below are not intended to offer an empirical refutation of the pedagogy recommended by Kahneman and others who have so far misappropriated the sciences of mind and brain. It is intended to spark our imaginations with the possibilities for a pedagogy that maintains the unity of the organism and its history. Education that works does so because it is harmonious with the integrated and embodied cognitive potentialities of human beings, and the frontier of neuroscience and other contemporary discourses sheds light on this harmony. We learn best when we cooperate in solving concrete problems. Let us turn to the evidence: first from anthropology, then from neuroscience.

Anthropological evidence

Kim Sterelny's study of anthropology presents wide-ranging evidence from over 1.5 million years ago that stone implements were made using an apprentice system.[69] He argues that the 'apprentice learning model' both fits the ethnographic data and also allows us to make further sense of it. Apprenticeship can develop slowly as a 'side effect of adult activity, without adult teaching and without adaptations for social learning in the young.'[70] Beginning as a side effect, apprentice learning can emerge gradually and can, gradually, become richer social learning incorporating more and more feedback loops in an increasingly complex context. Co-evolving with tool development, in what Sterelny calls the 'cooperation syndrome,' are foraging and hunting, reproductive cooperation, and sharing information about the environment in which the foraging and hunting take place. These are the 'first foundations of human sociality.'[71] The beginnings of trust are also present in this context. Evidence from chimps indicates that our ancestors were equipped with prosocial, affiliative emotions that accompany the coevolution of cooperative endeavor and the investment in relationships that is required. The subjective elements of trust follow on the 'interplay of joint action, shared experience, communication, and shared outcomes.' In the beginning, trust is 'built rather than displayed.' Trust 'is more interactive and interventionist, less a matter of signaling hidden preexisting psychological dispositions';[72] further, 'Systems of gesture, mime, and depiction would suffice to share important environmental information in ways that will be kept fairly honest by public signaling and by pooling data before action.'[73] And all of this before language is on the scene.

Within larger groups, individuals could become specialists in one task or another. As some information is picked up through 'unconscious processes of automatic imitation' in which 'neither model nor mimic is aware that the mimic is noticing and matching patterns in the model's actions,' with the appearance of specialists, apprenticeship intensified into more organized teaching and learning.[74] Specialist teachers had to reflect on their own competence to make what they do easier for others to learn. Learners and teachers engage in a joint endeavor. The teacher struggles to present a skill in a form that can be learned; learners, focusing on the skill, push teachers and themselves to a greater understanding of both teaching and learning. Those who use the product of a skill will also provide information about how well the teaching and learning are going. Feedback loops abound and become more complex. More opportunities for trust to be extended and deepened reside in this process. Trust involves anticipation – what is here today will be here tomorrow as well – and anticipation is akin to faith – 'personal faith in personal day-by-day working together with others,' as Dewey says.[75]

Normative considerations are already built into the mutual apprenticeship of teachers and learners. They extend to the anticipation of the tool users and to many other elements of the cooperation syndrome. As with skills, we absorb norms from our experiences with others. In all of this social learning, Sterelny relies on our plasticity, coevolution, and ever more complex feedback loops to explain what distinguishes even our earliest ancestors from the great apes.

There is no need for Sterelny to wait until a discussion of norms to introduce the importance of pattern recognition. Pattern recognition is an important part of skill development. As Sterelny points out when discussing apprentice toolmakers, the apprentices form their tools using the work of more skillful makers as a model. Apprentices also have to model their stone-striking gestures on the basis of more skillful makers so as to be able to effectively form a blade. In this situation and others, pattern recognition is shown to be a pre-linguistic phenomenon.[76]

Neuroscientific evidence

Our second area of evidence begins with the neuroscience of the mirror neuron system. Both Mark Johnson and Shaun Gallagher argue that neonate imitation shows that infants see others as like themselves and different from all the other items in their environment.[77,78] Gallagher says that the neonate 'is *always already* "coupled" with the other.'[79] Johnson says, 'Infants see other people as like themselves...infants are born with the capacity to discriminate animate expressive gestures of the sort characteristic of humans.'[80]

Studies of adult/infant interaction in which infants as young as 42 minutes imitate the facial expression of adults have led Gallagher to conclude that the infant imitates the adult not by copying the outward appearance of the adult's face but using the infant's proprioceptive capacities to replicate the intentional action and expression of the adult. Gallagher refers to the infant's ability to imitate facial gestures as 'body reading.'[81] He uses this locution as part of his interaction theory which sees subjectivity as arising from embodied practice.[82] In contrast to theory theory and simulation theory, which are about gaining access to other minds, interaction theory posits 'that in most intersubjective situations we have a direct understanding of another person's intentions because their intentions are explicitly expressed in their embodied actions, and mirrored in our own capabilities for action.'[83]

Following the work of Marco Iacoboni, we see the mirror neuron system as the foundation for what Gallagher calls 'body reading' and for Johnson's embodiment theory in which infants see others as themselves.[84] On the basis of his research, Iacoboni speculates that infants have mirror neurons which are further formed by the interaction between infants and adults.[85] Babies learn by imitating both adult behavior and the behavior of other youngsters.[86] This is a *'coordinated activity'* of interacting individuals – a *bidirectional* flow of information,' according to Iacoboni, within which language emerges.[87] This coordinated activity relies on several types of mirror neurons. Those that are 'strictly congruent' fire both when we act and when another performs the same act.[88] Those that are 'broadly congruent' fire when there is a similar goal, for example, getting food into one's mouth can be done by using one's hands to bring the food up to the mouth or it can be done by moving the mouth to the food and picking it up with the lips. The goal is to eat so a broadly congruent neuron will fire even if your eating activity is not the same as mine.[89] There are also 'logically related' mirror neurons. Iacoboni says:

> I see you grasping a cup with a precision grip and my precision grip mirror neurons fire...However, given that the context suggests drinking, the firing of other mirror neurons follows: these are my

'logically related' mirror neurons that code the action of bringing the cup to the mouth. By activating this chain of mirror neurons, my brain is able to simulate the intentions of others....Mirror neurons help us reenact in our brain the intentions of other people, giving us a profound understanding of their mental states.[90]

Observing the facial gestures of the adult, the neurons associated with those gestures are activated in the infant. Because 'perception and action are a unified process in the brain,' the infant, so far as it is capable, can make the same gestures.[91] Coordinated recognition blooms between adult and infant. There is both animacy and intentionality wrapped up in this recognition. The infant has received a signal and it signals back. Perception of action yields a connection with the actor. In making faces and imitating the face-making, and in switching places when the adult imitates the infant, two beings are discovering one another. Together they explore their motor possibilities, possible body postures, and body powers. Both adult and child can be transformed in the interaction. They are apprentices to one another in coordinated activity. Both Iacoboni and Gallagher call this coordinated activity primary intersubjectivity.[92, 93] Johnson calls it 'body-based intersubjectivity.'[94] For Johnson, the transaction between infant and care giver is a proto-conversation;[95] for Iacoboni, it is the beginning of the 'dance of dialogue,' which is easily observed once language is present.[96]

Against the background of the strictly congruent neurons, the activity of the broadly congruent and logically related mirror neurons begins to involve learning and habit formation. Iacoboni uses the example of getting food to one's mouth to illustrate broadly congruent neurons. There are a great many ways to do this. If I stab a piece of food with a fork held in my hand, you will have to know about forks and their use before your broadly congruent neurons for the goal of getting food to mouth will fire. You will have already learned a great deal about eating within the culture which we share. In the case of logically related neurons, Iacoboni himself comments on the significance of context in the quote we used above. As he says, when 'the context suggests drinking,' and when a cup is grasped with a precision grip, the neurons for bringing the cup to the mouth will also fire. The contexts that suggest drinking are, again, a matter of having experiences in the ways of one's culture.

Apprenticeship expands from the strictly congruent mirror neurons that facilitate human transaction from the very beginning as they also lay the ground, we argue, for the communal learning which Sterelny describes. The presence and activity of the mirror neuron system supports

Sterelny's view that apprenticeship learning develops without anyone's being directly concerned with creating it. Youngsters mimic the behavior of adults and of their peers giving rise to apprentice learning and the habits of life composing Sterelny's cooperation syndrome. Anthropology and neuroscience indicate that learning, and knowing have their roots in action taken in consort with others. Apprenticeship should be at the center of our thinking about education.

Narrative, a sense of unfolding time, runs through all the activities of apprenticeship. Anticipation is not only found in connection with matters of trust. It is found in the bodily comportment of a person just about to strike a stone to make an axe; in the actions of someone planting a seed; or in the hunters who watch spearheads being fashioned. Gallagher observes that even the simplest motor action 'involves the retention of previous postures and the anticipation of future action.'[97] Apprenticeship is an implicit and branching narrative with a very public dimension. We learn what to do first and how to continue in light of the preceding contingencies. As tasks grow more complex, apprentices need to remember how similar contingencies have been treated in the past. Memory, agency, and a sense of history develop together as patterns are recognized across time.

Jay Schulkin observes that 'the vast cognitive arsenal that underlies behavioral adaptation presupposes sets of timing and memory systems that create a sense of history.'[98] According to Schulkin, our sense of history is also an 'adaptation to inherent uncertainty.'[99] In narrative, we anticipate familiar patterns. When they do not materialize, there is a problem which must be solved. Early apprenticeship is the beginning of hypothesize-and-test. Schulkin reminds us that 'Knowledge is in part a contact sport, social in nature, tied to others, and emboldened by body sensibility and imagination.'[100] We learn from each other's experiences. Joint attention to tasks and objects, essential for collaborative work, is already present before we are two years old.[101] For Schulkin, we are natural-born problem solvers who are prepared to learn right from the start. Neural plasticity and neurogenesis play a major role in our ability to be apprentices. Schulkin notes that enriched environments can increase neuronal sprouting in several parts of the brain. This plasticity, which increases the number of synapses, is very important in learning, memory, and problem solving. Neurogenesis is also present in parts of the brain and has been linked to learning from others and to long-term potentiation.

James Zull appeals to a second area of evidence from neuroscience.[102] Zull's aim is to fashion education so that we learn to move from brain to mind – from being controlled by unconscious processes to making conscious decisions about how to behave. He distinguishes two pathways which sensory data follow upon exiting the thalamus. Some data go directly to the amygdala. This data are 'then passed directly to the brain stem and then out to the muscles of the body, generating rapid but crude actions somewhat like reflexes.'[103] The second path runs from thalamus to cortex, and then to the amygdala: 'The involvement of the cortex draws in more complex and potentially modulating data and memories, which allows for a less reflexive and slower response to the sensory data, and a deeper understanding of its significance and meaning.' These two pathways sound very like Kahneman's System 1 and System 2, or, at least, their neurological analogs. Zull goes on to argue that there is a reward system associated with the second pathway. This circumstance would make careful deliberation – the use of System 2 – pleasurable, not unpleasant or aversive as Kahneman asserts. Zull points out that the front integrative cortex uses working memory to solve problems, and recalls facts and stories in decision making.[104] This area of the brain is 'primarily' where dopamine is delivered.[105] Since, Zull argues, dopamine has been shown to be part of a reward system 'using working memory to solve problems, intentional recall of facts and stories, and decision making are all rewarding, satisfying, and even exhilarating.'[106] Schulkin's discussion of dopamine illustrates the complexity of this issue.[107] Dopamine is activated when an award is expected or perceived as probable. It is 'essential in the organization of movement [and] in the organization of thought and the prediction of events.'[108] But once a reward always occurs, dopamine is no longer activated.[109] It may be that dopamine is not so much associated with reward *per se* as it is with anticipation, expectation, prediction and calculation of the likelihood of reward. Even if this were the focus of dopamine activation, System 2 would be excited when reward is not a certainty. Given these considerations, it is unlikely that System 2 would always be lazy, as Kahneman suggests. Indeed, the opposite seems possible. With our penchant for learning and problem solving, System 2 would have to be habitually suppressed in order for it to become lazy.

A common and long-standing neuroscientific position, Schulkin points out, has been to establish a contrast, even a competition, between 'neocortical control and limbic expression.'[110] This view, he says, is 'not wrong, but overstated and under-tested.'[111] Schulkin notes that the brain has a chemical architecture that traverses all its physical structures, creating an interconnectedness that goes beyond local anatomy and making hierarchical approaches, like the distinction between Systems

1 (unconscious) and 2 (conscious), potentially misleading.[112] Chemical balances in the central nervous system play a crucial role in our lives. For example, the steroid hormone cortisol increases 'under conditions of adversity'[113] and increases the intensity of fear reactions to such things as loud noises, unpleasant smells and stimuli that cause pain.[114] It also increases with social isolation. People with low socioeconomic standing and members of minority groups tend to have higher levels of cortisol.[115] Long-term elevated cortisol impacts social behavior by making individuals more fearful, and more timid in unfamiliar contexts

Cortisol helps deal with short-term dangers like food or water deprivation, injury, or external[116] threats until they are resolved. However, the long-term presence of cortisol can have a negative impact on memory and on the neurogenesis which occurs with learning new things. Both of these effects are due to the impact of cortisol on the hippocampus, in which cells shrink or die or are not produced under the long-term impact of cortisol. People who have elevated levels of cortisol have smaller hippocampi.[117] Elevated cortisol can also have a negative impact on the prefrontal cortex, creating distractibility, undermining working memory and decision making, as well as executive functions. These deficits are particularly associated with the increased cortisol of depression characterized by 'a state of neural and psychological withdrawal in which the brain's ability to attend to, engage, and learn about the world is reduced.'[118]

Whether it is a question of the conditions under which broadly congruent and logically related mirror neurons will fire, under which dopamine will be released, or under which cortisol will be appropriately modulated, what happens 'inside' of us is very much regulated by what is going on 'outside' of us. Kahneman's discussion of two internal systems in conflict oversimplifies the complexity of who we are as historical beings that maintain and are maintained by a complex transaction with our surroundings.

Neuropragmatism's pedagogy

Pedagogy in the context of neuropragmatism challenges the common belief that the singular purpose of education is to impart information. It follows the lead of Dewey's pragmatism in proposing an education that equips individuals with dynamic habits for navigating complex situations. It promotes a delicate play between the acquisition of knowledge and norms and the development of problem solving capability that maintains resilience in the face of novelty. These are characteristics of apprenticeship, so we return to a study of apprenticeship to begin our exploration of neuropragmatic pedagogy. The bold claim that we are all apprentices, and we are all apprentices to one another, we have seen, is supported by evidence from anthropology and neuroscience. Beginning as a side effect of adult activity, Sterelny's apprentice learning model can become more complex, as he describes:

> As well as seeing expert practitioners in action and helping them, children often have a chance to listen to experts talk to one another and to more skilled apprentices about their expertise. Listening in helps in acquiring local lore as well as local practices. They learn from one another as well. The general picture is that much skill learning in forager society is accomplished by trial and error, but supervised and organized trial and error. Moreover, it is trial in an environment seeded with props and other cognitive tools. The specialist vocabulary to which children are exposed marks salient distinctions. Tools and artifacts – finished, half-finished, and broken – are available as sources of inspiration and comparison. In short, while the role of explicit teaching in traditional societies is often quite limited, adults can and do structure and engineer the learning environment, even without explicit teaching.[119]

Apprentice learning emerges gradually, and gradually becomes richer social learning, incorporating more and more feedback loops in an increasingly complex context. For Sterelny, it is about continuity, not sudden innovation: 'Coevolution, not a magic moment when a special light turns on inside, explains why we are so very different from the other great apes in our behavior, our cognition, and our social organization.'[120] As we have seen, foraging and hunting, reproductive cooperation, sharing information about the environment, and trust coevolved with tool development in the cooperation syndrome. Affiliative emotions which humans had already inherited from their ancestors were the backdrop for these developments.

Teaching and learning emerge as more specific activities. Those with special skills who now become teachers must reflect in a new way on what they do. The question of how one's skill can be learned by another requires self-reflection and a greater consciousness of what one does. Behavior that was previously done thoughtlessly from habit must now be scrutinized as the new skill of teaching appears on the scene. The presence of learners reinforces these demands on conscious activity and adds a new set of concerns about how to interact with the learners.

How is teaching best accomplished: I can show them what to do, but should I guide their hands when they try? If they don't pay attention, what do I do? Concomitantly, learning emerges as a skill. How is it best accomplished? Should I as a learner sit, stand or squat in front of, beside or behind the skilled one? Should I ask a lot of questions or be quiet and see what happens? As these behaviors emerge into specific roles, they feed back and forward into one another. Seeing-*with*-the-possible and seeing-*into*-the-actual are activated, and imagination comes into full play. In these activities, we move from bodily experience to more abstract matters. In learning and teaching, we ask whether we should 'tackle' a problem 'head on' or 'circle around it.' Mark Johnson has shown how abstract thinking is based upon such bodily experiences.[121] We talk about people we can trust as 'rock solid' or 'straight shooters.' Trust and anticipation involve seeing-*into*-the-actual with the possible – solid rock is here today and abides into the future – and anticipation is akin to the faith that teachers and learners have in one another when their interaction is working well. The presence of teaching and learning as distinct activities increase the complexity of individual subjectivity (or mind) that finds its ground in the primary intersubjectivity based in the mirror neuron system. Even as we are articulated as individuals, it is remarkable to realize the extent to which we are formed by those and that which surround us completely. Formed, that is, from the very beginning. We *are* all apprentices to one another, and we are apprentices all the way down.

We can see from the questions posed by the developing of activities of teaching and learning that norms appear right along with skills. Skill itself is normative, and so is apprenticeship. From the very beginning, with adult-infant interaction, apprenticeship calls for restraint, humility, an open willingness to engage, even wonder or awe, on the part of the adult. The spell that the infant casts, the call which it makes on us, finds its correlate in the apprenticeship of education in which the apprentices stand more and more on equal terms with one another. Increasingly, both parties are called on to exhibit the characteristics of the adulthood called forth by the infant. The relationship becomes more reciprocal as it matures while never losing its initial quality. Each party sees the potential embodied in the other. Goethe has a rich description of this apprenticeship in terms of a researcher approaching nature. Here he describes the apprentice as a person of lively intellect:

> When people of lively intellect first respond to Nature's challenge to be understood, they feel irresistibly tempted to impose their will

upon the natural objects they are studying. Before long, however, these natural objects close in upon us with such force as to make us realize that we in turn must now acknowledge their might and hold in respect the authority they exert over us. Hardly are we convinced of this reciprocal influence when we become aware of the twofold infinitude: in the natural objects, of the diversity of life and growth and of vitally interlocking relationships; in ourselves, of the possibility of endless development through always keeping our minds receptive and disciplining our minds in new forms of assimilation and procedure.[122]

We must distinguish this lively apprenticeship from another, more standard apprenticeship in which an apprentice has a master whose authority is established by a clear hierarchy. The norms of hierarchical apprenticeship are different from the lively variety. These apprentices are to master and control that which is required to accomplish the tasks of their trade. Whether they are tool and die makers, research scientists, bakers, brewers, chefs, or sociologists in the process of learning their trade, they become comfortable with hierarchy. It comes to seem quite natural that there are such things as apprentices and masters and that masters dominate and control both apprentices and the raw materials of their enterprise. The norms of domination and control are central to hierarchical apprenticeship; they are not so placed in lively apprenticeship. The difference between these apprenticeships is subtle but central. Certainly, lively apprentices have to master skills and knowledge. Yet their aim is not to move up in a hierarchy or to dominate some part of the world, but to prepare themselves to be receptive and flexible in order to be open to the diversity which surrounds them. Lively apprenticeship can be neuropragmatism's model for education because it is consistent with pragmatism's Darwinian vision that (1) we are evolved beings who, in order to survive, adapt, and flourish, had to become good at solving problems; (2) to be good at solving problems, we have to use our imagination to see-*with*-the-possible and see-*into*-the-actual; (3) to use our imagination well, we must take full advantage of our lively intellects; (4) to fully engage the capabilities inherent in (1), (2) and (3), we must exhibit flexibility in thought and action, which is not typically supported by hierarchical structures; and (5) the human and natural sciences are making discoveries that support an understanding of our ourselves as evolved problem solvers.

Much pedagogy uses positive reinforcement to reward proper answers about information which the student does not generate. This positive reinforcement keeps in place a cognitive-affective landscape in which the learner, as a relatively passive recipient of information, is encouraged to retain this information in a way that can be quickly retrieved with little distortion. Information has to be carefully conserved in relatively discrete units to be available to be combined in ways indicated by those in authority. Success in this situation depends on habits and attitudes whose smooth, automatic functioning is imperative and unobtrusive. Pattern recognition is surely part of this pedagogy, yet having to conserve information in discrete units cuts down on flexibility and creativity. Neural plasticity with the long-term potentiation of multiple pathways is more likely to be promoted when units of information are connected and rearranged in many and various ways. Using imagination in the process of seeing-*with*-the-possible and seeing-*into*-the-actual will accomplish this goal.

The pedagogy which renders the student a passive recipient of information and then tests for the retention of that information is often called 'the banking method.' It usually pits students against one another by creating competition among them for the highest grades. Working cooperatively is not a central component of this method, so it is usually rewarded infrequently. This means that the prosocial emotions, for example trust, which have coevolved with apprenticeship, are not potentiated by this method and habits of prosocial behavior are less likely to form. Due to neural plasticity, synaptic growth and branching which support habits of competition will be established more fully and firmly than prosocial habits. This means that when problems are encountered solutions which involve competition and produce winners and losers are more likely to appear than solutions which involve cooperation. The banking method may be appropriate in limited contexts, but it cannot form a central or foundational element in neuropragmatic pedagogy.

The central element of the pedagogy of neuropragmatism is a commitment to the apprenticeship model of learning. Sometimes this model is called 'learning by doing.' It might be more helpful to say doing-under-the-right-circumstances yields learning – awkward but accurate. Sterelny has already alerted us to the rich environments in which apprenticeship coevolved. So, the pedagogy of neuropragmatism demands environments rich in the tools, in the materials, and in the products of human endeavor – adjusted to account for the capacity of the learners. In turn, teachers must be apprentices to their students, engaging in a self-reflective attunement of their activities to the rhythms of student learning. By the same token, students need to be taught to learn. Lively apprenticeship is not widely encountered in contemporary contexts. It needs to be fostered not only by teachers who model it, but also by explicit explanation and practice. In all educational settings, participants are apprentices to one another, sharing and deepening the transaction brought about by their joint attention to a task.

Joint attention is not fostered by lectures, lesson plans, and exams which test the contents of short-term memory within the banking model of education. It is more likely to be fostered when all the apprentices take a journey together, establishing a joint narrative to which all can attend. In the process of the journey there will be problems to be solved that will require the use of imagination as well as knowledge of probabilities and the nature of the physical world. Imaginative problem solving could well lead to branches in the journey. Encountering various paths will yield disagreements and debates about which to follow. Questions of historical precedent may arise which lead to inquiry in that direction. Still, no one path may seem satisfactory to all, so more than one may be pursued. The journey becomes more richly complex as apprentices pursuing different paths exchange information. This information exchange will hold joint attention because all the apprentices are invested in the outcome. Listening is an important element of this attention. Students and teachers who learn to listen to the findings and experiences of those who are following different paths – paths which may conflict with their own – are learning *courageous* listening. It is not easy to listen respectfully and humbly to those who may challenge one's considered opinions and the completeness of one's ideas or experiences. Their statements or their paths in life may be deeply disturbing. *Courageous* listening is more profound than hearing a criticism of one's ideas or decisions from a person whom one respects or who feels like kith. It is being able to hear and to work at understanding those who seem foreign even when they speak your language and might even live in your neighborhood or town. The ability to listen courageously must be fostered early in our educational practices so that it may become second nature by the time we enter the public forums of democratic debate. Training in the arts could be an important aspect of this pedagogy. Arts training can enhance courageous listening by encouraging students' sensitivity to the materials of the art. Manipulation of physical materials – the drawing of the bow, the gesture in making a drawing, the experience of centering oneself to begin a dance, the feel of the clay, the interplay of colored pigments, papers, or waters – can, as Goethe suggests, make us aware of their endless complexity and make us aware of new capacities in ourselves which emerge in these transactions. When one becomes an apprentice, not only to the instruments of the art but to its history as well, one can become aware of how works which are foreign and deeply disturbing to the culture of one period can become well respected in another.

The challenge which artworks so often present brings to the fore the role of fearless speech within the context of lively apprenticeship. Natural and human scientists and cultural critics of all sorts, including philosophical liaisons, need to be ready and willing to speak fearlessly about the results of their inquiries. The presence of courageous listening allows fearless speech to flourish which, in turn, puts courageous listening to the test. As with courageous listening, fearless speech must be fostered early in our educational practices. Indeed, one cannot be practiced without the other. The practices we have been discussing are the very skills required to create lively democratic communities and a culture which will sustain them. We have described the careful attention which practitioners – whether a violinist, a medical professional, or a liaison – must pay to their objects of study and the sensitivity they must have to be apprentices to their practice. These considerations apply as well to those seeking social justice who must listen to the voices of oppressed and disenfranchised people in such a way that those voices, to quote Goethe again, 'close in upon us with such force as to make us realize that we in turn must now acknowledge their might and hold in respect the authority they exert over us.' Imagine in this process becoming aware of a 'twofold infinitude:' first, in the diversity of those voices and the vitality of the relationships among them, and, second, in the way in which we, the privileged, can undergo transformation as we stay receptive to what those voices are saying. For example, consider the long history of black people's speculation about what it is about white people that makes it so difficult for white people to recognize the humanity of black people. In 1829, David Walker, in his *Appeal to the Colored Citizens of the World*, asked and answered this question.[123] Walker's answer was that white people (in the United States, at least) were unable to recognize the humanity of black people because white people's 'secret monitor,' which God has placed in every human and which gives humans the capacity to recognize other humans, has been ruined by avarice and greed to the extent that it can no longer give accurate reports about which beings are human. Suppose white people were to courageously listen to Walker and, as Goethe observes, acknowledge the 'might' of his words and 'hold in respect' their authority. If white people were to undertake a lively apprenticeship with black people, whites would learn of the long history of Walker's question and the various answers which have been given in response. The diversity of these answers and the common themes among them could awaken in whites who listen courageously to this fearless speech, new visions of possibilities for white transformation and for social justice. Now, imagine that we all are apprentices to one another in this way: talking on street corners, exhibiting amicable cooperation, and not betraying democracy by 'intolerance, abuse, calling of names because of differences of opinion about religion or politics or business, as well as because of differences of race, color, wealth, or degree of culture,' for as Dewey says these 'are treason to the democratic way of life.'[124]

Adopting lively apprenticeship as a model for education and, by implication, as a model for citizenship in the United States, will not magically solve all of the national and international problems with which the United States is entangled. Our argument is that adopting lively apprenticeship will be an important step toward creating citizens who will courageously face these problems and, in response, will not defensively retreat from fearless speech or fraught complexity.

Notes

1. John Dewey, *Experience and Nature,* in *The Later Works of John Dewey,* vol. 1, ed. Jo Ann Boydston (Carbondale: Southern Illinois University Press, 1984), p. 224. Repeat citations of Dewey will use 'Dewey, LW1, p. x' following the volume of the collected works.
2. Daniel Kahneman, *Thinking, Fast and Slow* (New York: Farrar, Straus and Giroux, 2011).
3. Boba M. Samuels, 'Can the Differences between Education and Neuroscience be Overcome by Mind, Brain, and Education?' *Mind, Brain, and Education* 3(1) (2009): 45–55.
4. Lucia Mason, 'Bridging Neuroscience and Education: A Two-Way Path Is Possible,' *Cortex* 45 (2009): 548–549.
5. John T. Bruer, 'Education and the Brain: A Bridge Too Far,' *Educational Researcher* 26(8) (1997): 4–16.
6. Dewey, LW1, pp. 306.
7. Rebecca E. Martin and Jennifer S. Groff, 'Collaborations in Mind, Brain, and Education: An Analysis of Researcher–Practitioner Partnerships in Three Elementary School Intervention Studies,' *Mind, Brain, and Education* 5(3) (2011): pp. 115–120.
8. Usha Goswami, 'Neuroscience and Education: From Research to Practice?' *Nature Reviews Neuroscience* 7(5) (2006): 406–413.
9. Dewey, LW1, p. 305.
10. Ibid., p. 298.
11. Ibid., p. 306.
12. Ibid., pp. 29–37.
13. Dewey, *Art as Experience,* in *The Later Works of John Dewey,* vol. 10, ed. Jo Ann Boydston (Carbondale: Southern Illinois University Press, 1984), p. 345.

14. Thomas Alexander, 'Pragmatic Imagination,' *Transactions of the Charles S. Peirce Society* 26 (1990): 325–348.
15. Ibid., p. 341.
16. Dewey, 'Three Independent Factors in Morals,' in *The Later Works of John Dewey*, vol. 5, ed. Jo Ann Boydston (Carbondale: Southern Illinois University Press, 1988), p. 293.
17. Steven Fesmire, *John Dewey and Moral Imagination* (Bloomington: Indiana University Press, 2003), p. 126.
18. Ibid., p. 95.
19. James Campbell, 'Dewey's Conception of Community,' in *Reading Dewey: Interpretations for a Postmodern Generation*, ed. L. A. Hickman (Bloomington: Indiana University Press, 1998), pp. 23–42.
20. Jay Schulkin, *Cognitive Adaptation: A Pragmatist Perspective* (Cambridge and New York: Cambridge University Press, 2009), pp. 113–115.
21. Ibid., p. 103.
22. Dewey, *Democracy and Education*, in *The Middle Works of John Dewey*, vol. 9, ed. Jo Ann Boydston (Carbondale: Southern Illinois University Press, 1985), p. 313.
23. Samuels, 'Can the differences between education and neuroscience be overcome by mind, brain, and education?'
24. Daniel T. Willingham, 'Three problems in the marriage of neuroscience and education,' *Cortex* 45(4) (2009): 544–545.
25. Ibid., p. 544.
26. Ibid., p. 545.
27. Mark Tschaepe, 'John Dewey's Conception of Scientific Explanation: Moving Philosophers of Science Past the Realist-Antirealism Debate,' *Contemporary Pragmatism* 8(2) (2011): 187–203.
28. Ibid., p. 190.
29. Ibid., p. 196.
30. Donna Coch and Daniel Ansari, 'Thinking about Mechanisms Is Crucial to Connecting Neuroscience and Education,' *Cortex* 45(4) (2009): 546–547.
31. Martin and Groff, 'Collaborations in Mind, Brain, and Education.'
32. Coch and Ansari, 'Thinking about Mechanisms Is Crucial to Connecting Neuroscience and Education.'
33. Goswami, 'Neuroscience and Education: From Research to Practice?'
34. Larry A. Alferink and Valeri Farmer-Dougan, 'Brain-(not) Based Education: Dangers of Misunderstanding and Misapplication of Neuroscience Research,' *Exceptionality* 18(1) (2010): 42–52.
35. Samuels, 'Can the Differences between Education and Neuroscience be Overcome by Mind, Brain, and Education?'
36. Joanna A. Christodoulou and Nadine Gaab, 'Using and Misusing Neuroscience in Education-Related Research,' *Cortex* 45(4) (2009): 555–557.
37. Sashank Varma, Bruce D. McCandliss, and Daniel L. Schwartz, 'Scientific and Pragmatic Challenges for Bridging Education and Neuroscience,' *Educational Researcher* 37(3) (2008): 140–152.
38. Alferink and Farmer-Dougan, 'Brain-(not) Based Education.'
39. Marco Iacoboni, *Mirroring People: the New Science of How We Connect with Others* (New York: Farrar, Straus and Giroux, 2008).
40. Coch and Ansari, 'Thinking about Mechanisms Is Crucial to Connecting Neuroscience and Education.'

41. Susan J. Pickering and Paul Howard – Jones, 'Educators' Views on the Role of Neuroscience in Education: Findings from a Study of UK and International Perspectives,' *Mind, Brain, and Education* 1(3) (2007): 109–113.
42. Coch and Ansari, 'Thinking about Mechanisms Is Crucial to Connecting Neuroscience and Education.'
43. Eric Jensen, 'Brain-based Learning: A Reality Check,' *Educational Leadership* 57(7) (2000): 76–80.
44. Alistair Smith, *Accelerated Learning in the Classroom* (Bodmin, UK: Network Educational Press Ltd., 1996).
45. James P. Byrnes and Nathan A. Fox, 'The Educational Relevance of Research in Cognitive Neuroscience,' *Educational Psychology Review* 10(3) (1998): 297–342.
46. Alferink and Farmer-Dougan, 'Brain-(not) Based Education.'
47. Goswami, 'Neuroscience and Education: From Research to Practice?'
48. Coch and Ansari, 'Thinking about Mechanisms Is Crucial to Connecting Neuroscience and Education.'
49. Deena Skolnick Weisberg, Frank C. Keil, Joshua Goodstein, Elizabeth Rawson, and Jeremy R. Gray, 'The Seductive Allure of Neuroscience Explanations,' *Journal of Cognitive Neuroscience* 20(3) (2008): 470–477.
50. Goswami, 'Neuroscience and Education: From Research to Practice?'
51. Isabel Cohen and Marcelle Goldsmith, *Hands on: How to Use Brain Gym in the Classroom* (South Africa: Edu Kinesthetics, 2002), qtd. in Goswami, 'Neuroscience and Education: From Research to Practice?'
52. Kahneman, *Thinking, Fast and Slow*, p. 27.
53. Ibid., pp. 52–85.
54. Ibid., p. 28.
55. David Franks, *Neurosociology: The Nexus between Neuroscience and Social Psychology* (New York: Springer, 2010).
56. Michael S. Gazzaniga, *The Mind's Past* (Berkeley: University of California Press, 1998).
57. Franks, *Neurosociology*, p. 67.
58. Kahneman, *Thinking, Fast and Slow*, p. 44.
59. Ibid., p. 40.
60. Ibid., p. 45.
61. Ibid., p. 64.
62. Dewey, 'The Reflex Arc Concept in Psychology,' in *The Early Works of John Dewey*, vol. 5, ed. Jo Ann Boydston (Carbondale: University of Southern Illinois Press, 1972), pp. 96–109.
63. Dewey, 'The Influence of Darwinism on Philosophy,' in *The Middle Works of John Dewey*, vol. 4, ed. Jo Ann Boydston (Carbondale: University of Southern Illinois Press, 1980), pp. 3–14.
64. Joseph Henrich, Steven J. Heine, and Ara Norenzayan, 'The Weirdest People in the World?' *Behavioral and Brain Sciences* 33(2–3) (2010): 61–135.
65. Ibid., p. 64.
66. Ibid., pp. 70–72.
67. Tibor Solymosi and John Shook, 'Neuropragmatism and the Culture of Inquiry: Moving Beyond Creeping Cartesianism,' *Intellectica* 60 (2013): 137–159.
68. Dewey, LW1, p. 224.

69. Kim Sterelny, *The Evolved Apprentice: How Evolution Made Humans Unique* (Cambridge: MIT Press, 2012).
70. Ibid., p. 35.
71. Ibid., p. 122.
72. Ibid., p. 123.
73. Ibid., p. 140.
74. Ibid., p. 142.
75. Dewey, 'Creative Democracy – The Task Before Us,' *The Later Works of John Dewey*, vol. 14, ed. Jo Ann Boydston, (Carbondale: Southern Illinois University Press, 1991), pp. 224–230.
76. For example, Mark Johnson (*The Meaning Of The Body* (Chicago: The University of Chicago Press, 2007)) points out evidence which indicates pattern recognition takes place in infants. Cross modal pattern recognition is manifest when an infant recognizes the pattern of a pacifier on which it had been sucking (p. 42). In the case of object recognition, infants establish object permanence by recognizing repeated patterns (pp. 47–48).
77. Mark Johnson, *The Meaning Of The Body*.
78. Shaun Gallagher, *How The Body Shapes The Mind* (Cambridge: Cambridge University Press, 2005).
79. Ibid., p. 81.
80. Johnson, *The Meaning of the Body*, p. 40. Both Johnson and Gallagher draw on the work of Andrew N. Meltzoff and M. Keith Moore, 'Imitation of Facial and Manual Gestures by Human Neonates,' *Science* 198 (1977): 75–78.
81. Gallagher, *How the Body Shapes the Mind*, p. 227.
82. Ibid., p. 208.
83. Ibid., p. 224.
84. Iacoboni, *Mirroring People*.
85. Ibid., p. 134.
86. Ibid., p. 48.
87. Ibid., p. 95.
88. Ibid., p. 25.
89. Ibid.
90. Ibid., pp. 77–78.
91. Ibid., p. 17.
92. Ibid., p. 155.
93. Gallagher, *How the Body Shapes the Mind*, p. 225.
94. Johnson, *The Meaning of the Body*, p. 51.
95. Ibid., p. 39.
96. Iacoboni, *Mirroring People*, p. 98.
97. Gallagher, *How the Body Shapes the Mind*, p. 204.
98. Schulkin, *Cognitive Adaptation*, p. 69.
99. Ibid.
100. Ibid., p. 95.
101. Ibid., p. 98.
102. James Zull, *From Brain to Mind: Using Neuroscience to Guide Change in Education* (Sterling: Stylus Publishing, 2011).
103. Ibid., p. 59.
104. Ibid., p. 65.
105. Ibid., p. 64.

106. Ibid., p. 65.
107. Schulkin, *Effort: A Behavioral Neuroscience Perspective on the Will* (Mahwah: Lawrence Erlbaum Associates, 2007).
108. Ibid., p. 104.
109. Ibid., p. 105.
110. Schulkin, *Adaptation and Well-Being: Social Allostasis* (Cambridge: Cambridge University Press, 2011).
111. Ibid., p. 147.
112. Ibid., pp. 60–65.
113. Ibid., p. 80.
114. Joseph E. LeDoux, *Synaptic Self: How Our Brains Become Who We Are* (New York: Penguin Books, 2003).
115. Sheldon Cohen, Joseph E. Schwartz, Elissa Epel, Clemens Kirschbaum, Steve Sidney, and Teresa Seeman, 'Socioeconomic Status, Race, and Diurnal Cortisol Decline in the Coronary Artery Risk Development in Young Adults (CARDIA) Study,' *Psychosomatic Medicine* 68(1) (2006): 41–50.
116. Schulkin, *Adaptation and Well-Being*, p. 119.
117. LeDoux, *Synaptic Self*, pp. 278–279.
118. Ibid., p. 281.
119. Sterelny, p. 36.
120. Ibid., p. 29.
121. Johnson, *The Meaning Of The Body*.
122. Jeremy Naydler, *Goethe on Science: A Selection of Goethe's Writings* (Edinburgh: Floris Books, 1996).
123. David Walker, *Appeal to the Coloured Citizens of the World* (State College: The Pennsylvania State University Press, 2002).
124. Dewey, LW14, p. 227.

10
A Neuropragmatist Framework for Childhood Education: Integrating Pragmatism and Neuroscience to Actualize Article 29 of the UN Child Convention

Alireza Moula, Antony J. Puddephatt, and Simin Mohseni

Introduction

A 9-year-old Swedish girl, Milla Martin, was watching TV with her mother when she saw a film about starving African children. She became very sad and angry and asked, 'Why don't we do something to help these children?' Together with some other children, Milla and her mother decided to bake cakes and sell them in order to collect money to support the starving African children. Their responsible social action inspired other children and youth in Sweden who did the same thing. They succeeded in collecting hundreds of thousands of Swedish kroner for this human cause. As a result, Milla Martin was chosen as one of the few persons nominated as 'Swedish heroes of the year.' Consequently in 2011, on a popular nationwide TV program involving the Swedish prime minister, she received her prize: travelling with her family to Tanzania in a 'study-travel' to find out how her 'cake-baking-movement' could help these less fortunate children.[1]

In this inspiring story, a 9-year-old Swedish girl demonstrated how a group of children could create a social project and *organize their actions* to help needy children. Some of the most complicated human capacities, such as emotional-expression, thinking, decision-making, problem-solving, relationship-management, and responsible social action are all integrated in a nationwide example of how children from a welfare state can establish a bond with those of developing countries. This story helps us to generate assumptions that are at the core of our ideas, forming the basis of this chapter:

- Emotions are crucial for human action, for example in the creation of empathy, but are most effective when regulated by cognition.
- Reflective and goal-directed thought allows for creative problem solving.
- These human capacities develop in the context of relationships with others, and can develop into habits.
- Abstract philosophies about minded behavior are best understood when linked to the concrete science of cognition vis-à-vis neuroscience.

Through the integration of pragmatism and neuroscience, this chapter aims to provide a conceptual framework to help realize Article 29 of the Child Convention. The essence of Article 29 is focused on (a) developing children's mental abilities to their fullest potential, and (b) preparing them to live as responsible social actors. The crucial issue is that a child's mental capacity can develop in various directions. The Child Convention ties a child's capacity for development to their performance of responsible social actions. *Milla Martin's story (and the other children's who supported her) provides an illustrative case for how children can organize themselves in responsible social actions.* Following the Child Convention, all humans under 18 years of age are considered children; we use the terms children, or youth, to refer to this group.

Inspired by pragmatist philosophy and neuroscience, we argue that the enhancement of executive functions (vis-à-vis the prefrontal cortex of the brain) play a crucial role in developing important capacities such as thinking and problem solving in children. Pragmatist philosophy can be seen as the equipment for ameliorative social action.[2] Knowledge of the executive functions of the prefrontal cortex explains the biological make-up of these activities and contributes to the construction of our philosophical-scientific framework. This chapter will proceed by presenting: (1) Article 29 of the Child Convention, (2) the relevant aspects of social pragmatism, (3) a description of the executive functions of the prefrontal cortex, (4) a neuropragmatist conceptual framework that facilitates organizing the executive functions of the child in a responsible and goal-directed manner; and finally (5) a summary and discussion.

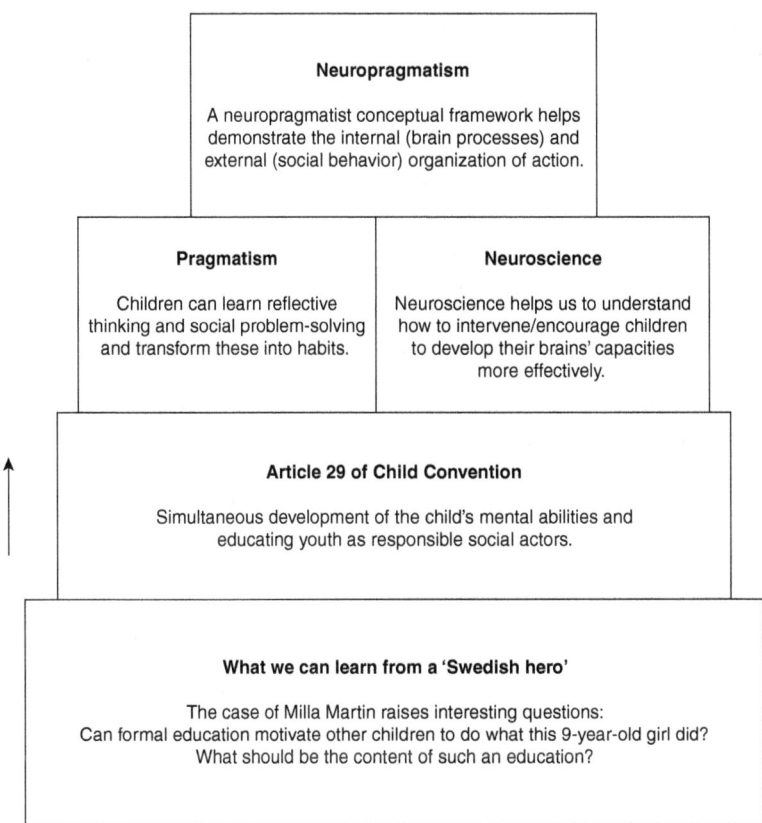

Figure 10.1 Neuropragmatism and the UN Child Convention

Figure 10.1 demonstrates three sources – The Child Convention, pragmatism and neuroscience – that we draw on to develop a neuropragmatist conceptual framework for realizing Article 29 of the Child Convention.

Article 29 of the Child Convention, mental development, and ameliorative social action

The United Nations' Child Convention is an international document that is ratified by almost all countries in the world, and demands the realization of children's rights. Built on varied legal systems and cultural traditions, the Convention is a universally agreed set of non-negotiable standards and obligations. They are founded on respect for the dignity and worth of each individual, regardless of race, color, gender, language, religion, opinions, origins, wealth, birth status, or ability and therefore apply to every human being everywhere.[3] In this chapter, we limit ourselves to Article 29 of this Convention, which refers to the direction of education and includes, briefly:

(a) The development of the child's personality, talents and mental and physical abilities to their fullest potential;
(b) The development of respect for human rights and fundamental freedoms, and for the principles enshrined in the Charter of the United Nations;
(c) The development of respect for the child's parents, his or her own cultural identity, language and values, for the national values of the country in which the child is living, the country from which he or she may originate, and for civilizations different from his or her own;
(d) The preparation of the child for responsible life in a free society, in the spirit of understanding, peace, tolerance, equality of sexes, and friendship among all peoples, ethnic, national and religious groups and persons of indigenous origin;
(e) The development of respect for the natural environment.

This list can be reduced down to two main conceptual points, which will guide our chapter: developing a child's mental abilities to his/her fullest potential, and educating children so that they learn respect for others and become responsible social actors. These points cover the fundamental issues at stake from Article 29, and connect more directly to what our neuropragmatist educational model aims to accomplish.

It is important to note that these two points are connected in an important way. That is, self-development occurs through the exercise of responsible social action, as individuals learn to respect other children's self-development in turn. Thus, self-development takes place in a social context. Children study and learn together, not in isolation from each other. As Dewey[4] has emphasized, children do not learn by merely sitting and listening to their teachers. Instead, acquiring information should be completed through experiential learning, or what Dewey refers to as learning by doing. We add a social component to this, emphasizing the importance of doing things together with others, in groups, in order to learn most effectively. To conceptualize this with reference to some recent research, we draw from Ryan and Deci's self-determination theory[5] and its three components: autonomy, relatedness, and competence. This theory assumes that all individuals have natural, innate, and constructive tendencies to develop an ever more elaborated and unified sense of self. People have a primary propensity to forge interconnections among aspects of their own psyche as well as with other individuals and groups in their social environments. Healthy development involves the complementary functioning of these two aspects of integration. Psychological growth and integration in personality should neither be taken as a given nor discounted. Rather, it must be viewed as a dynamic potential that requires nurturing. We hope that the conceptual framework we present in the following sections can help realize Article 29 of the Child Convention. Our neuropragmatist contribution places an emphasis on the child as an active biopsychosocial being. Our research project utilizes a holistic approach, taking into account the child's cognition, emotional states, social relationships, and the process and content of knowledge acquisition.

Pragmatism: developing positive habits of thought and action

In this section, we try to answer the *what, where, when, and how* of forming habits. We limit ourselves to three tightly related contributions of pragmatism: (1) the power of habits, (2) the importance of developing positive habits in children and youth through formal education, and (3) habits that are most central in regard to responsible social action. In so doing, we endeavor to show how the pragmatists envisioned many contemporary neuroscientific ideas prior to the 20th century, and are hence quite relevant to the conceptualization of these present-day issues. Let us start with William James, who can be considered as the founder of neuropragmatism.[6, 7, 8] The following quotation nicely captures his ideas about the power of habits in our everyday lives:

> All our life, so far as it has definite form, is but a mass of habits, – practical, emotional, and intellectual, – systematically organized for our weal or woe, and bearing us irresistibly toward our destiny, whatever the latter may be...I believe that we are subject to the law of habit in consequence of the fact that we have bodies. The plasticity of the living matter of our nervous system, in short, is the reason why we do a thing with difficulty the first time, but soon do it more and more easily, and finally, with sufficient practice, do it semi-mechanically, or with hardly any consciousness at all.[9]

Dewey[10] also had great respect for the power of habits: 'All virtues and vices are habits which incorporate objective forces. They are interactions of elements contributed by the make-up of an individual with elements supplied by the out-door world.' He added that habits form our effective desires, and they furnish us with our working capacities. They rule our thoughts, determining which shall appear and be strong and which shall pass from light into obscurity.

For Dewey, as the founder of progressive pedagogy,[11] the purpose of education is to prepare people to survive and flourish in a future that is by nature uncertain. Enabling children to take command of their own powers, rather than merely fashioning them to fulfill society's current needs, best provides this. At the same time, we have to be aware that Dewey's concept of self-development is a social project. He views the self as one whose actions and identity are governed by relations with others. William James, as not only a pragmatist but also a psychologist and a neurophysiologist, described the biological basis of habit formation in the brain:

> The great thing, then, in all education, is to make our nervous system our ally instead of our enemy. It is to fund and capitalize our acquisitions, and live at ease upon the interest of the fund. For this we must make automatic and habitual, as early as possible, as many useful actions as we can, and guard against the growing into ways that are likely to be disadvantageous to us, as we should guard against the plague.[12]

But *what* is the content or direction of this education, and *how* should it form good habits in children? The concept of intelligence is crucial for all pragmatists (for an excellent review see an article by Bhattcharya).[13] Dewey referred to intelligence as the leading-concept in forming habits in children:

> We are told almost daily and from many sources that it is impossible for human beings to direct their common life intelligently. We are told, on the one hand, that the complexity of human relations, domestic and international, and on the other hand, the fact that human beings are so largely creatures of emotion and habit, make impossible large-scale social planning and direction by intelligence. This view would be more credible if any systematic effort, beginning with early education and carried on through the continuous study and learning of the young, had ever been undertaken with a view to making the method of intelligence, exemplified in science, supreme

in education. There is nothing in the inherent nature of habit that prevents intelligence from becoming itself habitual; and there is nothing in the nature of emotion to prevent the development of intense emotional allegiance to the method.[14]

In an attempt to make his idea about the 'intelligent method' more concrete, Dewey indicated that the formation of ideas should be based on desire. This, he argued, is the driving force behind human beings' activities. Later, after the thoughtful consideration of concrete situations, we can develop these desires into purposes. Finally, we can plan, in some detail, in an effort to realize these purposes. In his book, *How We Think*, Dewey[15] described how children can learn to organize their process of reflective thinking. For Dewey, reflective thinking involves not simply a smooth sequence of ideas, but a series of consequences: 'a consecutive ordering in such a way that each determines the next as its proper outcome, while each in turn leans back on its predecessors. The successive portions of the reflective thought grow out of one another and support one another; they do not come and go in a medley.'

William James, in a lecture for teachers, also described how children can *organize their thinking processes*:

> Your pupils, whatever else they are, are at any rate little pieces of associating machinery. Their education consists in the organizing within them of determinate tendencies to associate one thing with another, – impressions with consequences, these with reactions, those with results, and so on indefinitely.[16]

But thinking should be actively organized for the sake of meaningful action. Dewey emphasized that children should learn to decrease their impulsive reactions and stop and think before acting. George Herbert Mead makes a very interesting comparison between how an individual *organizes her or his act* and how the central nervous system makes such an organization possible. When Mead was writing almost a century ago, there were some 'brain scientists,' but neuroscience as a discipline did not exist. Mead repeatedly referred to the role of the 'central nervous system' in our behavior. A long quotation from Mead illustrates his belief in what we can call *the internal (neurobiological) and external (social) dimensions of the organization of action*:

> Human intelligence, by means of the physiological mechanism of the human central nervous system, deliberately selects one from among

the several alternative responses which are possible in the given problematic environmental situation; and if the given response which it selects is complex – i.e., a set or chain or group or succession of simple responses – it can organize this set or chain of simple responses in such a way as to make possible the most adequate and harmonious solution by the individual of the given environmental problem. It is the entrance of the alternative possibilities of future response into the determination of present conduct in any given environmental situation, and their operation, through the mechanism of the central nervous system, as part of the factors or conditions determining present behavior, which decisively contrasts intelligent conduct or behavior with reflex, instinctive, and habitual conduct or behavior – delayed action with immediate reaction.[17]

Mead not only tells us that intelligent action requires deliberation and the selection of the best possible option to solve a problem, but he also indicates that this process is possible only through our brain's physiological capacity. This leads us to the next section, where we consider the role of the brain in organizing our actions.

The prefrontal cortex, executive functions, and the organization of thought and action

At birth, the average weight of the human brain is around 400 g, and it develops through childhood and adolescence to about 1400 g. These developmental stages are mostly under the control of genes, particularly in embryonic stages. However, they are strongly influenced by environmental factors, particularly after birth. For example, motor activities, sensory stimuli (auditory, visual, olfactory, and tactile), social experiences (especially the parental-child relationship), emotional input (love or hate), diet, hormones, and drugs, can influence the development of the brain at the cellular – and consequently the behavioral – level. The capacity of the brain to reorganize itself by forming new cellular connections due to new experiences or injuries throughout life is known as 'brain plasticity.'

Neuroscientists refer to the brain as the most complex structure in the universe. Nobel Prize winner Eric Kandel[18] described this structure:

> The brain is a complex biological organ of great computational capability that constructs our sensory experiences, regulates our thoughts and emotions, and controls our actions. The brain is responsible

not only for relatively simple motor behaviors, such as running and eating, but also for the complex acts that we consider quintessentially human, such as thinking, speaking, and creating works of art.

Today, we know that different parts of the brain are responsible for different functions. Vision, for example, is under the control of the posterior part of the brain close to the upper part of the neck (occipital lobe), while hearing is under the control of the temporal lobe located on the sides of the brain. The part of this complex biological organ which controls intelligence, concentration, self-judgment, planning and problem solving is under the control of the frontal lobe, particularly in its most anterior part known as the prefrontal cortex (PFC). This is the part of the brain we will focus on in this chapter.

As Goldberg[19] states, 'Like a large corporation, a large orchestra, or a large army, the brain consists of distinct components serving distinct functions. And like these large-scale human organizations, the brain has its CEO, its conductor, its general: The frontal lobes.' To be more precise, this role is vested in but one part of the frontal lobes, the prefrontal cortex (PFC). However, the term 'frontal lobe' is common shorthand for this. The Great Russian neuropsychologist, Alexander Luria, called frontal lobes the neurobiological site of civilization.[20] Fuster[21] emphasizes the role of the PFC in coordinating cognitive functions in the temporal organization of behavior: that is, in the formation of coherent behavioral sequences toward the attainment of goals. Studies of the PFC and its role in development and neuro-maturation demonstrate the pivotal role for this cortical region in the integration of cognition, social emotions, and behavior throughout childhood and adolescence. The acquisition of executive functions and self-regulation are vital for everyday expertise and social adaptation.[22] Squire[23] describes the PFC's functions in a simple way:

> What controls your thoughts? How do you decide what to pay attention to? How do you act appropriately while dining in a restaurant or listening to a lecture? How do you plan and execute errands? How do you manage to pursue long-term goals, like obtaining a college degree, in the face of the many distractions that can knock you 'off task'? In short, how does the brain manage to orchestrate the activity of millions of neurons to produce behavior that is willful, coordinated, and extends over time? This is called cognitive control, the ability of our thoughts and actions to rise above more reactions to the immediate environment and be proactive: to anticipate possible futures and coordinate and direct thought and action to them.

Neuropsychologists use the concept of executive functions to refer to the main roles of the prefrontal cortex. However, Anderson, Jacobs, and Anderson[24] warn that although 'the prefrontal cortex clearly plays a crucial role in executive function, it appears to be increasingly accepted that this region does not act in isolation, but is part of a broader functional system, which involves other regions.' Executive functions include mental capacities necessary for formulating goals, planning how to achieve them, and carrying out these plans effectively. Thus executive function is a collection of interrelated functions, or processes, which are responsible for goal-directed or future-oriented behavior, and has been referred to as the conductor which controls, organizes, and directs cognitive activity, emotional responses, and behavior.[24]

But how and where are children supposed to learn executive functions? This is precisely the question that a group of educators raise:

> If parents do not engage children in thinking, teachers tell them only facts but do not tie the facts to the children's own experiences, television does not ask them to think and analyze, and video games provide excitement without reflections, where exactly do we expect them to develop the kinds of skills that help develop executive functions and prepare students to become responsible, thinking adults?[25]

In the last two decades, educational theorists, neuropsychologists and neuroscientists have organized themselves through a series of groupings, conferences and publications to develop the new field of neuroeducation.[26] Much attention is paid to the role of executive functions in formal education. For example, Meltzer, Pollica, and Barzillia[27] emphasize the importance of teaching executive function processes such as planning, organizing, prioritizing, and self-editing in schools. They argue that even though school is the ideal place for children to learn executive functions, these skills go far beyond school:[28]

> As the demands of our school curricula increase, students are expected to use executive processes for more and more assignments in order to prepare for high school, college, and beyond. The primary goal for teachers has been to prepare students by teaching them the content and skills valued by our highly literate society, such as reading, writing, spelling, math, history, and science. While the end product of learning is important, it is evident that students do not retain all the content they are taught from year to year. Therefore, it is even more important to teach students the executive function processes

that will carry over from elementary school to middle school, high school, college, and even into the real world.

As the world becomes more and more complex and demanding, life success depends increasingly on the mastery of executive processes such as goal setting, planning, organizing, prioritizing, memorizing, initiating, shifting, and self-monitoring.[29] In other words, executive functions provide the ability to regulate behavior within a fluctuating and unpredictable environment. This entails the integration of what the individual wants to achieve with what one can do based on the situational context, and the individual directs this energy and creates plans accordingly.[30] Zelazo et al.[31] have conceptualized executive functions as a problem-solving framework with distinct phases. This problem-solving framework is a kind of macro-construct that illustrates the way in which distinct executive processes operate in an integrative way to solve a problem or achieve a goal. These four phases are: (1) problem representation, (2) planning, (3) execution, and (4) evaluation. As Anderson[32] indicated, this approach to executive functions emphasizes an integrative functional approach, considering how different executive processes work together to achieve a goal. This framework is also appreciated by social psychologists working from a biosocial perspective: that is, to see the brain as functioning within a social environment. In line with this, Martin and Failows[33] suggest an approach with a holistic functioning of persons within sociocultural contexts and practices of problem solving and goal-directed, strategic activity. In line with Zelazo et al., and with Martin and Failows,[34] we present a problem-solving model that youth can learn and use to help organize their thoughts and actions, thus mastering their executive functions.

The prefrontal cortex, the executive functions, and the internal and external organization of the act: a neuropragmatist tool for guiding reflective thought and goal-directed action

> Thinking is important to all of us in our daily life. The way we think affects the way we plan our lives, the personal goals we choose, and the decisions we make. Good thinking is therefore not something that is forced upon us in school: it is something that we want to do, and want others to do, to achieve our goals and theirs.[35]

In the above quotation, Jonathan Baron indicates how thinking is important for decision making, achieving our goals, and planning our daily life. In this section, we introduce Alireza Moula's model of the 'daily-life machine.' This model contains eight concepts organized through four major human capacities: (1) cognition (including thinking, decision-making, problem solving, and life-management), (2) emotion, (3) relations with others, and (4) metacognition (integrating previously acquired knowledge with a psychological tool to guide life-management). All of these are necessary in order to meet the challenges of daily life in a goal-oriented way (see Figure 10.2).

An individual is the conscious (and many times the unconscious) driver of this machine, that is, the owner of her or his body, brain and mind. This analogy of the machine is inspired by the familiarity of using machines in our everyday lives. For example, we can drive a car while doing something else at the same time, like talking on a cell phone or daydreaming. We often drive very effectively without any conscious focus on the car or the road. Such action takes place after the habit of driving has become a routine act. But driving is not an exception; we do many other things through our habits and routines without reflective thinking or deliberative decision making. The analogy of the machine warns us that there are moments in our lives where we should not rely solely on habits, driving our proverbial car in a half-conscious daze. Rather, in this daily life-machine analogy, we should wake out of our unconscious habits and deliberate toward higher goals. We should know where we are going, and we should think about how to get there. The 'motor of the machine' represents the power of cognition, and reveals how thinking is at the heart of decision making, which is in turn at the heart of problem solving, central to life-management.

Much like a car needs a motor, it cannot move without wheels. In our symbolic 'daily-life machine,' the 'wheels' are the emotions, which are essential for daily life functions. Because of the vitality of the emotions, we describe their role in connection with cognitive functions such as decision-making. Ratey[36] writes that emotion is 'messy, complicated, primitive, and undefined because it's all over the place, intertwined with cognition and physiology.' He hopes that with the help of science, our knowledge of emotional functioning will expand, such that practical lessons for daily emotional management might become available. Ratey's own casework with his patients indicates that 'lack of emotion leads to poor reasoning and ultimately to poor social judgment, even when factual intelligence is still intact.'[37] As Franks[38] observes,

> One can know cognitively that others regard killing as a serious crime, but without empathetic emotions, the private compulsion

making such behavior automatically undesirable to the individual is lacking...Mastery of any culture's right and wrong must contain an emotional component. Socialization is not just a process of stuffing heads with shared and disembodied symbols; it involves the body as well as our thoughts. Without physiologically grounded constraint of emotions, role taking would just as well produce a society whose 'understanding' of others would be used only for their manipulation.

There is no doubt about the vitality of emotions in our lives. However, as Damasio[39] notes, an emotional signal is not a substitute for proper reasoning. It has an auxiliary role, increasing the efficiency of the reasoning process and makes it speedier. Damasio[40] also describes different levels of life regulation, seeing emotions as complex, stereotyped patterns of responses. Feelings, too, are described as sensory patterns signaling pain or pleasure. Emotions and feelings are under the line of consciousness, while higher reasoning is above the line of consciousness. Reasoning is described as complex, flexible, and customized plans of response that are formulated in conscious images and can be executed as behavior.

In line with this, Franks[41] states that most rational business choices are laced with emotions, and the anticipation of regret, patience in making well thought out decisions, and expectations of long-term success involve emotions. Franks criticizes ideas of 'pure logical determinacy,' meaning that rational decision making and social control depend on their own supportive embodied emotions. Emotions often accompany and color our memories of previous experiences, shaping our cognitive decisions. As Damasio writes, the emotional signal can operate entirely under the radar of consciousness. However, new situations may be like previous ones or may not. On the one hand, an individual once bitten by a dog may feel frightened as soon as he or she sees the next dog. On the other hand, reasoning may reassure him or her that the majority of dogs do not bite, leading to overcoming the fear. Therefore, strong emotional experiences may make one more cautious but should not be seen as a complete barrier to the future action of individuals. It is quite possible that an individual has no time or motivation to apply reasoning toward making a decision. Then, one trusts one's feelings because they accompany similar previous experiences. What is proper is that, whenever emotions 'come up,' the individual tries to reflect on them. The person can apply reasoning strategies to what one's emotions demand, and carefully think about what to do, making decisions on the basis of awareness of consequences.

The gas tank symbolizes part three of the 'machine'; our relationships with other human beings are our fuel for life. In the language of neuroscience, a brain cannot adequately develop without contact with other brains. As Wexler[42] explained, human behavior cannot be determined only by its biological factors. Social factors play a crucial role as well, and researchers should discover new ways to study how the social and biological, through interaction with each other, determine our behavior. Dick and Overton[43] emphasize the role of social interaction in the executive functions of the prefrontal cortex. They indicate a need to bridge cognitive and neuro-physiological emphases reflected in the contemporary executive functioning literature with the social aspects of concern in the literature about theories of mind and social understanding. Dick and Overton[44] clarify their holistic perspective in the following:

> Holism encourages a view of the psychological organism and brain as interpenetrating part systems of the embodied agent actively engaged in the world. These part systems, like other part systems of the embodied agent, are dynamic and self-organizing in character, and not decomposable into foundational elements. Explanation is not found in independent causes, but in formulating models that account for the system functioning at each level of analysis, further exploring the systematic relations among levels and, within this context searching for specificity of conditions.

Thus, focusing on the social aspects of the individual's life situation is key if a complete understanding of their self-development is to be realized.

Part four, the 'lights of the machine,' symbolize the human metacognitive capacity, and how, through knowledge and skills, we can find our route. This includes (a) knowledge a child usually acquires in school, and (b) a psychological tool (mental model) for interpreting and/or using this knowledge to guide cognition in the processes of life-management. 'Metacognition' refers to cognition about cognition or thinking about thinking. Thus, this mental model's final aim is facilitating thoughtful and goal-directed life-management. Schools usually provide general knowledge through teaching lessons from many diverse disciplines. However, they usually do not provide metacognitive lessons for how youth might organize their behavior in daily life. Kozulin[45] draws on Vygotsky to define psychological tools as those symbolic artifacts – signs, symbols, texts, formulas, graphic-symbolic devices – that help individuals master their own 'natural' psychological functions of perception, memory, attention, and so on. Kozulin emphasizes that psychological tools offer a fresh perspective in classroom learning, teaching, and creative problem solving.

> One of the essential characteristics of learning based on psychological tools is its ability to use models, that is, schematized and generalized representations of objects, processes, and their relationships. Moreover, students' own thinking and problem-solving activity can be represented as a model with the help of psychological tools, thus becoming an object of the students' conscious deliberation, planning and decision making.[46]

Considering the relation between these four major parts of the machine, we suggest that by teaching the child (or youth) a mental model – a psychological tool – the child can organize their other capacities in a goal-directed manner. Therefore, we present a model that can work as a psychological tool for organizing youths' thinking, decision-making, problem-solving, emotional regulation, and relationship management. Our construct, the machine, will draw attention to these concepts, helping youth to be conscious about these capacities and their relation to each other. As a psychological tool, the model suggests that by learning and using this model, an individual can organize these capacities.[47, 48, 49, 50, 51]

Figure 10.2 shows Moula's 'daily life machine,' which is constructed to support individuals in becoming more conscious about their capacities and how they can organize their internal acts (brain processes) and external acts (social behaviors). This fictive machine has four parts that includes eight concepts. Part one is the motor of the machine and symbolizes human cognitive power, including thinking, decision-making, problem solving, and life-management. The second part symbolizes the wheels of the machine and illustrates the importance of emotions in our daily life performances. Part three, the fuel tank, symbolizes the vitality of our social relations in our lives, and the fact that we cannot function in a society without them. Part four, the lights of the machine, includes (a) the general knowledge that the individuals gain through school, family, Internet, and so on, and (b) a specific skill that youth, as the driver of the machine, can learn to organize their internal and external acts in daily life. The advantage of this symbolic machine is that it conceptualizes the individual as a bio-psycho-social being and avoids dualism between thinking and acting, emotions and cognition, and the internal (neurobiological) and external (social).

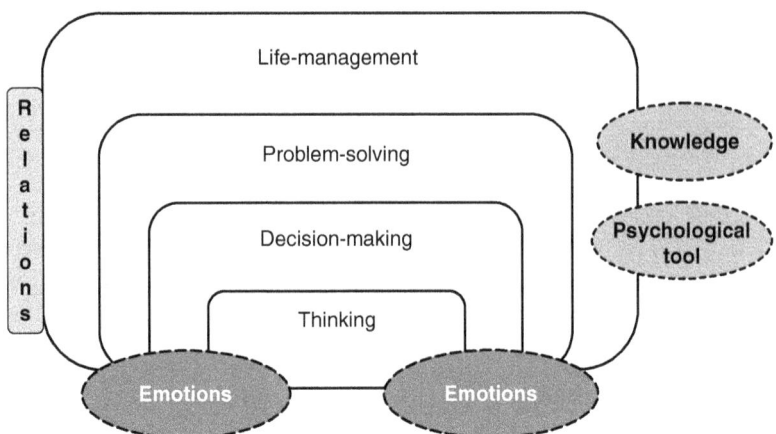

Figure 10.2 Moula's daily-life machine and internal and external organization of action

A mental model for organizing thought and action

> Most of our behavior is anchored in experience. Much of it consists of learned habits, sequence of more or less automatic responses to internal and external stimuli. Some of it, most characteristically in our species, depends not only on habits but also on educated choice, and leads to purposeful creation of new changes in the environment and new relationships between us and the environment. This form or part of our behavior can appropriately be termed deliberate, for it is guided by the cognitive deliberation of alternatives, expected risk and value, and deliberative purpose.[52]

After having presented the four capacities necessary for daily life through the analogy of the machine, we complete our conceptual framework by presenting a model – a psychological tool – that can help youth to organize these capacities. Learning to use a problem-solving model to achieve goals can be an effective tool for the internal and external organization of action. As Dewey[53] emphasized, although human beings have the necessary biological capacities in place, they are not born with ready-made 'rudiments of mental discipline.' Intelligent habits are not a gift of nature, but 'the main office of education is to supply conditions that make for their cultivation.' Mesulam[54] echoes this, indicating that while the prefrontal cortex gives human beings the capacity for reflection, change, and choice, the PFC cannot predetermine the content of these reflections, the direction of change, or the choices we make. As Mead[55] would expect, these are qualities that a child acquires in social interaction with others. The Rahyab model has been used in social work practice, teaching, and research in Sweden and Iran since 1997.[56] In this chapter, we develop the model so it can be used as a psychological tool for use in schools. Our experience using Rahyab shows that it encourages people to think more deliberately, exchanging reactive and impulsive action for careful, reflective thought to guide action more effectively. Below, we present the steps of this model:

Step one

Thomas and Thomas[57] wrote that if individuals define their situations as real, they are real in their consequences. This theorem connects people's understanding of situations to their construction of action and the subsequent consequences. Others may help to shape the definition of the situation, but individuals are at the center of this interpretive phase. We can compare this dialogical process of mutual definition – dialogue with self and others – with a funnel. Thoughts that may be general or ambiguous in the beginning become more and more specific. Somewhere near the bottom of the funnel, we arrive at the final definition of a concrete situation or problem.

Step two

Dialogue (with self and others) continues, and the individual tries to imagine the desirable situation. The desirable situation is connected to the present situation or problem as defined in step one. It is possible that the dialogue (in step one) cannot lead to definition of the problem, but when the individual talks about the desirable situation (in step three) and what she/he wants, then, it becomes clearer what she/he *does not* want, thus providing the definition from step one. Thus, practitioners need not adhere to the chronology of these steps too strictly and should remain flexible about their order. Further, an individual may think of several desirable situations. In this case, the next step would ask which desirable situation is most possible from a practical point of view.

Step three

Watzlawick has said the best help that we can give someone is to find a new alternative.[58] First, all possible alternatives should be found without any attempt to rank them. Then, youth are asked to consider barriers and resources for each alternative. Finally, *the consequences of each alternative should be examined very carefully.* Only at this stage can the individual rank the alternatives. Sometimes we have three or four alternatives, and for each alternative we have to look at barriers, resources, and consequences. This is the step of the model that demands a great deal of thinking and patience.

Step four

Often, it is not easy to choose an alternative. Youths may put pressure on others to recommend one. The practitioner should encourage help-seeking individuals to study the alternatives themselves and choose one. After selecting, he or she can plan for action with the necessary details in place.

Step five

The youth can look back and decide whether or not they are satisfied with the course of action.

A few general thoughts on the model

Rahyab is a humanistic model that recognizes the individual as the expert of her/his own life. It is an effective 'help to self-help' and seeks to avoid creating dependency on experts. It encourages youth to put reflective thinking at the heart of their daily performances. Saleebey[59] believes that the greatest good you can do for another is not just to share your own riches, but also to reveal people's own strength to them. Dewey believes that the cultivation of our own reflective intelligence is of the utmost importance to our development and well-being, and that nothing can prevent us from using this intelligence in our daily lives. For Dewey, intelligence has to do with remaking of the old through union with the new. The Rahyab model is an intelligent decision-making and problem-solving model for two reasons: (a) it connects the past and present to the future, and (b) it compares different alternatives in a deliberative manner before acting. These habits of continual comparison, reflection, evaluation, and choice increase the intellectual capacity of individuals in their daily-life performances. There are often no easy or quick solutions in the problems presented in life, and this model helps to raise the consciousness of youth about the complexity posed in life across various situations.

A good example of this model comes from an educational project that was launched by Alireza Moula in Sweden in September 2009. This three-year project involved the participation of a group of students who were beginning secondary school in grade 7 (13-years-old), and followed these children until they had completed grade 9 (15-years-old). Since the results are still being analyzed, we briefly draw on this project to better illustrate and provide support for our model. The project had three phases: The first had students learning Rahyab and applying it to solve various fictive personal and family problems. The second phase involved the youth applying the model to try and solve international issues, such as poverty in developing countries. The third and final phase involved the youth engaging in practice exercises to enhance their executive functioning. A preliminary analysis suggests that these youth completed the first phase with success, learning to apply Rahyab to solve fictive problems. In the second year, they used Rahyab in an effort to meet international challenges such as poverty in the world, which culminated in these youth writing a fictive letter to UNICEF. In this letter, the youth outlined several potential programs that could be implemented to reach out to youth in developing countries. Some of the proposed solutions were to enroll famous Swedes, and fundraise using a variety of methods, in order to sponsor them. Further, the students wanted to take on a number of strategies to better understand the experiences of youth in developing regions by setting up interviews, visits, and exchanges. This project has been quite inspiring, and shows that youth have the ability to apply Rahyab successfully in group-based formats, which expands their empathy toward others, and enables them to creatively and effectively solve social problems.[60] This in turn helps shape habits to maximize their use of cognitive functioning in daily life, thus producing responsible and engaged citizens as required by Article 29 of the UN Child Convention.

Discussion

The aim of this chapter is to construct a conceptual framework that can facilitate the realization of Article 29 of the Child Convention. At least two main points can be seen in this article: (1) to develop children's mental capacities, and (2) to create educational spaces where children can learn to become responsible social actors. In the introduction, we presented the case of Milla Martin, a Swedish girl who was chosen as one of the nations' heroes of 2011 due to her sympathetic and creative social project. Is self-development through active engagement in responsible social action the only characteristic of 'heroes'? Our hypothesis – currently being tested in a school in Sweden – is that any child has the capacity to learn social problem solving through responsible social action and make this a habit. The youths' symbolic letter to UNICEF suggests that they could learn a problem-solving model and use it in a creative way to suggest how youth in welfare states can help needy children. Milla Martin's real act – which resulted in gathering money and sending it to African children – cannot be compared with what children *can* learn. Indeed, such lessons have the potential for action but cannot guarantee the act will take place. But that is the nature of the Child Convention's Article 29. It is about what children should learn at school, and like much other learning, in itself it does not guarantee these lessons will be used in reality. At the same time, it is quite reasonable to think that if only a few percent of the children put their learning into practice, they could create many responsible social projects. Milla Martin was only one 9-year-old, yet she made a difference. The ideal situation would be to create projects that children in school not only learn about but also practically engage in to realize. Such projects may need the support of the pupils' parents, but these would have the capacity to transform children's learning into real, responsible social actions and lasting life lessons.

Daily life is complex, and hence, psychological tools can play an important part in organizing our thoughts and actions. As LeDoux[61] indicated, our brains have not evolved 'to the point where the new systems that make complex thinking possible can easily control the old systems that give rise to our needs and motives, and emotional reactions.' However, he immediately added that this does not mean that we're simply victims of our brains and should give in to our urges. It means that for the prefrontal cortex to be in the leading position, sometimes hard work is necessary. What we suggest, on the basis of pragmatist philosophy, is that if youth can learn to think systematically when their brain is at a very plastic state of development, then systematic thinking can become a lifelong habit. Once this is established, they have a good chance to use systematic reasoning as adults with less effort. Our world is becoming increasingly complex, and youths' access to Internet is also increasing. As such, there are countless attractive options for youth, who need to use their own power to judge these options and make good choices. For example, we live in a world of many religions, political ideologies and parties, and diverse social projects such as saving animals, saving the natural environment, working for peace, and so forth. It is crucial for young people to be able to reflect on all of these options before making choices. Our suggested model empowers youth by helping them to think systematically, encouraging goal-orientation, and improving the organization of their thoughts and behavior.

As contemporary neurophilosopher Patricia Churchland[62] stated, mere philosophical ideas are 'in peril of floating on a sea of mere, albeit confident, opinion,' so there is a need to connect abstract theories of the human condition to the 'hard and fast,' or that which is concrete. An example of such empirical knowledge is what we know about the position of the PFC, its connection with other brain's structures, and the leading role that the PFC plays in the brain.[63] Neuropragmatism – as an attempt to connect neuroscience to pragmatist philosophy – is still very young. In order for it to grow and become strong, it is in need of support and creative contributions. If neuropragmatism succeeds in finding a place in the academic world, it has the potential to combine a practical philosophy with a practical science. This makes it a suitable option for meeting complicated challenges that need a philosophic-scientific knowledge base with both depth and breadth.

More than a century ago, William James[64] wrote:

> The fact that the brain is the one immediate bodily condition of the mental operations is indeed so universally admitted nowadays that I need spend no more time in illustrating it, but will simply postulate it and pass it on. The whole reminder of the book will be more or less of a proof that the postulate was correct.

Thus, James shows an early awareness of the centrality of brain for the mind, and spends the rest of his more than 1000-page book establishing this. James foresaw that 'the psychologist is forced to be something of a nerve-physiologist.' His prediction is now a reality and neuropsychology has a stable place within the discipline of psychology. More than a century after James's insight, the current distinguished brain scientist Michael Gazzaniga wrote:[65]

> My college, Dartmouth, is constructing a magnificent new building for psychology. Yet its four stories go like this: The basement is all neuroscience. The first flour is devoted to classrooms and administration. The second floor houses social psychology, the third floor, cognitive science, and the fourth cognitive neuroscience. Why is it called the psychology building?

What Gazzaniga meant is that psychology has really embraced neuroscience, and some other disciplines have started to pay attention to neuroscience. Here we refer to some interesting examples; psychiatrists,[66, 67] philosophers,[68, 69] education theorists, and even a few neurosociologists.[70, 71] This chapter is an attempt to integrate knowledge gained from neuroscience with the philosophy of pragmatism, to present a neuropragmatist conceptual framework. As demonstrated in our example of an educational study in Sweden, the Rahyab model is an effective way to foster empathy and creative problem solving among youth, bolstering executive functioning, and providing the end-product of responsible, reflective, and socially-aware citizens. Using this model can help to provide a deeper conceptualization of, and hopefully, the realization of, Article 29 of the Child Convention. Simply, we wish to empower youths to become more reflective thinkers and responsible social problem-solvers.

Notes

1. *Aftonbladet* (Swedish daily newspaper, December 17, 2011).
2. David Hildebrand, *Dewey, A Beginner's Guide* (Oxford: Oneworld, 2008).
3. UNICEF's homepage, http://www.unicef.org/crc.
4. John Dewey, *Experience and Education: The 60th Anniversary Edition* (Bloomington and Indianapolis: Indiana University Press, 1998).
5. Edward Deci and Richard Ryan, 'Overview of Self-Determination Theory,' in *Handbook of Self-Determination Research*, ed. E. Deci and R. Ryan (Rochester, NY: The University of Rochester Press, 2002).
6. William James, *The Principles of Psychology* (Cambridge, MA: Harvard University Press, 1890/1981).
7. William James, *Talk to Teachers on Psychology; and to Students on Some of Life's Ideals* (Charleston: BiblioBazaar, 1899/2007).
8. Alireza Moula, Toomas Timpka, Anthony Puddephatt 'Adult-Adolescent Interaction and Adolescent's Brain Development: Integrating Pragmatism/Interactionism and Neuroscience to Develop a Platform for Research on Adolescent's Life Regulation,' *Sociology Compass*, 3(1): 118–136.
9. James, *Talk to Teachers on Psychology*, p. 49.
10. John Dewey, *Human Nature and Conduct* (New York: Dover Publication, 1922), p. 16.
11. Arne Malten, *Vad är Kunskap?* (Swedish book: *What Is Knowledge?*) (Malmö: Liber, 1981).
12. James, *The Principles of Psychology*, p. 126.
13. Nirmal Bhattcharyya, 'The Concept of "Intelligence" in John Dewey's Philosophy and Educational Theory,' *Educational Theory* 19 (1969), pp. 185–195.
14. Dewey, *Experience and Education*, p. 100.
15. John Dewey, *How We Think* (New York: Dover Publication, 1910/1997), pp. 2–3.
16. James, *Talk to Teachers on Psychology*, p. 59.
17. George Herbert Mead, *Mind, Self and Society from the Standpoint of a Social Behaviorist* (Chicago: The University of Chicago Press, 1934), p. 98.
18. Eric Kandel, *In Search of Memory* (New York: Norton and Company, 2006), p. xii.
19. Elkhonon Goldberg, *The New Executive Brain: Frontal Lobes in a Complex World* (Oxford: Oxford University Press), p. 23.

20. Goldberg, *The Executive Brain*.
21. Joaquin Fuster, *The Prefrontal Cortex*, Third Edition (Amsterdam: Academic Press, 2008).
22. Paul Eslinger and Kathleen Biddle, 'Prefrontal Cortex and the Maturation of Executive Functions, Cognitive Expertise and Social Adaptation,' in *Executive Functions and the Frontal Lobes*, eds V. Anderson, R. Jacobs, P. Anderson (New York: Taylor and Francis, 2008), pp. 299–316.
23. Earl Miller and Jonathan Wallis, 'The Prefrontal Cortex and Executive Brain Functions,' in *Fundamental Neuroscience*, ed. Larry Squire et al., (Amsterdam: Academic press, 2008), pp. 1199–1222.
24. Peter Anderson, 'Toward a Developmental Model of Executive Function,' in *Executive Functions and the Frontal Lobes*, pp. 3–21.
25. Renate Caine, Geoffrey Caine, Carol McClintic, and Karl Klimek, *Brain/Mind Learning Principles in Action* (California: Corwin Press, 2009), p. 8.
26. *The Jossey-Bass Reader on the Brain and Learning* (California: John Wiley and Sons, 2008).
27. Lynn Meltzer, Laura Pollicia, and Mirit Barzillia, 'Executive Function in the Classroom,' in *Executive Function in Education: From Theory to Practice*, ed. L. Meltzer (New York: The Guilford Press, 2007), pp. 165–193.
28. Ibid., pp. 186–187.
29. Meltzer, et al., *Executive Function in Education*, 2007.
30. Seana Moran and Howard Gardner in Meltzer, *Executive Function in Education*, pp. 19–38.
31. P. Philip Zelazo, et al., 'Early Development of Executive Function: A Problem-Solving Framework,' *Review of General Psychology*, 1(2) (1997), pp. 198–226.
32. Anderson, in Anderson, Jacobs, Anderson, *Executive Functions and the Frontal Lobes*, pp. 3–21.
33. Jack Martin, and Laura Failows, 'Executive Function: Theoretical Concerns,' in *Self and Social Regulation: Social Interaction and the Development of Social Understanding and Executive Functions*, ed. B. Sokol et al. (Oxford: Oxford University Press, 2010), pp. 35–55.
34. Ibid.
35. Jonathan Baron, *Thinking and Deciding* (Cambridge: Cambridge University, 2000), p. 5.
36. John Ratey, *A User's Guide to the Brain: Perception, Attention, and the Four Theaters of the Brain* (New York: Vintage Books, 2001).
37. Ibid., p. 292.
38. David. Franks, 'Emotions: The Importance of Emotions to Symbolic Interaction,' in *Handbook of Symbolic Interactionism*, ed. Larry Reynolds, Nancy Herman-Kinney (New York: Altamira 2003), pp. 787–809.
39. Antonio Damasio, *Looking for Spinoza: Joy, Sorrow, and the Feeling Brain* (New York: Harvest Book, 2003).
40. Antonio Damasio, *The Feeling of What Happens: Body and Emotion in the Making of Consciousness* (New York: Harcourt Brace & Company, 1999).
41. David Franks, 'Mutual Interest, Different Lens: Current Neuroscience and Symbolic Interactionism,' *Symbolic Interaction* 2 (2003): 613–630.
42. Bruce Wexler, *Brain and Culture: Neurobiology, Ideology, and Social Change* (London: MIT Press, 2006).

43. A. S. Dick and W. F. Overton, 'Considering Mechanisms of Executive Function,' in *Self and Social Regulation: Social Interaction and the Development of Social Understanding and Executive Functions*, ed. J. Carpendale, G. Iarocci, U. Müller, B. Sokol, & A. Young (New York: Oxford University Press, 2010), pp. 7–34.
44. Ibid., p. 19.
45. Alexander Kozulin, *Psychological Tools: A Sociocultural Approach to Education*, (Cambridge: Harvard University Press, 1998).
46. Ibid., p. 160.
47. Alireza Moula, *Population Based Empowerment Practice in Immigrant Communities*, Doctoral dissertation, (Linkoping, Sweden: Linköpings universitet, 2005).
48. Alireza Moula, *Population Based Empowerment Practice in Immigrant Communities* (Published doctoral dissertation with minor changes) (Saarbrucken: Lambert Academic Publisher, 2010).
49. Alireza Moula, 'An Invitation to Empowerment-Oriented Neurosociology,' in *Learning from Memory: Body, Memory and Technology in a Globalizing World*, ed. Bianca Maria Pirani (Cambridge: Cambridge Scholar Publishing, 2011), pp. 234–266.
50. Alireza Moula, Toomas Timpka, and Antony Puddephatt, 'Adult–Adolescent Interaction and Adolescents' Brain Development: Integrating Pragmatism/Interactionism and Neuroscience to Develop a Platform for Research on Adolescents' Life Regulation,' *Sociology Compass* 3(1) (2009): 118–136.
51. Hamideh Addelyan-Rasi, Alireza Moula, Antony J. Puddephatt, Toomas Timpka, 'Empowering Single Mothers in Iran: Applying a Problem-Solving Model in Learning Groups to Develop Participants' Capacity to Improve their Lives,' *British Journal of Social Work* (2012), pp. 1–20.
52. Fuster, *The Prefrontal Cortex*, p. 336.
53. Dewey, *How We Think*, p. 28.
54. Marsel Mesulam, 'The Human Frontal Lobes: Transcending the Default Mode through Contingent Encoding,' in *Principles of Frontal Lobe Function*, ed. D. T. Stuss & R. T. Knight (Oxford: Oxford University Press, 2002), pp. 8–30.
55. Mead, *Mind, Self and Society*.
56. Alireza Moula, *Empowermentorienterat socialt arbete* (Swedish book: *Empowerment Oriented Social Work*), (Lund: Studentlitteratur, 2009). See also notes 8, 47, 48, 49, and 50 above.
57. William Thomas and Dorothy Thomas, *Child in America* (New York: A. A. Knopf, 1928).
58. Anders Engquist, *Konsten att samtala* (Swedish Book, *The Art of Conversation*) (Lund: Studentlitteratur, 2013).
59. Dennis Saleebey, *The Strength Perspective in Social Work Practice*, vol. 3 (Boston: Allyn and Bacon, 1992).
60. Alireza Moula, forthcoming.
61. Joseph LeDoux *Synaptic Self: How our Brains Become Who We Are* (New York, Viking, 2002), p. 323.
62. Patricia Churchland, *Braintrust* (Princeton: Princeton University Press, 2011).
63. Goldberg, *The New Executive Brain*.
64. James, *The Principles of Psychology*, p. 18.

65. Michael Gazzaniga, *The Mind's Past* (Berkeley: University of California Press, 1998), p. xi.
66. Eric Kandel *Psychiatry, Psychoanalysis, and the New Biology of Mind* (Washington, D.C.: American Psychiatric Publishing, Inc., 2005).
67. David Brendel, *Healing Psychiatry: Bridging the Science/Humanism Divide* (Cambridge: MIT Press, 2006).
68. Churchland, *Braintrust*.
69. Paul Churchland, *Neurophilosophy at Work* (Cambridge: Cambridge University Press, 2007).
70. Franks, 'Emotions.'
71. Franks, 'Mutual Interest, Different Lens.'

Part IV
Ethics, Neuroscience, and Possibility

11
Pragmatism and the Contribution of Neuroscience to Ethics
Eric Racine

There have been several claims that neuroscience will inform, and even transform, morality and, more precisely, ethics as a discipline. In this chapter, I first briefly review such claims and underscore how they are often supported by a positivist (or strong naturalistic) epistemology.[1] I argue that this epistemology is at odds with the evolution of practical ethics and its integration of empirical research, as exemplified in the field of bioethics and the sub-field of neuroethics. One likely explanation for these prevailing positivist claims is that interpretations of neuroscientific knowledge are founded on a ubiquitous belief in the epistemic supremacy of neuroscientific explanations and their ability to provide 'foundations' for ethics. Such positivist claims, however, have triggered a strong backlash by raising assertions against the relevance of neuroscience in ethics. Taking a middle-ground perspective, I posit that a critical assessment of the contribution of neuroscience to ethics should not necessarily lead to a flat-out rejection of the potential for neuroscience to make important contributions to ethics, that is, anti-naturalism. One caveat is that neuroscience's role in ethics may need to be situated within a comprehensive framework that underscores the role of interdisciplinary and empirical research as well as the ways evidence informs practical judgment in ethics. Thus, the second part of this chapter is dedicated to exploring a moderate form of naturalism inspired by the thinking of American philosopher John Dewey, in particular his views on the role and nature of evidence in ethics. I also discuss how common claims about neuroscience's contribution to ethics are modulated within a pragmatic (moderately naturalistic) theoretical context. Readers should keep in mind that this chapter is not by any means the final word on this complex topic, but hopefully frames the nature of a significant tension point in the literature and suggests some pitfalls to avoid as well as novel areas to investigate further.

The epistemic supremacy of neuroscience explanations

Several scholars have claimed that neuroscience will inform and even transform morality and, more precisely, ethics as a discipline. Some of these claims are more comprehensive and encompassing, while others are nuanced, reflective, and mindful of the challenge of integrating evidence from neuroscience into the framework of normative and practical ethics.

Neurobiologist Jean-Pierre Changeux has long argued that ethics should be based on neuroscientific knowledge and that such an ethics would foster happiness in individual lives as well as yield the foundation of evidence-based ethical norms and behaviors, creating greater social wellbeing.[2,3] More recently, cognitive neuroscientist Michael Gazzaniga has written that neuroethics 'is – or should be – an effort to come up with a brain-based philosophy of life.'[4] Gazzaniga contends that neuroscience provides a novel way to examine ethical beliefs and assumptions and that these may be revised based on developments in neuroscientific knowledge. For example, he argues that, 'while certain beliefs may have made sense when they were formed, science has now taught us a few things about how the brain works and we need to be willing to change those beliefs.'[5] Some of Gazzaniga's claims attribute specific features to a neuroscience-based ethics; he asserts:

> I believe, therefore, that we should look not for a universal ethics comprising hard-and-fast truths, but for the universal ethics that arises from being human, which is clearly contextual, emotion-influenced, and designed to increase our survival. This is why it is hard to arrive at absolute rules to live by that we can all agree on. But knowing that morals are contextual and social, and based on neural mechanisms, can help us determine certain ways to deal with ethical issues.[6]

The mandate of 'neuroethics' is thus, according to Gazzaniga, 'to use our understanding that the brain reacts to things on the basis of its hardwiring to contextualize and debate the gut instincts that serve the greatest good – for the most logical solutions – given specific contexts.'[7]

In many ways, strong assertions like these suggest an epistemic supremacy of neuroscientific knowledge in its ability to provide foundations for ethics or to disconfirm unfounded ethical beliefs. This impact of neuroscience is part of what some have called a 'neuroscience revolution.'[8] Accordingly, neuroscientists have been called on to 'recognize that their work may be construed as having deep and possibly disturbing implications' outside of academia.[9] Likewise, cognitive neuroscientist Martha Farah and colleagues have stated that, '[t]he question is therefore not whether, but rather when and how, neuroscience will shape our future.'[10] It is important to consider how some fundamental aspects of the idea that neuroscience will have fundamental implications for ethics are perhaps not novel, but part of an ongoing debate on the contribution of neuroscience to the humanities and general culture.

Taking a historical perspective reveals that in other epochs, the supremacy of neuroscientific explanations also generated great interest and debate. In the early 19th century, for example, neuroanatomist Franz Joseph Gall was known for advancing localizationism and advocating for broad, and often controversial, applications of phrenology, such as within the criminal justice system.[11] Some decades later, popular American phrenologists, the Fowler brothers, claimed the unique ability of phrenology to explain the human character. The Fowlers professed that phrenology was a mind-reading technique with implications for child-rearing, medicine, and even morality, deeming it 'a Powerful Lever in Self-Improvement, in Moral and Intellectual Advancement.'[12] Later in the century, leading neuroanatomists and neuroscientists like Theodor Meynert, August Forel, Paul Fleschig, and Oskar and Cécile Vogt put forward the notion that the future of humanity and morality could be guided by neuroscientific findings.[13] Premonitory of some contemporary writing on consciousness,[14] Meynert suggested that a primitive 'primary ego' dealt with selfish functions of pain, hunger, and warmth while a 'secondary ego' (located in the associative fibers and developing gradually with cortical maturation) was 'associated with the ideas of mutuality, reciprocity, brotherhood.'[15] According to Michael Hagner, Fleschig thought that, 'brain research would replace philosophy as the dominant science of cultural orientation' and called for a 'physiological theory of morality/moral teaching' or a 'moral physiology' to establish a 'true culture.'[16] In the same period, Forel wrote:

> [A]s a science of the human in man, neurobiology forms the basis of the object of the highest human knowledge which can be reached in the future. It will and must find increasing numbers of workers, ever increasing recognition, if our culture is to move forward and not backward. It will however, also have to provide the correct scientific basis for sociology (and for mental hygiene, on which a true sociology must be based).[17]

The Vogts, continuing in this tradition, propelled the view that man was a 'brain animal,' and argued that, 'a fortuitous future of our species depended significantly on the expansion of brain hygiene.'[18] These statements drift into the realm of biological reductionism and are ostensibly emblematic of the eugenic and public hygiene programs and developments of the time.

The supremacy of neuroscientific explanations surfaced in the writings of some physicians and ethicists in the 1960s and 1970s as well. Physician and neuroscientist Paul MacLean, famous for the triune brain theory but also deeply interested in the neuronal bases of empathy, wrote in 1967 that a neuroscientific understanding of empathy could shed light on moral decision making and empathy in medicine.[19] Furthermore, in one of the first issues of the *Journal of Medicine and Philosophy* (a special issue dedicated to neuroscience with a contribution from Roger Sperry), bioethicist and philosopher Tristram Engelhardt sagaciously pointed out that neuroscience 'has revised our view of ourselves and of our moorings in this world.'[20]

However, while over the past century several neuroscientists have articulated adamant perspectives supporting a substantial contribution of neuroscience in ethics, many others have been more reserved in their statements, remaining sensitive to the epistemic and practical problems of sweeping claims. For instance, Churchland writes, '[m]y sense is that the details of decision making, of choice, of acquisition of character and temperament, and of development of such things as moral character are going to elude us until we have made more progress on certain fundamentals of neuroscience – namely the dynamic properties of neural networks.'[21] Notably, Churchland has explored in detail how folk psychological explanations of morality could be vetted by neuroscientific ones. While whether her neurophilosophical perspective also provides a sound theoretical framework has been debated, it clearly concedes significant epistemic weight to neuroscientific explanations without necessarily closing the door on interdisciplinary understandings of ethics.

Other contemporary philosophers including Roskies, Casebeer, and Greene have explored these questions and attempted to articulate how neuroscience's contributions can navigate a range of conceptual and methodological problems. One idea formulated within the field of neuroethics contends that '[a]s we learn more about the neuroscientific basis of ethical reasoning and self-awareness, we may revise our ethical concepts.'[22] Accordingly, Casebeer has presented the argument that Aristotle's moral philosophy best reflects what we understand about moral psychology based on neuroscience research, including the contribution of affect to moral cognition and the nature of abstract moral reasoning.[23] Joshua Greene on the other hand, claims that, '[s]ocial neuroscience is, above all else, the construction of a metaphysical mirror that will allow us to see ourselves for what we are and, perhaps, change our ways for the better.'[24] His point is rather descriptive since, in his view, neuroscience contributes to the empirical basis of meta-ethics as opposed to the normative ethics seen in Casebeer's perspective. Through neuroscience, 'understanding where our moral instincts come from and how they work can...lead us to doubt that our moral convictions stem from perceptions of moral truth rather than projects of moral attitudes.'[25]

The point of alluding in a cursory fashion to history is not in any way an attempt to disqualify the interest of the question or to suggest unanimity amongst neuroscientists about the contribution of neuroscience to ethics. Indeed, some neuroscientists like Charles Sherrington strongly resisted the introduction of neuroscience into the humanities for fear of reductionism and the undermining of human values.[26] Rather, my point is that various historical contexts have shaped the interest of neuroscientists and other scholars by promoting a neuroscientific understanding of ethics – based implicitly and sometimes explicitly – on the belief of the supremacy of knowledge about the brain in the domain of ethics. This positivist, neurocentric interpretation has percolated in different public discourses, typically exhibiting rather crude forms of reductionism. Recent work on the public understanding of contemporary neuroscience by others and myself has suggested that neuroscience, especially areas of neuroscience which bear significance for behaviour, personality, and morality, are interpreted following a similar positivist trend. For instance, studies on the media coverage of fMRI have shown a wide-ranging proclivity to equate the self to the brain, such as claiming that 'we are our brains,' which I have described elsewhere as *neuro-essentialism*.[27, 28, 29] Moreover, neuroscience techniques like fMRI and PET scans are vested with the power to offer objective and definitive evidence. They have been claimed to provide an ultimate visual proof, which outweighs other types of evidence because fMRI, for example, shows neuronal changes associated with task performance. This phenomenon, which I originally characterized as *neuro-realism*,[30] has been described in more depth in other works by myself and colleagues.[31, 32, 33] Some studies examining the impact of neuroimages and neuroscientific explanations on the evaluation of scientific explanations have also suggested the epistemic supremacy of neuroscience. Subjects in experiments rated the validity of explanations with neuroscientific explanations[34] and neuroimages[35] higher than those without such explanations or images.

In a context of strong claims and media clamour, some ethicists and philosophers have voiced early criticism against the introduction of neuroscientific perspectives to ethics, especially given the inclination to interpret this body of research along the lines of moderate to strong naturalism[36,37] (see left column of Table 11.1 in the next section). Claims reviewed in this first section exhibit features of a strong naturalist perspective, particularly that (1) ethics is an empirical discipline, ideally a kind of applied field of neuroscience (ethical knowledge is reducible to empirical knowledge); (2) neuroscientific knowledge could provide a foundational perspective and guide ethical behavior such as, for example, potentially universal natural laws and principles could be identified to guide conduct; and (3) ethics is empirical in the sense that empirical knowledge trumps normative ethics and ethical reflection (no clear distinction between 'is' and 'ought'; ethics predicates are natural properties). In response to the strong naturalist approach to ethics, some common counterarguments have been presented (see Table 11.1 for a quick review), which are not further discussed here.[38] Indeed, it may seem that many neuroscientists of ethics suffer from what Steve Morse has called the 'brain over-claim syndrome' – '[a] cognitive pathology...that often afflicts those inflamed by the fascinating new discoveries in the neurosciences.'[39] Their interpretations of neuroscience's contribution to ethics reflect the same belief in the epistemic privilege of neuroscience that others have had in the past. I have myself critically reviewed common anti-naturalist arguments against the use of neuroscience knowledge in ethics elsewhere and have come to the conclusion that none of the main arguments are entirely compelling, although they do bring wisdom to modulate strong positivist and naturalistic claims (see Table 11.1 below for sample claims) and help us better understand potential contributions of neuroscience.[40] In general, anti-naturalistic arguments temper strong naturalistic claims and call for an alternate framework to incorporate neuroscience in ethics.

Following anti-naturalists, I take the perspective that skepticism about the epistemic supremacy of neuroscience is healthy, well-placed, and even founded given the historic precedent and the potential implications of biological reductionism in health policy and other areas of public life. However, I reject the claim that ethics will not benefit from the fruits of neuroscience research because of the idea that ethics is impermeable to empirical evidence, and stress that the anti-naturalist proponents of this perspective are dangerously wrong. Apropos, Changeux writes that, 'the search for a "common ethics" cannot sidestep the recent immense contribution of the humanities, anthropology, history of cultures and of law, psychology and neuroscience, and evolutionary naturalism' (*author's translation*).[41] Indeed, critics have often been more radical than

Table 11.1 Common conceptual arguments against the neuroscience of ethics

Neurological determinism
Contribution of neuroscience to ethics appears to jeopardize beliefs in free will and to support forms of determinism.

Naturalistic fallacy
Contribution of neuroscience conflates an 'is' with an 'ought'; operates a slide from an 'is' to an 'ought'.

Semantic dualism*
Brain properties are different from mind properties precluding any sound integration of neuroscience on ethical thinking because brain properties are confused with mind properties.

Biological reductionism and eliminativism
Neuroscience reduces ethical concepts to the point of examining only their trivial components, which are theoretically and practically irrelevant. Eliminativism would err in triaging folk psychological explanations of ethics not reducible to lower-level explanations.

Neuroscientism and the threat to ethics
Fear that neuroscience will damage human values and beliefs about free will, honesty, personhood and so on or further contribute to bureaucratic models of organization of care, impersonal and overly objective medicine, and disrespect for persons.

*Substance dualism is in our eyes a moot position in the current neuroscience context, hence its exclusion from this table.

Source: Based on Racine, Bar-Ilan, and Illes, "fMRI in the Public Eye," pp. 159–164 and Racine, 'Pragmatic Neuroethics: Improving Treatment and Understanding of the Mind-Brain.'

those who have put forward arguments in favour of a neuroscience of ethics. When taken in extreme forms (which is not the case for all authors cited above), both perspectives reflect what Dewey called a 'quest for certainty,' in which absolute universal principles are considered to be true irrespective of context, time, and geographical location. What I offer is a middle-ground perspective, inspired by Dewey's writing, which (1) recognizes and values the contribution of empirical research and evidence to ethics (including neuroscience) but (2) avoids different forms of reductionism and crude ethico-logical fallacies (Table 11.1) within a view of practical ethical judgment that stresses the role of individual perspectives and experience in moral deliberation.

Pragmatism and the contribution of evidence in practical judgment

Tempering claims about the contribution of neuroscience to ethics *does not*, or at least from the gist of this chapter *should not*, automatically lead to a flat-out rejection of the potential for neuroscience to make an important contribution to ethics. Debates over the epistemic supremacy of neuroscientific explanations in ethics, while in need of some corrections and nuances, should not overshadow the possible contributions of neuroscience to a more comprehensive understanding of the role of empirical research in ethics and to informing practical ethical judgment. The second part of this chapter explores pragmatic ethics, a paradigm of thought based largely, but not solely, on Dewey's thinking about the nature and role of evidence in ethics. I present how common claims about the contribution of neuroscience are modulated in a pragmatic theoretical context. By drawing upon Dewey's work on ethics, I wish to provide an alternative framework in which to consider neuroscience evidence in ethics.

Dewey's ethical thought and general pragmatism are notoriously hard to define or synthesize, and it is beyond the scope of this chapter to present Dewey's general moral philosophy. Table 11.2 below provides an overview of key features of Dewey's thinking on ethics. These concepts are not principles in the traditional sense, but provide fundamental meta-ethical scaffolding for ethics inspired by pragmatism. Some of these concepts relate more specifically to the use of empirical evidence in ethics and are explained briefly in the following paragraphs, which will hopefully clarify the pragmatic framework presented.

In the first section of the chapter, I reviewed some strong naturalistic claims in the neuroscience of ethics. Table 11.3 below shows how moderate pragmatic naturalism offers different perspectives on the nature of ethics, the role of ethical principles, and empirical knowledge grounded in the basic features of pragmatic ethics described in Table 11.2. In the following paragraphs I elucidate some of the divergences between strong and pragmatic (moderate) naturalism with respect to the neuroscience of ethics.

Ethics is a form of reflexive, creative, and social deliberative enterprise seeking the 'good(s)'

Strong naturalism presents an understanding of ethics as applied neuroscientific knowledge, which actually usurps some of the elementary normative aspects of ethics. A pragmatic naturalistic approach, on the other hand, considers ethics as a reflexive, creative, and social deliberative enterprise seeking the 'good(s),' which comes in plural forms. Ethics only makes sense in relation to other individuals because of its intrinsic social nature. As Figure 11.1 captures, in contrast to linear

Table 11.2 Key features of Dewey's ethics

Non-foundationalism
Reluctance to posit *a priori* sources of ethics such as foundational ethical principles or the authority of religious ethics or natural order.

Deliberative, adaptive, and creative
Ethics and ethical norms (in contrast to morality based on revelation and religious belief) are developed by humans and are generated as the outcome of human ethical deliberation and creativity consistent with the 'method of democracy,' otherwise said collective deliberation, without falling prey to blunt relativism.

Situationalism
A general emphasis on the importance of context to understand human behavior and to assess if an act is ethically appropriate for a specific and concrete situation or context while also keeping in mind the possible broader implications beyond a specific situation.

(Radical) empiricism and experientialism
Empirical knowledge and experience not only *inform* our understanding of ethics but have a potential *transformative* role in shaping views on the nature of ethical behavior and the assessment of real-world consequences of acts, including consequences on the development of an agent's moral habits and character. Knowledge of these consequences yields '(social) intelligence' through what Dewey described as a process of 'inquiry.'

Social nature of ethics
The recognition that the scope of ethics is broader than the individual because individual behavior can only be understood as being shaped by social networks and systems.

Interdisciplinarity
Ethics is the act of applying knowledge to human situations rather than a special province of knowledge or expertise in and of itself; contributions of a wide range of disciplines can inform on 'human nature' as it relates to moral decision making and moral behavior.

Practice- and action-oriented
Ethics is practice- and action-oriented and cannot be solely an academic or scholarly endeavor; ethics is a reflective process in service of action but which integrates an iterative process of deliberation and action.

Instrumentalism of ethical principles and ethical theory
Ethical principles have worth inasmuch as they help guide ethical conduct and generate good outcomes; they are 'hypotheses' which should be assessed based on their consequences.

Moderate consequentialism
Consequences of ethical attitudes and of actions cannot be ignored from an ethical standpoint. However, moderate consequentialism does not equate with utilitarianism: that is, the belief that we should maximize utility to achieve the best decisions; 'consequences' have broader and potentially more transformative effect on ethics.

Continued

Table 11.2 *Continued*

Pluralism
Diversity contributes to generating richer and more insightful deliberations and inquiries; in rapidly changing societies, monolithic and rigid belief systems are underpowered with respect to the ethical challenges created by science and technology.

Source: First published in Eric Racine, 'Pragmatic Neuroethics: The Social Aspects of Ethics in Disorders of Consciousness,' *Handbook of Clinical Neurology* 118 (2013), pp. 357–372.

Table 11.3 Interpretations of neuroscience's contribution to ethics following strong and moderate (pragmatic) naturalisms

	Strong naturalistic interpretation	Moderate (pragmatic) naturalistic interpretation
Nature of ethics	Ethics is an empirical discipline, ideally a kind of applied neuroscience; ethical knowledge is reducible to empirical knowledge.	Ethics is inherently a normative discipline (it tries to identify the good thing to do and generate wisdom or knowledge of how to use knowledge), which needs to rely on factual understandings of ethical situations, outcomes of actions as well as on the nature of moral judgment and moral deliberation as such.
Ethical principles and foundations	Neuroscience knowledge could provide a foundational perspective and guide ethical behavior such that universal natural laws and principles could be identified to guide conduct.	Foundations (in a strict sense) in ethics are elusive and most often reflect a debatable quest for certainty. The task of ethics is not solely to provide foundations or universal principles but to propose scenarios of actions based on contextualized moral deliberation, which take into account knowledge of the natural world and should be assessed in specific circumstances and contexts.

Continued

Table 11.3 Continued

	Strong naturalistic interpretation	Moderate (pragmatic) naturalistic interpretation
Empirical knowledge	Ethics could be empirical in the sense that empirical knowledge trumps normative ethics and ethics reflection (no clear distinction between 'is' and 'ought'; ethics predicates are natural properties). Knowledge from neuroscience vets other knowledge.	Distinction between 'is' and 'ought' is acknowledged as a continuum, but ethics is a deliberative process which integrates an understanding of matters of facts as well as an understanding of outcomes of actions. Knowledge from neuroscience could enrich other bodies of knowledge and empower moral agents.

Source: Defined in Racine, 'Which Naturalism for Bioethics? A Defense of Moderate (Pragmatic) Naturalism,' pp. 92–100.

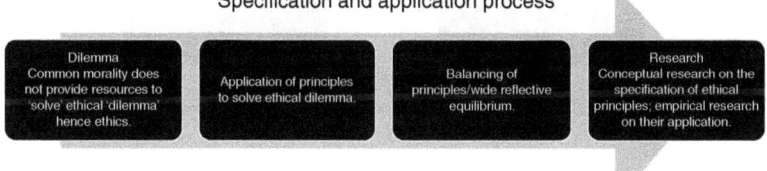

Figure 11.1 Comparison of the processes implied in applied ethics and pragmatic ethics. Process of applied ethics

Source: Racine, 'Pragmatic Neuroethics: The Social Aspects of Ethics in Disorders of Consciousness,' pp. 357–372.

models of applied ethics, in which principles are summoned to respond to ethical dilemmas and then specified, balanced, and entered into a wide reflective equilibrium process (Figure 11.1), pragmatism suggests an iterative and deliberative process (Figure 11.2). However, to be fair, within modern ethics, and in particular bioethics, the value of empirical research and experience is now well recognized and broadly accepted.[42,43] Most bioethicists have come to acknowledge the challenge of anti-naturalistic arguments and strong interpretations of the naturalistic fallacy.

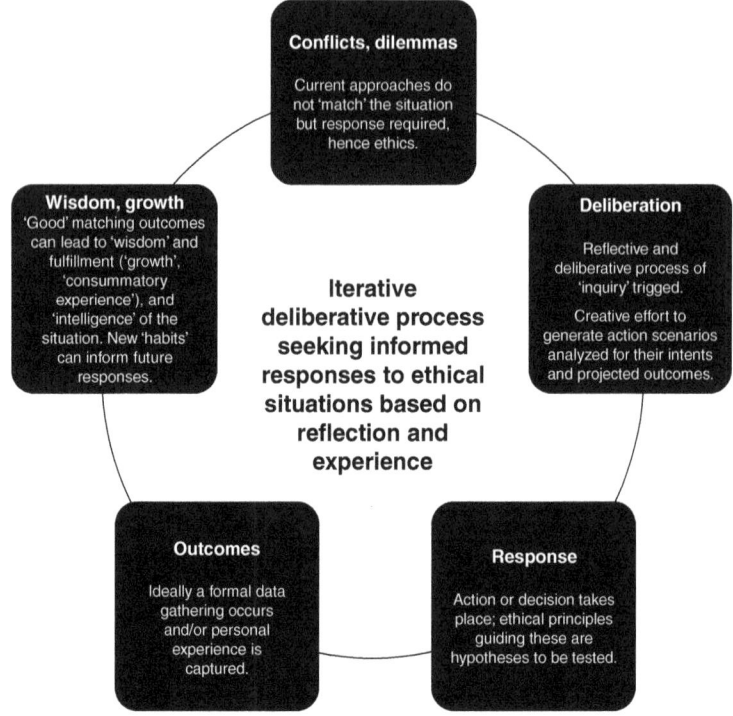

Figure 11.2 Comparison of the processes implied in applied ethics and pragmatic ethics. Process of pragmatic ethics

Elucidating this challenge, bioethicist, Daniel Callahan, comments that, '[s]ince "is" is all the universe has to offer, to say that it cannot be the source of an "ought" is tantamount to saying *a priori* that an ought can have no source at all – to say that is no less than to say there can be no oughts.'[44]

Other contemporary authors like Beauchamp and Childress also recognize the value of experience and empirical evidence in specifying ethical principles:

> The abstract rules and principles in moral theories are extensively indeterminate; that is, the content of these rules and principles is too abstract to determine the acts that we should perform. In the process of specifying and balancing norms and in making particular judgments, we often must take into account factual beliefs about the

world, cultural expectations, judgments of likely outcome, and precedents previously encountered to help fill out and give weight to rules, principles, and theories.[45]

A foundationalist interpretation of ethical principles is contrary to the very nature of ethics

In a pragmatic framework, ethical dilemmas and the like spark the need for an inquiry (see Figure 11.2) similar to common approaches in bioethics. For instance, pragmatism entails a process by which scenarios are proposed and rehearsed in deliberation to foresee their outcomes as integral to ethics. Responding in some form to ethical dilemmas and situations as well as capturing outcomes through personal experience or formal data gathering (although Dewey himself was less clear on this aspect) are also parts of ethics. If the response to the dilemma *matches* the problematic situation, then the individual involved experiences *growth*, and *wisdom* is generated from a deliberative process based on an *intelligence* of the situation (italicized terms have specific meanings in Dewey's writings which cannot be reviewed here). Moderate pragmatic naturalism offers an understanding of ethics which navigates between strong naturalism and anti-naturalism, circumventing some significant problems found in these two perspectives. For example, pragmatic naturalism does not linger on the question of whether ethics is an entirely empirical discipline with no specificity or a unique normative discipline without connections to empirical disciplines. Rather, the debate centers on how empirical knowledge can be brought to bear meaningfully in the pursuit of moral good(s) and how ethics as a discipline should be dedicated to these tasks. Dewey writes:

> Whether the goal be thought of as pleasure, as virtue, as perfection, as final enjoyment of salvation, is secondary to the fact that the moralists who have asserted fixed ends have in all their differences from one another agreed in the basic idea that present activity is but a means. We have insisted that happiness, reasonableness, virtue, perfecting, are, on the contrary, parts of the present significance of present action. Memory of the past, observation of the present, foresight of the future are indispensable. But they are indispensable *to* a present liberation, and enriching growth of action. Happiness is fundamental in morals only because happiness is not something to be sought for, but is something now attained even in the midst of pain and trouble, whenever recognition of our ties with nature and

with fellowmen releases and informs our action. Reasonableness is a necessity because it is the perception of the continuities that take action out of its immediateness and solution into connection with the past and future.[46]

In summary, ethics pursues the goals of the good, wisdom, and happiness. The neuroscience of ethics could re-describe these goals in the language of neuroscience but could never do away with them. In ethics, empirical knowledge serves to inform practical judgment in fulfilling these goals and intents.

In contrast to strong naturalism, it is antagonistic to the nature of ethics to rely on established principles following a deductive or foundational interpretation of ethical principles or of a singular piece of knowledge. Dewey writes, apropos of this, 'The attempt to set up ready-made conclusions contradicts the nature of reflective morality [ethics].'[47] His writings also suggest an inherently creative role in practical judgment in ethics. There is a sense that ethical approaches are 'founded' inasmuch as this foundation process leaves room for creative thinking in which imagination is a key ingredient to identify the 'good' approach(es) in a specific context. In other words, reflection and deliberation are processes through which empirical evidence is integrated into an ethical response; however, it does not become foundational in any conventional sense. Accordingly, ethical principles are considered to be hypotheses – part of a flexible and creative moral enterprise that adjusts responses based on the circumstantial nature of every given situation:

> But in morals a hankering for certainty, born of timidity and nourished by love of authoritative prestige, has led to the ideal that absence of immutably fixed and universally applicable ready made principles is equivalent to chaos. In fact situations into which change and the unexpected enter are a challenge to intelligence to create new principles. Morals must be a growing science if it is to be a science at all, not merely because all truth has not yet been appropriated by the mind of man, but because life is a moving affair in which old moral truth ceases to apply.[48]

Ethics is inherently situational or context-dependent

Some strong naturalistic interpretations of the neuroscience of ethics suggest that universal laws or principles will be discovered by neuroscience and therefore guide human behavior or radically inform other discourses on ethics. From a pragmatic standpoint, however, ethics is inherently situational or context-dependent. This does not mean that answers and ethical approaches are relative; rather, just that their 'truth' and 'fit' can only be assessed in specific sets of environmental constraints. As McGee describes, 'Dewey found moral investment in the existential context of the social situation itself, not in narrow notions of acceptability and condemnation that we bring to every problematic area. In this way Dewey articulated not relativism, but a careful and subtly contextual ethics.'[49] He also emphasized the constitutive function of habits and social context in shaping and making possible ethical behavior. For instance, he argued that one of the most important sophisms of philosophical thinking is what he called the 'philosophical fallacy,' the neglect of context, which accordingly leads to 'the supposition that whatever is found true under certain conditions may forthwith be asserted universally or without conditions.'[50]

Experience and knowledge of facts offer various contributions to a deliberative pragmatic ethics but do not provide foundations

Both moderate and strong forms of naturalism acknowledge the value of factual knowledge, with the specification that pragmatism also calls for the recognition of the value of first-person experience and intersubjective knowledge – a departure from a strict positivist epistemology. Experience and knowledge are intimately connected to each other and have various contributions at different steps of the deliberative pragmatic ethics process (Figure 11.2). These contributions include a better understanding of the ethical situation or dilemma, providing evidence to support deliberation, informing the feasibility of different action scenarios, supporting data gathering and reflecting on personal experience, and understanding how outcomes match (or fail to match) a situation. They are also now well acknowledged in bioethics, which is often reliant on, as others and myself have argued, an implicit form of pragmatism or naturalism.[51,52] Dewey summarizes it eloquently: '[l]ack of insight always ends in despising or else unreasoned admiration...What cannot be understood cannot be managed intelligently.'[53]

The nuances of this approach are important to delineate. Firstly, facts do not bring certainty, and no single kind of knowledge possesses an inherent supremacy. Additionally, the acquisition of knowledge does not override the specificity of the 'knowledge of using knowledge' wisdom, which is a key goal of ethics. Dewey, for example, clearly outlines how empirical knowledge would not 'solve' ethical dilemmas:

> [I]t is not pretended that a moral theory based on moral realities of human nature and a study of the specific connections of these realities with those of physical science would do away with moral struggle and defeat. It would not make the moral life as simple a matter as wending one's way along a well-lighted boulevard ... But morals based upon concern with facts and deriving guidance from knowledge of them would at least locate the points of effective endeavour and would focus available resources upon them.[54]

Knowledge liberates and empowers the agent to act on the world

In a strong naturalistic epistemology, empirical knowledge could somehow guide behavior. Within a pragmatic and moderately naturalistic stance, empirical knowledge liberates and empowers the agent to act on the world. For example, knowing about biases in decision making (such as neuroscience suggesting a hard-wired personhood network in the brain could be tricked by non-persons like patients in a vegetative state or fetuses, according to Farah and Heberlein[55]) does not support determinism or a call to conform to these biases. It allows the agent to exercise control of these factors through training or technology, such as understanding how a personhood network is hard-wired and can be triggered by non-persons. Dewey stresses the liberating nature of knowledge that offers the moral agent to be active and free within the conditions of real-world existence:

> [w]e are told that seriously to import empirical facts into morals is equivalent to an abrogation of freedom. Facts and laws mean necessity we are told. The way to freedom is to turn our back upon them and take flight to a separate ideal realm. Even if the flight could be successfully accomplished, the efficacy of the prescription may be doubted. For we need freedom in and among actual events, not apart from them.[56]

Dewey also pointedly highlights the conceptual flaws which equate knowledge of facts to determinism, asserting that 'no amount of insight into necessity bring with it, as such, anything but a consciousness of a necessity. Freedom is the "truth of necessity" ... '[57] Hence, no amount of neuroscientific knowledge about ourselves will explain away our capacity to use this knowledge in the pursuit of practical goals. In this process, different interpretations and uses of this knowledge will unavoidably surface. One can hope this knowledge will breed enlightenment, but absolute foundations in any substantial understanding of this term are elusive.

Non-reductionism and interdisciplinary scholarship and practices are required

Strong naturalism may suggest that neuroscience could provide foundational evidence for ethics in contrast to 'flimsy' disciplines. A moderate approach argues that a broad range of disciplines is needed to integrate the first-person perspective through relevant methods and disciplines. As Dewey argues,

> Morals is the most humane of all subjects. It is that which is closest to human nature; it is ineradicably empirical, not theological nor metaphysical nor mathematical. Since it directly concerns human nature, everything that can be known of the human mind and body in physiology, medicine, anthropology, and psychology is pertinent to moral inquiry. Human nature exists and operates in an environment...Moral science is not something with a separate province. It is physical, biological and historic knowledge placed in a human context where it illuminates and guides the activities of men.[58]

Dewey stresses how ethics is not necessarily a discipline confined to definite academic and institutional boundaries – it is much more fluid. Ethics is knowledge garnered from both empirical evidence and theoretical insights, mobilized in service of individual happiness and reciprocal enrichment. The neuroscience of ethics can offer a contribution to ethics knowledge, but it must fit within a comprehensive, that is, non-reductionist, framework. In this framework, the first contribution of empirical disciplines is to enrich and broaden perspectives on the understanding of the very nature of moral situations rather than quickly disposing of unsupported moral views and suggesting incontestable foundations for ethics. The latter is, in fact, a disservice to the goals of ethics.

Conclusion

This chapter reviewed general claims about the possible contribution of neuroscience to ethics and underscored how there are implicit – and sometimes explicit – references to the epistemic supremacy of neuroscience explanations in contemporary writings in ethics and related fields following a strong naturalistic (or positivist) trend. Unfortunately, prophetic language from neuroscientists and others have triggered alarmist responses entrenched in anti-naturalistic epistemological assumptions and conceptual debates that fail to capture the role of evidence and empirical research in ethics. Inspired by Dewey's thinking, I delineated a pragmatic account of how evidence and research serve ethics in practice and in scholarship. Within this framework, neuroscientific explanations will certainly have important contributions to make – but only if they are situated within a broader framework that recognizes the value of empirical research and brings neuroscientific explanations into dialogue with perspectives from other disciplines as a means to support and inform the deliberative processes. More simply put, a humble interpretation of the neuroscience of ethics needs to coalesce within an interdisciplinary understanding of ethics.[59] In conclusion, I acknowledge Dewey for this compendious final point of attention: whenever we think we are on the verge of an ultimate understanding or explanation, a pragmatic approach importantly reminds us that '[h]umility is more demanded at our moments of triumph than at those of failure.'[60]

Acknowledgments

Research and writing for this chapter was possible thanks to a grant of the Social Sciences and Humanities Research Council of Canada (SSHRC). Thanks to members of the Neuroethics Research Unit and in particular Ms. Megan Galeucia, Ms. Allison Yan, and Ms. Victoria Saigle for feedback on a previous version of this manuscript. This chapter has been first published as a journal article in *Essays in Philosophy of Humanism* and content of this chapter overlaps with the following previous publications: Eric Racine, 'Why and how take into account neuroscience in ethics: Toward an emergentist and interdisciplinary neurophilosophical approach,' *Laval théologique et philosophique* 65(1) (2005): 77–105; Eric Racine, 'Interdisciplinary approaches for a pragmatic neuroethics,' *AJOB-Neuroscience* 8(1) (2008): 52–53; Eric Racine, 'Which naturalism for bioethics? A defense of moderate naturalism,' *Bioethics* 22(2) (2008): 92–100; Eric Racine, and Emma Zimmerman, 'Pragmatic neuroethics and neuroscience's potential to radically change ethics,' in *Neuroturn in the Humanities and the Social Sciences*, ed. Melissa Littlefield and Jenell Johnson (in press); Eric Racine. *Pragmatic Neuroethics: Improving Treatment and Understanding of the Mind Brain* (Cambridge, MA: MIT Press, 2010).

Notes

1. Eric Racine, 'Which Naturalism for Bioethics? A Defense of Moderate (Pragmatic) Naturalism,' *Bioethics* 22 (2008): 92–100.
2. Jean-Pierre Changeux, 'Le point de vue d'un neurobiologiste sur les fondements de l'éthique,' in *Cerveau et psychisme humains: Quelle éthique?*, ed. Gérard Huber (Paris: John Libbey Eurotext, 1996), pp. 97–109.
3. Jean-Pierre Changeux, 'Les progrès des sciences du système nerveux concernent-ils les philosophes?,' *Bulletin de la Société française de Philosophie* 75 (1981): 73–105.
4. Michael S. Gazzaniga, *The Ethical Brain* (New York/Washington, D.C.: Dana Press, 2005).
5. Ibid.
6. Ibid.
7. Ibid.
8. Paul Root Wolpe, 'The Neuroscience Revolution,' *The Hastings Center Report* 32 (2002): 8.
9. Anonymous, 'Does Neuroscience Threaten Human Values?,' *Nature Neuroscience* 1 (1998): 535–536.
10. Martha J. Farah, Judy Illes, Robert Cook-Deegan, Howard Gardner, Eric Kandel, Patrick King, Eric Parens, Barbara Sahakian, and Paul Root Wolpe, 'Neurocognitive Enhancement: What Can We Do and What Should We Do?,' *Nature Reviews Neuroscience* 5 (2004): 421–425.
11. Gérard Huber, ed. *Cerveau et psychisme humains: Quelle éthique?*, Collection *Éthique et Sciences* (Paris: John Libbey Eurotext, 1996).
12. Eric Racine, 'Pragmatic Neuroethics: Improving Treatment and Understanding of the Mind-Brain,' *Basic Bioethics*, ed. Glen McGee and Arthur Caplan (Cambridge: MIT Press, 2010).
13. Michael Hagner, 'Cultivating the Cortex in German Neuroanatomy,' *Science in Context* 14 (2001): 541–563.
14. Antonio R. Damasio, *The Feeling of What Happens: Body and Emotion in the Making of Consciousness* (San Diego: Harvest Book Harcourt, 1999).
15. Hagner, 'Cultivating the Cortex in German Neuroanatomy,' pp. 541–563.
16. Ibid., pp. 541–563
17. Ibid., pp. 541–563
18. Ibid., pp. 541–563
19. P.D. MacLean, 'The Brain in Relation to Empathy and Medical Education,' *The Journal of Nervous and Mental Disease* 144 (1967): 374–382.
20. Tristram Jr Engelhardt, 'Splitting the Brain, Dividing the Soul, Being of Two Minds: An Editorial Concerning Mind-Body Quandaries in Medicine,' *Journal of Medicine and Philosophy* 2 (1977): 89–100.
21. Patricia Smith Churchland, 'Neuroconscience: Reflections on the Neural Basis of Morality,' in *Neuroethics: Mapping the Field, Conference Proceedings*, ed. Steven J. Marcus (San Francisco: The Dana Foundation, 2002), pp. 20–26.
22. Adina Roskies, 'Neuroethics for the New Millenium,' *Neuron* 35 (2002): 21–23.
23. William D. Casebeer, 'Moral Cognition and Its Neural Constituents,' *Nature Neuroscience* 4 (2003): 841–846.

24. Joshua D. Greene, 'Social Neuroscience and the Soul's Last Stand,' in *Social Neuroscience: Toward Understanding the Underpinnings of the Social Mind*, ed. A. Todorov, S. Fiske and D. Prentics (Oxford and New York: Oxford University Press, 2006).
25. Joshua D. Greene, 'From Neural "Is" to Moral "Ought": What Are the Moral Implications of Neuroscientific Moral Psychology?' *Nature Reviews Neuroscience* 4 (2003): 847–850.
26. Roger Smith, 'Representations of Mind: C.S. Sherrington and Scientific Opinion, C. 1930–1950,' *Science in Context* 14 (2001): 511–529.
27. Eric Racine, Ofek Bar-Ilan, and Judy Illes, 'fMRI in the Public Eye,' *Nature Reviews Neuroscience* 6 (2005): 159–164.
28. Eric Racine, Sarah Waldman, Jarrett Rosenberg, and Judy Illes, 'Contemporary Neuroscience in the Media,' *Social Science and Medicine* 71 (2010): 725–733.
29. Eric Racine, 'Pourquoi et comment doit-on tenir compte des neurosciences en éthique? Esquisse d'une approche neurophilosophique émergentiste et interdisciplinaire,' *Laval théologique et philosophique* 61 (2005): 77–105.
30. Ibid., pp. 77–105.
31. Racine, Bar-Ilan, and Illes, 'fMRI in the Public Eye,' pp. 159–164.
32. Racine, Waldman, Rosenberg, and Illes, 'Contemporary Neuroscience in the Media,' pp. 725–733.
33. Racine, 'Pourquoi et comment doit-on tenir compte des neurosciences en éthique? Esquisse d'une approche neurophilosophique émergentiste et interdisciplinaire,' pp 77–105.
34. Deena S. Weisberg, Frank C. Keil, Joshua Goodstein, Elizabeth Rawson, and Jeremy R. Gray, 'The Seductive Allure of Neuroscience Explanations,' *Journal of Cognitive Neuroscience* 20 (2008): 470–477.
35. David P. McCabe, and Alan D. Castel, 'Seeing Is Believing: The Effect of Brain Images on Judgments of Scientific Reasoning,' *Cognition* 107 (2008): 343–352.
36. Gunther S. Stent, 'The Poverty of Neurophilosophy,' *Journal of Medicine and Philosophy* 15 (1990): 539–557.
37. Alasdair Macintyre, 'What Can Moral Philosophers Learn from the Study of the Brain?' *Philosophy and Phenomenological Research* 58 (1998): 865–869.
38. Racine, 'Which Naturalism for Bioethics? A Defense of Moderate (Pragmatic) Naturalism,' pp. 92–100.
39. Stephen Morse, 'Brain Overclaim Syndrome and Criminal Responsibility: A Diagnostic Note,' *Ohio State Journal of Criminal Law* 3 (2006): 397–412.
40. Racine, 'Pragmatic Neuroethics: Improving Treatment and Understanding of the Mind-Brain.'
41. Changeux, 'Le point de vue d'un neurobiologiste sur les fondements de l'éthique,' pp. 97–109.
42. Mildred Z. Solomon, 'Realizing Bioethics' Goals in Practice: Ten Ways "Is" Can Help "Ought",' *Hastings Center Report* 35 (2005): 40–47.
43. Pascal Borry, Paul Schotsmans, and Kris Dierickx, 'The Birth of the Empirical Turn in Bioethics,' *Bioethics* 19 (2005): 49–71.
44. Daniel Callahan, 'Can Nature Serve as a Moral Guide?,' *Hastings Center Report* 26 (1996): 21–22.
45. Tom Beauchamp, and James Childress, *Principles of Biomedical Ethics*, 5th ed. Vol. 9 (Oxford: Oxford University Press, 2001).

46. John Dewey, *Human Nature and Conduct: An Introduction to Social Psychology* (New York: Holt, 1922).
47. John Dewey, and James H. Tufts, *Ethics* (New York: Holt, 1932).
48. Dewey, *Human Nature and Conduct*.
49. Glenn McGee, ed. *Pragmatic Bioethics* 2nd ed. (Cambridge, MA: MIT Press, 2003).
50. Dewey, *Human Nature and Conduct*.
51. Jonathan Moreno, 'Bioethics Is a Naturalism,' in *Pragmatic Bioethics*, ed. Glenn McGee (Nashville/London: Vanderbilt University Press, 1999), pp. 5–17.
52. Racine, 'Which Naturalism for Bioethics? A Defense of Moderate (Pragmatic) Naturalism,' pp. 92–100.
53. Dewey, *Human Nature and Conduct*.
54. Ibid.
55. Martha J. Farah, and Andrea S. Heberlein. 'Personhood and Neuroscience: Naturalizing or Nihilating?' *American Journal of Bioethics* 7 (2007): 37–48.
56. Dewey, *Human Nature and Conduct*.
57. Ibid.
58. Ibid.
59. Racine, 'Pragmatic Neuroethics: Improving Treatment and Understanding of the Mind-Brain.'
60. Dewey, *Human Nature and Conduct*.

12
Pragmatist Ethics: A Dynamical Theory Based on Active Responsibility

Markate Daly

Introduction

The main focus of moral theories has long been the ethics of rules and principles. However, another very different ethics complements and competes with it in American society. This chapter provides a theoretical framework for that ethics, locating it within the American pragmatist philosophical tradition and supporting its claims with data from the cognitive and behavioral sciences. That framework is a complex dynamical systems approach to a moral situation. A systems approach locates a moral action in an agent's attunement to the complex of relationships among particular persons at a particular time when a choice takes place. A skillful moral judgment aims through its action to conserve such community-building values as trust, respect, kindness, forgiveness, compassion, and generosity.

In current society, 'responsibility' is the central concept used to discuss this relational ethics, in the forward-looking sense of 'taking responsibility' in an evolving situation, rather than the sense of 'being held responsible' for violating a rule. Since interpersonal relationships are the central variables in this moral system, the argument starts with a discussion of the social ontology of an interpersonal relationship and proceeds to its moral dimensions. This supports the description of the relational ethics framed in terms of prospective responsibility. Because this ethics is based on relations between persons rather than between parts within a person, the traditional problem of a free mind moving a mechanical body is not applicable. An alternative metaphysics is proposed, based on the philosophy of George Herbert Mead and the research of Daniel

Kahnemann. The final section illustrates how moral judgments look in a systems approach and considers systemic evil. The proposed theory is an extension into interpersonal relationships and morality of recent embodied enactive approaches to mental activities.

Because the decision process of a moral agent figures prominently in moral theories, information on that process from the cognitive and behavioral sciences has altered the groundwork of moral theories. Rapid cultural change in America has also contributed to a shifting moral landscape, not only in how people choose what to do, but the language they use to describe it, and their judgments of what is good. In the last 50 years, that language has come to center on the concept of responsibility in its active and forward-looking sense of taking responsibility, shirking it, growing into it, abandoning it, sharing it, and so forth.[1] We try to raise our children to be responsible adults in this sense of the term. The goal of a responsible action is an unspecified positive outcome in an interlocking network of social relationships and within a societal structure. This chapter explores the living morality in use behind this language, provides a theoretical framework for it, and supports it with a modern scientific understanding of human choice and action.

A morality of active, forward-looking responsibility has many features that differ from those described in traditional moral theories. To begin with, it presumes a conception of a person as related to others in a web of interpersonal relationships, each with its own history, and enmeshed in a cultural and social milieu. A moral agent then needs to be connected with related others both cognitively through a 'theory of mind' and affectively through empathic understanding of the other. An agent's action is a move in an ongoing series of actions and reactions with other agents and within the containing social entity and can rarely be viewed as an incident isolated from those pressures and tensions. Because the action itself involves particular people in a particular time and place, a moral judgment of what to do is made to fit that particular situation with sensitivity to the complexities of each relationship involved, considering likely reactions, and alert to possible societal fallout. To function in this moral climate of particulars, an agent needs the skills of social perception and moral judgment. Social perception requires empathy, an understanding of the personal interests each party has at stake were the situation to change, and the courage to acknowledge what is in fact the case. A judgment of what to do is a skill much like that exercised by Aristotle's 'man of practical wisdom'; it requires practice and cultivation. To be able to do these things and act in this way is what it means to be a mature, responsible adult. Because a moral situation has constantly changing potentialities and tensions among interested parties, it is best conceived of as a multilevel, complex dynamical system. The moral good toward which the system would ideally tend would be to increase the interpersonal virtues that are the foundation of a flourishing society, such as: trust, kindness, and mutual respect. Changes in the situation brought on by an agent's action are given a positive judgment when these result. This moral theory is a version of consequentialism that is relational, particular, and dynamic.

The other morality practiced in America, principle- or rule-based ethics, has been exhaustively described and studied by moral theorists. Here the moral agent is a solitary individual acting through rational deliberation in accord with a universal principle to choose the right or good action. Among the many versions of such a moral principle are: obey the commands of the moral law found in scriptures or in reason itself, do only what you could want everyone to do, secure the greatest happiness for the greatest number of people, do whatever would further self-interest, honor the contracts you have made, and keep yourself pure. An action is judged morally good when in accord with the principle. Because this morality is modeled on a legal system, failure to be in accord with the principle is accounted blameworthy and punishable, much like a crime or sin. Currently in American society, most people use a principle-based ethics together with the morality of active responsibility. The languages of these two systems are a constant feature both of public discourse and of interpersonal discussions of what to do.

At the center of responsibility theory is the integrity of an interpersonal relationship and, accordingly, the first section argues that not only is it an existent, but that its participants can perceive it. The second section explores the sociobiological structure of this relationship and its moral dimensions, showing how it can be a conduit for moral behavior. In the third section, I offer an ethical theory built on the moral dimensions of an interpersonal relationship. In a morally sensitive situation, one agent's action calls forth a response from others, changing the calculus of what is owed or must be given between the parties. Added to these changing tensions within the situation are pressures from higher levels of organization – social groups, society and culture, and pressures from individuals and groups only tangentially linked to participants. This complex of influences is easily visualized as a complex dynamical system. This chapter is an attempt to theoretically underwrite this common ethical system as it appears in American culture. I will show how it complements and competes with principle-based ethics. The fourth section explores moral agency in a relational ethics using a systems approach.

Here I rely on a way of understanding human decisions and action independently developed by George Herbert Mead and Daniel Kahnemann. Mead and Kahnemann conceive of an agent as having two modes of thinking and decision-making conceptualized as a spontaneous self and a deliberative self. Those are Daniel Kahnemann's terms, ones he derived from decades of empirical research into human decision-making. This modern, empirically derived duality from Mead and Kahnemann comes with its own explanatory power and set of problems for ethics, and it is these that I consider in this section. First among them is the fact that so much of what any person chooses to do comes from that vast repository of intelligence outside the light of conscious awareness. An agent's spontaneous and unaware self acts to achieve personal goals just as the conscious deliberative self does. I consider the possibility that what we admit into our arena of conscious thought can be a choice. The whole person would then choose all purposive actions whether conscious, pre-reflective, or performed without conscious awareness. In the last section, I will describe how a relational ethics of responsibility works using the language of complex dynamic systems.

A Metaphysics of ethics

If we are to have an ethics based on relationships between an agent and other members of a moral community, relations must be more than similar thoughts in the minds of the participants. Otherwise we would have another version of individualistic ethics with moral agents choosing and acting through consultation only with their own thoughts, whether principled or self-interested. Furthermore, for a relationship to have an effect in interpersonal interactions it must be perceptible, to be sensed more or less accurately. That relations in general are real and have causal impact is the less contentious of these claims. Physics provides abundant evidence for the existence of force fields at the cosmic level and entanglement at the quantum level. That human beings can directly perceive relations in their ordinary world, especially interpersonal relationships, is less widely accepted. But for a moral theory to be founded on integrity in interpersonal relationships, these must be as real as objects and perceptible.

That relations are real and perceptible is not a new idea. Both William James[2] and J. J. Gibson[3] among others have given arguments for the direct perception of relations, although neither suggested how this was possible. This is a particularly fraught question because relations cannot be visually represented the way objects can be. How can you prove that one is there if you cannot see what it is, describe it, or measure it? Arguably, what you can't prove has no claim to being real in science. There are two good answers to this. The first is that metaphysical assumptions usually can't be proven in this sense. They can only be validated by the robustness of the theories they support. But they are always vulnerable to replacement with a different metaphysical assumption capable of supporting a better theory. That only objects, in this case other persons, are real is one of these vulnerable assumptions, not a scientific fact.[4] The second answer offers evidence that animals can perceive without phenomenal awareness; perception of relations would fall into that category. The capabilities of humans and other species suggest that non-phenomenal perception is not uncommon. For example, a recent experiment has found the neural correlates of a magnetic sense in the brainstem of pigeons.[5] In another example from animal studies, monkeys responded to a decrease in entropy in an array of perceived icons.[6] Many studies on human subjects have shown that a visual cue, flashed too quickly to be consciously noticed, nevertheless affects their cognitive processes. The phenomenon of 'blindsight' is another well-documented case of unconscious, non-representational perception of distance, direction, and perhaps mass.[7] Blindsight also offers a mechanism for how such vision is possible. Visual signals from the retina travel via the ventral pathway to the primary visual cortex giving a conscious representation of 'what' is out there. And they travel via the dorsal visual pathway, an older visual pathway, to the brainstem and then to the parietal lobe that senses distance and direction to 'we don't know what.' The dorsal visual system guides action through spatial relationships without consciousness or pictorial representations. This system is brought into relief by damage to the primary visual cortex. Blindsighted people who are able to navigate without pictorial awareness of what is out there have been the subject of intense study and conflicting interpretations. But this secondary visual pathway most likely operates behind and with the primary visual pathway in all sighted animals.[8] This mechanism may also underlie the perception of Gibsonian affordances understood by ecological psychologists as a relation of benefit or harm between an animal and some feature of its environment.[9]

When we come to consider a social relation, there is so much more going on than the mere existence of spatial relations and the benefits or harms another person might afford oneself. That would be a good definition of a psychopath's perception. In a normal social relation, each party has some access to the thoughts and emotions of the other, can empathize the other as a conspecific, brings to spontaneous interactions cultural mores and moral imperatives, and can develop an ongoing relationship with the other person with its own emotional valence and strength. The question then arises as to how much of this complexity is directly perceived or sensed in some way and has a claim to being real, and how much is calculated from perceived clues and therefore is more of an inference.[10] In a human social relation, an affordance offered by the other person very likely contains many of the complexities of a normal 'theory of mind': ideas, emotions, beliefs, desires, imaginings, and intentions. However, the neuroscience that impinges on this question is currently in intense development, and there isn't enough information to support a confident assertion at this time. As an example, a study of socially coordinated finger wagging shows that when the two participants can see each other, they spontaneously switch from independent movements to in-phase synchronization and back when vision is blocked. Synchronization was correlated with a distinctive oscillatory pattern in the participants' EEG's 'as if' there were a connection between their brains.[11] Might this be the physical trace of an interpersonal relationship?

A picture emerges from the forgoing discussion of a socially active person who is the nexus and origin of relations with many elements in the environment, including other persons, each having meaning for the agent and are likely to respond to any initiative with one of their own. I will use this view as the basis for a theory of responsibility ethics. A moral agent is linked through interpersonal relationships in a web of relations with others, who all have some access to each other's mental states and are conditioned with some level of empathy for them in a system of mutual adjustments. Then, depending on the agent's neurophysiological endowment and practice in decision making, he or she acts with some level of skill in moral choices. Responsibility ethics sits very comfortably on this metaphysics of a moral agent. I propose to show that a morality based on relations is more realistic, explains more, and gives better practical guidance than one based on the actions of a solitary rational agent.

Moral dimensions of an interpersonal relationship

Ethical systems function to maintain social order. People's expectations of each other must be fulfilled for the most part to avoid potentially ruinous social conflict. One way of achieving such order is to use a set of rules together with an enforcement mechanism to ensure obedience, such as shamans, chiefs, gods, or social censure. Only rarely will a society forego the use of such rules even though it requires an intensive use of resources to enforce. But in addition to rules that are meant to be obeyed, all settled societies have been found to have developed some version of the Golden Rule: 'Do unto others as you would have them do unto you.' This is not a rule to be obeyed but a heuristic to help people to empathically connect with the person they are about to do something to. It asks, 'How would *you* like it?' If the members of a community used this heuristic regularly, social order would be maintained with little cost and effort, trust among the members would increase, cooperation and investment would increase, and the general wealth and well-being of the community would increase. The Golden Rule serves as a bridge between care for children and family members to friendship with unrelated others in the community. This primeval and ubiquitous morality centers on the personal relationship between an agent and another human being. Patricia Churchland has recently offered an extended argument for the hypothesis that morality evolved among mammals from bonds of care and affinity with offspring and mates, later extended to a wider social group.[12]

Support for the idea that morality evolved from within an interpersonal relationship also comes from game theory in evolutionary biology. I am assuming that the evolution of a moral practice along the lines of the Golden Rule could be modeled by the evolution of cooperation. Since the 1970s, the prisoner's dilemma has served as a model for the conflict between cooperative and exploitative behavior between two players. Scoring awards five points if you defect on a cooperator who gets zero, three points each to mutual cooperators, and one point each for mutual defectors. Tournaments of an iterated prisoner's dilemma pitting the same two parties against each other used different mathematical algorithms of cooperation and defection. The contests were held to see which strategy of cooperation and defection would be the most advantageous to the player. Those tournaments ended when a simple algorithm consistently won. Nicknamed 'Tit-For-Tat,' it instructs the player to cooperate until the opponent defects, then defect on the next move, and switch back to cooperate on the following move.[13] The winning strategy is: be kind, enforce reciprocity, forgive, offer trust, and repeat. It was found that cooperation evolved between parties even in this computer simulation of an interpersonal relationship.[14] But in real world trials, allowing choice and refusal of partners accelerates the emergence of mutual cooperation.[15] Anthropologists studying world cultures and evolutionary biologists have both focused on an interpersonal relationship as a central element in moral systems. And even in mass societies where impersonal relations are ubiquitous, what we would do in an interpersonal relationship serves as our model of a morally acceptable action and forms our intuitions about how we should treat distant others. The responsibility ethics described below is an American version of this naturally occurring ethics distinctive of human interpersonal relationships.

If we accept that an interpersonal relationship is a real biological structure and central to moral agency, what are its properties and how does it function? Since our perceptions of relationships are non-representational, descriptions of our experience of them must use metaphors. Among the more common are line, channel, thread, bond, tie, connection, and link.[16] Interpersonal relationships have been explored in exquisite detail for hundreds of years by literary artists and for the last century by the counseling professions. Much scientific research has been done in the past few decades on specific issues concerning human relationships, but a general theory of the structure and function of interpersonal relationships has yet to appear.[17] Because it will be necessary for the rest of this theory, I will attempt a hasty sketch of a relationship between two parties at the biological level of human conspecifics.

A relationship forms a line of communication between two humans with these features:

It forms spontaneously on contact,

Its endurance can be fleeting or lifelong,

It transmits an awareness of each other's mental states,

It recognizes the other as another center of human consciousness,

It stirs empathy on recognition,

It carries an emotional valence either positive or negative or indifferent,

It has elasticity, a pull or repulsion,

It can survive separation,

It can open or close to the other on many of these dimensions.

What follows will examine three ethical dimensions of an interpersonal relationship: reciprocity, care, and openness. These dimensions, I will suggest, are crucial to understanding the forces shaping ethical decision-making.

An interpersonal relationship functions to transmit social goods and harms from one party to the other. On the affective side, respect and affection, as well as resentment and disapproval, pass to and are received by the recipient. A response from the recipient nearly always follows. Among transactions, gifts and services are offered, and gratitude accepted; injuries are caused, and apologies expressed. Again, the recipient will respond, and the dance of an ongoing interpersonal relationship continues. In everyday social interactions, people are very skillful in their perceptions of interpersonal dynamics, and their moves are so swift as to be pre-reflective. The governing standard people use for these interactions is reciprocity, but not merely as an ideal. Parties to a relationship keep a careful accounting and act to right any imbalance. If you do me a favor, I will feel in your debt and will try to return the favor in some appropriate way. If I can't do that, I may feel diminished as a person. If I won't do it, you may feel exploited. If you injure me in some way and do not make amends, I might try to punish you. Failing to achieve that balance can also diminish me. Anthropologists and other social scientists have generated a vast and complex literature on reciprocity. For our purposes, it is enough to note that people seek fairness and justice in their relationships. Kant would describe this in terms of not using others as a means but honoring them as ends in themselves. We enforce reciprocity regarding other people's treatment of oneself and even their treatment of others. This latter is a form of moral altruism when an individual does it, because enforcing a moral standard can incur a personal cost, much like the case of a whistle-blower.[18] Governing bodies usually adjudicate serious cases of an imbalance in a personal relationship over areas such as loans, stealing, injuries, defamation and the like as a part of a civil or religious justice system.

When there is a great power disparity between parties to a relationship, the ruling ethic is normally one of care for the vulnerable, rather than reciprocity. This is the second moral dimension of an interpersonal relationship. Interestingly, the golden rule covers this case too, 'if I were in a position of weakness, would I want someone to help me?' When we see a vulnerable person in an emergency situation, our urge to help is immediate and pre-reflective. I see an untended toddler run into the street, my feet start moving. I stumble on a curb, the stranger next to me grabs my arm. A man in a wheelchair approaches the door, I reach to open it. People even put themselves in danger following a spontaneous helping impulse.[19] We would think there was something strange about someone who lacked these impulses. The case of a vulnerable family member or friend also exerts a strong pull of moral obligation to provide help. We provide care for our children, our frail elderly, and our sick family members and friends. To respond appropriately in these cases is thought to be a test of character and failure to respond provokes strong social censure. The much harder cases involve the myriad dependencies incurred among ordinary people in our modern world. I depend on others for nearly every necessity from my food, clothing, and housing to my Internet service. I trust other people to stop for a red light as I drive through on green, and I trust business people not to sell my credit card numbers. Robert Goodin wrote a book-length treatise arguing in great detail how in each kind of relationship the moral imperative to provide help lies in the vulnerability of the weaker party even when an implied or explicit contract has been formed.[20]

The line of communication between parties to an interpersonal relationship can be opened and closed, sometimes at will but more often as an unconscious choice. This is the third ethical dimension of an interpersonal relationship. Our openness to others is a finely tuned instrument that discriminates closely in each particular relationship or kind of relationship and can vary moment by moment. If I am walking down a busy Manhattan street, I will be closed to all contact, as will everyone else, unless someone catches my eye. Then I can immediately make myself available to that person with some degree of openness. At the other end of the spectrum are intimate relationships, but, even in these, variability in openness to the other is part of normal life and not always under voluntary control. Between these two extremes lie the relationships we all have with business associates, friends, acquaintances, professionals, neighbors, and passersby. When we are tired or injured, we close ourselves off. Each person has an economy of internal strength to manage; in that economy, opening is costly, and closing off other people conserves energy. By fine-tuning our openness to others, we control the direction and degree of empathy, trust, and willingness to cooperate or provide assistance. To be open is like unilateral disarmament, a vulnerable state that is often used as a bargaining chip to elicit trust and openness from the other person. Moral systems have been built on each of these three moral dimensions of an interpersonal relationship. Reciprocity serves as the moral standard in tribal cultures using the rule 'an eye for an eye; a tooth for a tooth.' A feminist approach to moral theory uses the dimension of care at its center, usually presented as a morality used by women.[21] The dimension of openness is celebrated by the Buddhist concepts of compassion and mindfulness. Modern science has found that the neuropeptide, oxytocin, facilitates the expression of trust and empathy and is sometimes called the 'moral molecule.'[22]

In a morally sensitive situation, many of the interested parties are connected in relationships of various degrees of openness, with attractions and aversions, differing degrees of power or vulnerability, and each with its own history of interactions that exerts pressures to give or to take something from the other. Any action in this web of relationships – a benefit conferred or extracted, a harm imposed or suffered – alters the balance sheet of each party through which debts are calculated. It is a fluid dynamic system that evolves and changes recruiting one person to give up something, another to benefit, and another to take the responsibility to strike a fair balance in making a needed change. But this is only one move in an ongoing series of interactions that make up everyday human life.

A relational ethics of responsibilities

The foregoing section described the moral dimensions of an interpersonal relationship: how that biological structure allows, forms, and transmits actions of a moral nature. An interpersonal relationship conditions a range of possible actions, interpretations, and responses. And each person relies on their intuitive understanding of these possibilities to navigate their social and ethical world. This section examines the characteristics of an ethical system that has developed to fit the contours of this social world. I propose that interactional sequences generating trust between the parties are morally good, while those that produce anger, fear, and paranoia are not. This allows for evolution of morality without relativism, without licensing slavery or genital mutilation, for example. Trust is a natural world good produced through open, reciprocal, caring interactions in interpersonal relationships. Engaging others with integrity means being a trustworthy person in this sense. Trust between people benefits not only the parties involved but also the whole social group as well. If I find that I can trust you, I will be more willing to trust another person. When that trust is rewarded, I will trust even more people. In this way, trust expands throughout a society, forming the basis of a civil society as a systemic good; peace and prosperity follow.[23] Notice that this moral assessment attaches to the quality of an interactional sequence rather than to a solitary agent considered in isolation from his or her historic, systemic, and relational connections. Diagnosing what went wrong and correcting it is as important for a relational ethics as for any other morality. But rather than rely on blame and punishment of an offender, this ethics would focus instead on all the systemic influences involved in producing a harmful action, including the agent's own contribution. A corrective would be tailored accordingly; it would school the sensitivities and judgment of offending parties to bring their moral skills more in line with the complex demands of their situation. Sometimes applying the shock and pain of punishment may be necessary, but usually this causes a decrease in trust and undermines the motivation to act well.

Because the language of principle-based ethics dominates scholarly discourse, I will describe relational ethics and its language of prospective responsibility in terms of their differences. The first difference is the character of a moral situation. In principle-based ethics, the current situation is an instance of a type that is regulated by universally applicable abstract principles. Moral deliberation seeks to determine the type of situation at hand and what universally applicable rule should govern an agent's action. An agent then applies the rule, tempered by any relevant exceptions. In a relational ethics, the moral situation exists only at a particular time as a relationship among particular individuals in a particular context, and its resolution must be designed on the spot to fit those people. John Dewey's moral philosophy is similar; because the problematic situation is unique, the weight and burden of morality is transferred to intelligent judgment.[24] The location of moral decision-making is a 'concrete situation' rather than in an ethical ideal. The kind of judgment needed here depends on knowledge of what each party has at stake with an idea of the way the world works to choose what would be best; what Aristotle called 'a man of practical wisdom,' we would call 'being a responsible person.'

The relative weight given to the rational versus emotional input to moral agency is another difference between these two moralities. Traditional moral theory celebrates rational deliberation and conscious choice as the seat of moral action. Rawls's reflective deliberation is an attunement between the agent's moral judgments and the more abstract principles thought to contain the essence of morality.[25] In Kohlberg's theory of moral stages, moral maturity means understanding the most abstract and general form of moral principles.[26] A counterexample to this view of moral agency is the case of psychopaths. Those who exhibit a psychopathic personality are able to do moral reasoning and can learn the difference between right and wrong as well as normal people, but they have no motivation to act accordingly. When you ask a psychopath, 'How would you like it if someone did that to you?' He might answer, 'I wouldn't, but what does that have to do with what I am going to do to him?' A normal person would be able to mentally switch roles and imaginatively feel the harm of the incipient action. Imagining the pain that will be produced triggers feelings of fear and empathy for that victim and leads to second thoughts in normal people. Psychopaths, however, feel neither fear of pain nor the social emotions.[27] Since the psychopath is the model of an amoral person, and the defect is a lack of emotional impulse, morality could not be just a matter of using rational deliberation in applying general principles. Emotional engagement is a necessary condition for moral agency. In a relational ethics of responsibility, this engagement is an equal partner with intelligent judgment in choosing well.

Both relational ethics and principle-based ethics use the term 'responsibility,' although in different senses. In rule or principle-based ethics, the appropriate sense of responsibility is one of 'being held responsible.' This sense is often called 'moral responsibility.' An agent who obeys a moral rule is rewarded with praise, and one who disobeys is found guilty; blame and punishment follow. Since an agent is passively being held responsible by the authorities upholding the rule, its form is retrospective and passive. This contrasts with its prospective and active form in a relational ethics of taking responsibility. The two systems overlap in important ways. Being irresponsible when one's role or situation requires taking responsibility can retrospectively bring blame and punishment: for example, a negligent parent or sleeping security guard or inattentive chief financial officer. Another way the two senses of responsibility overlap is in moral learning. Social disapproval after violating a norm trains the intuitions of the miscreant, channeling future behavior along more acceptable paths. The content of moral rules and principles are most often prohibitions. Then the course of action is very clearly a single action with the instruction not to do it. When a moral rule enjoins a duty to take positive action, it is often of a general sort that is similar to having a prospective responsibility. Take the commandment, for example, 'Honor thy father and mother.' When regulating moral conduct in this interpersonal relationship, the duty is an open-ended responsibility to serve their parents' interests in some important ways. That duty is identical to taking prospective responsibility. In principle-based ethics, the obligation comes from a commandment either from God, from moral authorities, or from one's own rational deliberations; in a relational ethics, the obligation comes from the concrete moral situation in a filial relationship.[28]

These two forms of responsibility and their different ethical systems are often in conflict, however. In current American society, both the system of moral rules that demands obedience and the system of open-ended responsibility for acting well coexist, yet they tend to undermine each other. Moral rules are imposed most strictly when moral authorities assume that allowing personal judgment would license harmful, chaotic behavior. But in the interest of social stability, moral agency is then reduced to acquiescence, and the idea of taking responsibility and using one's own judgment carries the suspicion of moral cheating. Then, too, since the moral authority resides in the rules, and the authorities that enforce them, if one can get away with breaking the rule, this is only doing what is expected. Rule-breaking behavior then confirms the idea that human nature is inherently selfish and justifies the strict imposition of rules. However, in order to develop the moral judgment necessary in a system of active responsibility, an agent must be entrusted with the freedom to exercise it. Developing the social intelligence, the sensitivity, and judgment to choose well when taking active responsibility requires practice and perhaps training. A moral agent needs the freedom to experiment and learn, social support while doing so, and role models who are honored for their good judgment. While moral rules provide guidelines to acceptable behavior, the fear that relaxing the pressure of rules could license moral chaos curtails the freedom to develop the judgment and the moral maturity needed to take prospective responsibility in a morally sensitive situation.

Both of these moralities are under stress in current American society. Our system of moral rules relies on monitoring by authority figures, peer groups, and ultimately on an all-seeing God for enforcement. In a highly mobile, increasingly digital, and less religious society, it is not surprising that surveillance on all these fronts is in serious decline. Then, too, a celebration of the renegade since the Beat generation has given rule-breaking cultural cachet, leading to a more individualistic understanding of our obligations to each other. This focus on the self and autonomy tends to undercut the moral claims of relationships, putting pressure also on the relational ethics of responsibility that has been doing a large share of the moral work all along. If these trends are assessed from the standpoint of a rule or principle-based ethic, a moral deterioration of American society is clear and obvious; scholars and public commentators have called for a reinvigoration of morality, interpreted from the perspective of principle-based ethics. But, assuming that this trend cannot be reversed, it would be wise to compensate for a weakened system of moral rules by deliberately cultivating the personal judgment required by our relational ethics. Since this is traditionally framed in terms of individual intelligence and sensitivities, it takes advantage of modern needs for autonomy and control. That those actions serve social obligations and satisfy human prosocial desires – for affiliation, belonging, and helping – adds to their attraction. Rather than moral decline, a more sanguine interpretation of our current moral landscape would hold that people are gaining practice in using their own judgment in a responsibility ethics of relationships. Should this be true, moral rules and principles would continue to be used to develop intuitions and form habits, and they would continue to serve as ideals. But this role is very different from their traditional role as imperatives to be obeyed.

A systems approach to moral theory must include what Robert Jackall has called systemic evil.[29] Systemic evil refers to an environment inimical to the operation of personal judgment and empathy in interpersonal relationships. Aristotle considered a corrupt polis to be such an environment; it was impossible for a citizen there to be a good man. Jesus thought that simply being rich impeded entry into the kingdom of heaven. The modern version, Jackall analyzed, is a hierarchically organized social institution, a bureaucracy, because it requires each of its members to cede their agency to their immediate superior. Corporate goals and values originate outside human moral systems and are often antisocial and rapacious. The most notorious example of this is an army, but the same can be true of corporations, governments, and many smaller affiliation groups. The commands from a superior to an inferior are 'the moral law' in a corporation, writes Jackall and constitute 'a vast system of organized irresponsibility.'[30] Some hierarchical organizations, however, endorse the practice of interpersonal ethics or substitute its own code of ethics designed to mimic its results; for example, religions honoring love and compassion or service-oriented business ventures.[31] These counterexamples suggest two ways to combat systemic evil: either use a superior power to legislate an alignment of corporate goals with the common good, or cultivate the moral judgment of individuals throughout the corporate hierarchy in a relational ethics of active responsibility. A critical mass of people with whistle-blower morals might be a corrective. This is one of the most serious ethical problems in contemporary society.

Traditional moral theory assumes that the mental activity leading to a moral action is undertaken by a solitary, rational, and conscious mind. I argued above that the rational element is insufficient for a moral action; an emotional element is also required. I will argue in the next section that the conscious mind and the stories it tells itself account for only a fraction of moral activity. Here I want to challenge the idea that an individual mind is solitary, and mental activity is isolated from a person's surrounding social milieu. Given the duality of human experience between its solitary and social aspects, is there any reason to privilege one over the other, or is the proposition that mental activity is solitary just another metaphysical assumption? Certainly sensations and the stream of consciousness are experienced alone; no one feels my itch or hears my internal chatter. But morality is necessarily about other people, who are not just considered as inert features of the environment but as living persons with their own goals and agency.[32] Interacting with them requires a mutual attunement, both affective and cognitive, in order to perceive the moral characteristics of a situation. This requires a theory of the person that is social in its constitution.

George Herbert Mead provides such a theory of the person with his dual methods of the self's engagement with the world dubbed a 'Me' and an 'I.'[33] In Mead's social philosophy, the self, interacting in a system of relationships, forms a concept of a generalized other. Then, taking the standpoint of that generalized other, the self reflects back upon oneself and understands oneself from that perspective as 'Me.' As the self interacts in many different social systems, it constructs other versions of oneself. These various versions of 'Me' coalesce into a person's self-concept that can then be used to interpret past happenings and imagine future actions. Mead identifies the 'I' with the spontaneous action of the self in the instantaneous present moment in response to specific particular happenings. The self, who as an 'I' has no consciousness of acting, observes its own action after the fact and reconfigures its self-understanding as a new version of 'Me' in that social system. Mead argues that the spontaneous action of the self as an 'I' is the leading edge of novelty in an evolving and changing self-system. The development of intelligence in Mead's theory depends on the self's engagement in multiple dynamic systems with other people, alternately acting in a novel particular circumstance from a previous and static idea of 'Me' and then reconstituting that conception of 'Me' to take into consideration what the self actually did. Using something close to Mead's theory, we will turn next to a consideration of moral agency for a socially constituted self.

The source of moral agency

Moral agency is a fraught topic even for traditional principle-based theory, since it assumes an agent can freely choose which rule to apply and then follow through with it. This seemed possible when the mind/soul/spirit of a rational agent was thought to be separate from its mechanically determined physical body. A very serious challenge to this whole program has come from modern scientific identification of the mind with the brain as part of the mechanical physical world; voluntary choice and moral responsibility would seem to be mere illusions. There have been many creative responses to this theoretical move all trying to save moral responsibility; some charge scientific overreach, some try to fine-tune the science to find evidence of free choice, and many philosophers accept the compatibility of free will and determinism. For principle-based ethics, free will is crucially important for holding an agent responsible for a past action, assigning blame, and meting out punishment. But a relational ethics of prospective responsibility is based on the relations between persons rather than between parts within a person – a free mind and a mechanical body. A systems approach based on relations between people needs instead a metaphysics of a person that is unified in action but uses different methods of thinking and choosing what to do. Mead has provided an appropriate one with his social conception of a person as a complex of interrelated systems within a varied social landscape. A modern, empirically derived way to organize our understanding of decision-making and moral agency comes from Daniel Kahnemann's decades of research into the psychology of decision-making.[34] Actions that have an ethical dimension share decision and activation mechanics with all other actions, so his theory can easily be adapted to moral agency.

Kahnemann uses the terminology of 'System 1' for intuitive, emotional, and fast thinking, and 'System 2' for deliberative, logical, and slow thinking and invites us to think of these thought systems as two selves, named 'the experiencing self' and 'the remembering self,' 'the spontaneous self' and 'the deliberate self.' This dichotomy not only arises out of scientific research but meshes well with Mead's philosophical psychology. 'System 1' resembles Mead's 'I,' the spontaneous, unreflective self who acts, and 'System 2' his 'Me,' the conscious, deliberative self. Each style of thinking is intelligent, purposive and effective, although each makes errors undermining the person's best interests, and presumably moral errors, also. The spontaneous or experiencing self is the repository of cultural and experiential learning, is emotionally sensitive and expressive, discriminates the good-for-me from the bad-for-me, can understand and respond to complex verbal repartee, can judge the distance and trajectory of many moving bodies simultaneously, can navigate with subtlety in complex social situations, can read and speak in whatever languages it has learned, and has stored at the ready the complex rules of any game it has mastered: stock market, chess, sports, business, cooking, whatever. All of this is done instantaneously and effortlessly while adjusting to variable and unexpected input from the environment. The remembering or deliberative self, System 2, is the corrective when System 1 gives results that don't compute. The deliberative self marshals attention and trains it on the weak link. While the spontaneous self operates effortlessly, the deliberative self's efforts are labor-intensive and, since attention is a very limited resource, tune out much of the rest of the world. The deliberative self can orient attention towards a part of the environment and hold it there to scan for new data, integrates new information into the existing routines of the spontaneous self, can logically compute what is too complex for that self, and is the seat of conscious choice.

As is clear from the list of its accomplishments, the experiencing self is very intelligent, pursuing the person's goals with insight and agility, even though most of this is submerged below the light of conscious awareness. System 1 manages most of the dimensions of an interpersonal relationship; the experiencing self instantly categorizes other people and events as they unfold. I see this person as a friend, a con artist, someone else's child, a potential mate, one of 'them,' an authority figure, or an instance of other categories and stereotypes that carry with them templates for how they should be treated.[35] This is the cognitive background to moral agency. Even though categorizations frame and justify moral actions, they usually elude conscious scrutiny. System 1 opens or closes a relationship, keeps a running score of reciprocal benefits and harms, and judges whether the current situation requires care of another person. Personal habits and character, virtues and vices, operate through the spontaneous self. It carries the main cognitive burden in moral decision making, for good or ill. Traditional moral philosophy would identify System 2, the deliberative self, as the seat of moral choice. In this mode of operation, the agent withdraws from active engagement to consider the moral value of various options and to critically evaluate his or her spontaneous urge to choose one way or another. This self notices that I have applied an unfair stereotype and withholds action, or it notices that I have caved into my boss's demeanor of authority and reconsiders my obedience. Unlike the spontaneous self, this self operates in the full light of conscious awareness, or does it? The same deliberative process can be used to select which principle should guide my action. For example, after considering all the relevant factors, I may decide that a duty to my shareholders outweighs any claims from customers, employees, or the community; my bonus and promotion are irrelevant. Uncovering one's own self-deception is very hard work and the deliberative self receives no help from my spontaneous self.

The capacity to choose is critically important if we are to have responsibility of any kind. Everyday human experience testifies to the ability to choose a novel action. We have all had the experience of managing our own weaknesses by inventing constraints: I can't make myself exercise, so I get a dog that I will have to walk; I'll eat that whole package of chocolate crisps, so I buy cookies I don't like as much; I can't get much art work done at home, so I rent studio space. Another argument in support of the ability to choose comes from dynamical systems theory that has been used to model moral decision making in this theory. In a closed system that allows for control of all the variables, the output of the system is completely determined by the input. In an open system, however, with an unlimited number of variables impinging on the action in question, unusual and abrupt changes are common. You could argue that this is just an epistemological problem; if we knew the speed and trajectories of all those influences, the action of the system would be seen to be just as completely determined. But this speculation has no compelling factual support, and we are free to assume the opposite, especially since it accords with experience and is necessary for social order. I suggest we do that.

The experiencing and the deliberative selves each have its own pattern of constraints and possibilities for novelty. For the experiencing self, the unpremeditated decision to act occurs at the moment action is initiated. As we have seen, Dewey and Mead thought that choice occurred in the lived moment when a person responds to novel, unique, particular occurrences with a repertoire of many different templates and skillsets in order to achieve either long-term goals or an immediate advantage. No action or set of circumstances is exactly like any other, and all animals must be able to adjust their actions to extract from the current circumstances some treasured goal. This is an instance of a radically open system and allows for novelty. Can the deliberative self freely choose a novel way of doing things, of solving a problem? Personal accounts of finding a creative way of doing something highlight its origin just outside of conscious effort – 'It just came to me in the shower,' is typical. But it is also conceded that without a conscious struggle, turning over all the alternatives, it probably wouldn't have come to me at all. Dynamical systems typically oscillate before coming to rest; in this case, the deliberation process behaves like a dynamical system seeking a trough.[36] Whether this deliberative process produces a novel outcome or just leads up to it so that the spontaneous self can take over is just one of the many unknowns in the field of decision theory. Creativity and how novelty is produced are very active research topics and more will be known in coming decades.

How, then, is a moral agent to be understood? If both the experiencing self and the deliberative self are intelligent, are each a seat of moral motivation, and both can initiate moral actions, then the whole person is the moral agent and can take or refuse responsibility in a moral situation. An objection arises, though: a person could not take responsibility in a situation they are not even aware of. There is much to recommend this view, especially if that person were incapable of the needed awareness, as from a neurological defect or simply exhaustion. Inattention to the effects of one's action can also come from concentrating a limited supply of attention on an area of intense interest. System 1 can have all its attention captured by some startling emergency. Or System 2 can consciously allocate attention to only a particular area of life, neglecting all else. This inattention can be a reflection of my priorities and chosen for the advantages it delivers. In the short term, a person can often get away with negligently causing harm, but it is considered irresponsible not to notice the harm one is causing. If this behavior persists people notice and count it as a debt to be repaid, avenged, or forgiven, but they do not dismiss as morally irrelevant those interactions that are not deliberately chosen.

In the moral psychology of interpersonal relationships, people feel responsible for the welfare of closely related others. This is true even though an agent's action is not causally connected with what happened. When something bad happens to someone close, an adult will think/feel, 'I should have...' or 'If only I had...' Even very young children hold themselves responsible for the welfare of an important other in a relationship. If a parent falls ill or dies or just goes away, a child will most often feel responsible for making that happen. For children their failure was magically enacted. An adult will know that she or he did not cause the untoward occurrence, but a conviction of having failed someone with whom one shared a relationship often lingers in the subconscious mind. Much therapy aims to bring this 'failure' to consciousness and relieve those guilty feelings. The perception of failure in a relationship works in the other direction also, as when anger for abandonment is directed toward a spouse who died. This nearly ubiquitous sense of responsibility for related others, irrespective of intentionality or causality, shows a psychological conviction that the whole person is the moral agent, not just the conscious mind, and a moral failure attaches to the relationship rather than to individual acts. That our moral psychology leads us to moral illusions – moral mistakes – is an important and serious consideration for this theory of ethics. Since the vast majority of our moral actions flow from our character and the cultural habits of System 1, how can we tell the difference between our gut feeling of responsibility and our real moral obligations? Judging from Kahnemann's research, detecting an illusion from within an action is difficult. Using an open systems approach to morality would preclude the option of stepping outside the natural world for a regulatory anchor. Such a complex topic is outside the scope of the sketch presented here of a relational moral theory, but it deserves to be pursued. The moral situation depicted here is fluid, dynamical and almost infinitely complex as is true of open systems in general. For all practical purposes it is also non-quantitative, so I will present a simplified qualitative description of relational ethics and how it can be used to interpret a moral situation.

A moral situation viewed with dynamical systems theory

In a moral situation, the participants' choices will result in some allocation of benefits and harms for the participants, and perhaps distant others, groups, or the environment. It is a dynamical system, because the tensions and interactions between the interested parties fluctuate and change over time as each seeks an action that fits the situation from their point of view. Furthermore, each participant is a small dynamical system interacting with many non-participating others even through a separation in space and time. Besides relations with external people, other influences on the participants' choices include religious and cultural ideals, pop culture, historical conditioning, resemblances with other people, and analogies with similar situations.[37] A moral situation fits the description of a non-linear dynamical system, since each potential agent has a unique motivational structure, and these cannot be summed. Such systems are inherently chaotic exhibiting sudden reversals and unpredictable behavior. This accords with everyday experience and also with great literature, which portrays moral situations with intricate detail. However, some regularity in moral life can be observed. For example, a person's character and habitual ways of acting in a kind of situation could form a pattern that the system tends to settle into, an orbit or attractor basin. In this section, I want to explore the idea that the characteristics of dynamical systems condition the behavior of the agents within them. A consideration of how dynamic systems work might lead to a deeper understanding of human moral life and suggest ways to bend the system toward a better society.

Complex systems theory was designed to predict an outcome given the previous state of the system and the trajectories of its variables within the state space of possible outcomes. However, with an ethical system, we seek to evaluate an outcome rather than predict it, and also to evaluate the process leading to an outcome, since that process will continue as the other participants in the situation react. For a relational ethics, this means tracking the level of trust between the participants in an evolving moral situation or the level of trust in the whole system, since this is the moral good. The level of trust between people and within a group is a psychological fact that can be sensed as the degree of openness between participants. Because this is in theory quantitative, it can in principle be measured, even though no appropriate technology currently exists.[38] The usual way to compensate for this lack would use instead the external markers for trust and openness, such as, the level of social ease, mutual respect, lightness, intimacy, mutual understanding, lack of hostility, instances of caring, and support movements. Once these parameters are chosen, they could be applied to the relevant social system. Different levels of social organization will often display contradictory indications of trust, as in the case of honor among thieves but not including their victims within the society. Bonnie and Clyde were tight, for example, but only with each other. But this is common wherever an in-group abuts an out-group, as in racial and sexual discrimination, school children bullying, and exploitative business practices.

The question here is whether an agent's immoral or injurious actions result more from dynamical patterns in the containing social system or from the individual's moral psychology and character. Both our legal system, and morality interpreted in the legalistic terms of rules and principles, acknowledges the extenuating circumstances of a containing social system. But both focus on an individual action as the cause of a moral event and mitigate the blame or punishment to fit the environmental pressures on the agent. But in the metaphysics and ethics proposed here, each person is enmeshed in and constituted by multiple systems of relationships, some overlapping and some at different levels of organization ranging from family to ecological habitat. Both the agent and each of these systems are causally efficacious. Alicia Juarrerro presents a strong argument supporting the causal role of systems at each level of organization.[39] Tracing out causal influences serves the goal in a relational ethics of either changing the pressures and tensions in the participating systems, educating the agent in better judgments, or both. In traditional ethical systems based on principles and rules, the goal is to assign blame and apply punishment to correct the agent's non-compliance, sin or crime, through pain and fear. The contrast can be seen through how we judge Hamlet's actions in Shakespeare's play.

The only open, trusting relationship in the play links Hamlet with his friend Horatio, who is not involved in the action. Hamlet is completely isolated in his dilemma. Popular judgments of Hamlet's failure to kill the king derive from the dominant morality of our culture based on principles and rules. The usual assessment of Hamlet's actions and inactions accuses him of being weak and vacillating; that is his tragedy. But a systems assessment would consider his back and forth deliberation process to be normal. This is how a dynamic system seeks a trough or approaches a basin of attraction that constitutes a decision to act. And according to a relational theory of responsibility, Hamlet's reluctance to avenge his father's murder until his uncle, the king, incriminates himself, shows respect for all concerned and admirable restraint of his passions. Seeking a clandestine way to carry out his duty to avenge his father also shows a mature sense of responsibility, since an outright murder would produce a popular uprising with serious consequences for his country. His tragedy is that there is no way for him to honor all of these responsibilities.

In a modern real-world scenario, we find large American banks that are chartered to serve the public good stripping assets from individual citizens, other corporations, and the public treasury. This is all legal, since the legislature has cooperated in removing regulations that would protect the rest of society. As their profits increased, the banks' influence increased, until they were able to virtually write their own rules. Since this behavior does not contribute to economic development or the prosperity of either citizens or businesses, it conflicts with the public benefit rationale of their charter. All this is well known and the subject of much analysis. Much could be said about the character of individual bankers, but let's look first at the systemic influences on their actions. Rationality is a highly esteemed ideal in our culture. On the conventional definition of rationality only those actions that increase personal gain are rational; for bankers to do anything that doesn't maximize monetary gain is then irrational, even if it promotes economic development and increases the wealth of the nation. Rewards for employees at the managerial and technical levels depend on how much money each brought into the firm, again focusing narrowly on getting money by any legal means. Neither background morality nor public benefit is an issue here. Laws regulating banking practices tend to identify growth of a bank with the growth of the economy, and by extension with the public interest, justifying deregulation of banks. These assumptions are also written into our appropriation legislation and tax code. To correct for this redistribution of assets to the banks, should we try to reform each banker and their regulatory enablers, retool their value systems, jail them, or give them courses on ethics? Or should we change the system parameters to encourage integrity, honest behavior, respect, and care for public good?[40]

The literature on dynamic systems contains many examples of phase changes where an established system of attractor basins bifurcates to a new system with different attractors. The same description can equally apply to human social systems. If a man announces to his family that he is moving in with his mistress, the level of trust in the extended system will plummet, channels of communication close, and new alliances may form. On a larger scale, a revolution causes a massive breakdown of a social system. It can result in utter chaos, or it can produce new forms of social organization. Times of natural disasters similarly cause a bifurcation in the trajectory of a system. The immediate aftermath can bring chaos with looting and crimes against persons, lowering the level of trust in the system that will be hard to recoup. But disasters can also dissolve the walls each person erects against others, leading to extraordinary acts of compassion and rescue as people spontaneously band together in mutual support. In this case, the loss of established routines and customs throws people back on their humanity, relying on the ethics of interpersonal relationships. This situation is usually short-lived as people recover their boundaries and cultural habits.

The view presented in this chapter describes and provides a theoretical underpinning for a living morality used in American society alongside a morality of principle-based ethics. This morality is based on an interpersonal relationship, a biological structure linking human beings that shapes moral psychology among members of our species. The characteristics of this link include a resonance or attunement to the cognitive and affective contents of others' mental states, the capacity for shared intentionality, empathy, and a bias toward cooperative behavior. While none of these is unique to our species, the combination probably is.[41] In a complex moral situation involving several agents, the locus of moral agency can shift from one to another as each person reviews his or her options, makes their position known, or simply takes the initiative to do something. When someone takes responsibility to make a change, the others must then consider what can or should be done to support or rectify the situation. Each person in this scenario brings a history of needs, desires, hurts, and commitments and is enmeshed in a web of relationships with the others, each with its own history of reciprocity, care, and openness. An agent achieves the moral good when his or her contributions to the moral situation conserve the integrity of relationships interpreted as fostering trust among all parties who have something at stake in its resolution.

Notes

1. Winston Davis, in *Taking Responsibility*, ed. Winston Davis (Charlottesville: University Press of Virginia, 2001), p. 1. Moral philosophers have portrayed the ethical systems of their own cultures and times with their theories. I am doing this for current American culture.
2. William James, *Essays in Radical Empiricism, A Pluralistic Universe* (Glouchester, MA: Peter Smith, 1967), p. 42 and p. 326.
3. James J. Gibson, *The Ecological Approach to Visual Perception* (Boston: Houghton Mifflin, 1979), p. 127.
4. Traditionally Western philosophy has classified relations as abstract objects where 'abstract' can be interpreted as neither concrete nor mental.
5. Le-Qing Wu and J. David Dickman, 'Neural Correlates of a Magnetic Sense,' *Science* 1216567, published online April 26, 2012.
6. Described in Anthony Chemero, *Radical Embodied Cognitive Science* (Cambridge, MA: MIT Press, 2011), pp. 127–134.
7. Paul Azzopardi and Alan Cowey, 'Is Blindsight Like Normal, Near-Threshold Vision?' in *Proceedings of the National Academy of Sciences of the United States of America*, 94 (1997): 14190–14194. Here is a stunning video demonstration of non-phenomenal perception: http://www.scientificamerican.com/video.cfm?id=navigating-by-blindsight-2010-04-21.
8. S. E. Leh, H. Johansen-Berg and A. Ptito, 'Unconscious Vision: New Insights into the Neuronal Correlate of Blindsight Using Diffusion Tractography,' *Brain* 129(7) (2006): 1822–1832. They have shown that the vision in blindsight travels first to the superior colliculus and then to the dorsal visual pathway. In humans, 10% of axons from the retina lead to the superior colliculus, whereas in non-vertebrate animals this pathway through the analogue organ, the optical tectum, is the main visual system.
9. Anthony Chemero gives an in-depth analysis of the many interpretations of Gibson's concept of affordances in his *Radical Embodied Cognitive Science*. An affordance on his view is a relation between an organism's ability and some feature of the environment, ibid., pp. 142–145.
10. Perhaps direct perception versus inference is another one of those dichotomies that make for rousing arguments but obscure the complexities of the world, spread confusion, and foreclose new avenues of investigation.
11. Emmanuelle Tognoli, Julien Lagarde, Gonzalo C. DeGuzman, and J. A. Scott Kelso, 'The phi complex as a neuromarker of human social coordination,' in *Proceedings of the National Academy of Sciences*. May 8, 2007, pp. 8190–8195.
12. P. S. Churchland, *Braintrust: What Neuroscience Tells Us about Morality* (Princeton: Princeton University Press, 2011).
13. Robert Axelrod, 'The Evolution of Strategies in the Iterated Prisoner's Dilemma,' in *The Dynamics of Norms*, ed. Bicchieri, R, Jeffrey, B. Skryms (Cambridge: Cambridge University Press, 1997), pp. 1–17.
14. L. McNally, S. P. Brown, and A. L. Jackson, 'Cooperation and the Evolution of Intelligence,' in *Proceedings of the Royal Society of London B*, 279 (2012): 3027–3034.
15. E. Ann Stanley, Dan Ashlock, and Leigh Tesfatsion, 'Iterated Prisoner's Dilemma with Choice and Refusal of Partners,' in *Artificial Life III*, Volume

XVII, *Santa Fe Institute Studies in the Sciences of Complexity*, ed. C. Langton (Reading: Addison-Wesley, 1994), pp. 131–175.
16. I will use the concepts of a line or a link, because they do not connote extension or mass.
17. Ellen Berscheid, 'Help Wanted: A Grand Theorist of Interpersonal Relationships: Sociologist or Anthropologist Preferred,' *Journal of Social and Personal Relationships*, 12 (1995): 529–533.
18. As described in Daniel Kahnemann, *Thinking, Fast and Slow* (New York: Farrar, Straus, and Girous, 2011), p. 308. Using fMRI studies, an increased activity in the pleasure centers accompanies punishing a stranger for being unfair to another stranger.
19. J. A. Piliavin, J. F. Dovidio, R. Gaetner, R. and R. Clark, *Emergency Intervention* (New York: Academic Press, 1981).
20. Robert E. Goodin, *Protecting the Vulnerable* (Chicago: University of Chicago Press, 1985).
21. Nell Noddings, *Caring: A Feminine Approach to Ethics and Moral Education* (Berkeley: University of California Press, 1984).
22. Michael Kosfeld, Markus Heinrichs, Paul J. Zak, Urs Fischbacher and Ernst Fehr, 'Oxytocin Increases Trust in Humans' *Nature* 435 (2005): 673–676.
23. Robert Putnam, *Bowling Alone: The Collapse and Revival of American Community* (New York: Simon and Schuster, 2000).
24. John Dewey, *Reconstruction in Philosophy* (New York: Henry Holt, 1920), p. 163.
25. John Rawls, *Theory of Justice* (Cambridge, MA: Harvard University Press, 1971).
26. Lawrence T. Kohlberg, 'Moral Stages and Moralization: The Cognitive-Developmental Approach,' in *Moral Development and Behavior: Theory, Research and Social Issues*, ed. T. Lickona (New York: Holt, Rinehart and Winston, 1976), pp. 31–53.
27. The psychopathic defect is an abnormality in the amygdala and paralimbic system, the neural centers mediating emotional responses. K. A. Kiehl 'A Cognitive Neuroscience Perspective on Psychopathy: Evidence for a Paralimbic System Dysfunction,' in *Psychiatry Research* 142 (2006): 107–128. See also A. Glenn, A. Raine and William Laufler, 'Is it Wrong to Criminalize and Punish Psychopaths?' *Emotion Review*, 3 (July 2011): 302–324.
28. John Dewey said something very similar of a problematic situation, 'It thinks' is a truer psychological statement than 'I think,' in *Human Nature and Conduct* (New York: Henry Holt, 1922), p. 287.
29. Robert Jackall popularized the concept of systemic evil in his study of ethics in corporate bureaucracies. Jackall, *Moral Mazes: The World of Corporate Managers* (New York: Oxford University Press, 1988, 2010).
30. Ibid., p. 95.
31. Kaiser Permanente Medical Group mandates empathy training for its employees to more closely align their care with the needs of the patient.
32. Elise Springer describes an ethics of interaction with a moral agent deciding what to do in communication with others who are also moral agents in Springer, *Communicating Moral Concern: An Ethics of Critical Responsiveness* (Cambridge, MA: MIT Press, 2013).

33. George Herbert Mead, *Mind, Self, and Society* (Chicago: University of Chicago Press, 1934), p. 178. That human intelligence is also socially constructed has received support in the last few decades from research in evolutionary biology on the development of the primate brain. See Louise Barrett and Peter Henzi, 'The social nature of primate cognition,' *Proceedings of the Royal Society B* (2005), pp. 1865–1875.
34. Daniel Kahnemann, *Thinking: Fast and Slow*. I will use Kahnemann's terminology for its descriptive ability, even though his subject is a solitary and static agent.
35. Paul Churchland provides a theoretical background for such prereflective categorization in Churchland, *The Engine of Reason, The Seat of the Soul: A Philosophical Journey into the Brain* (Cambridge, MA: MIT Press, 1995).
36. James T. Townsend and Jerome Busemeyer, 'Dynamic Representation of Decision-Making,' in *Mind as Motion: Exploring the Dynamics of Cognition*, ed., Robert F. Port and Timothy Van Gelder (Cambridge, MA: MIT Press, 1995).
37. M. Dehghani, D. Gentner, K. Forbus, H. Ekhtiari, and S. Sachdeva, 'Analogy and Moral Decision Making,' *Proceedings of the 2nd International Analogy Conference*, (2009) Sofia, Bulgaria.
38. The coupled oscillations described in the finger wagging experiment may lead to a technology that can detect openness between two parties. See note 12.
39. Alicia Juarrero, *Dynamics in Action: Intentional Behavior as a Complex System* (Cambridge, MA: MIT Press, 1999).
40. See Juarrero, *Dynamics in Action*, and Richard H. Thaler and Cass Sunstein, *Nudge: Improving Decisions about Health, Wealth, and Happiness* (New Haven: Yale University Press, 2008).
41. Michael Tomacello, *A Natural History of Human Thinking* (Cambridge, MA: Harvard University Press, 2014). Tomacello provides evidence that the key to human cognitive uniqueness is our ability for cooperative social interaction.

13
Moral First Aid for a Neuroscientific Age
Tibor Solymosi

The rise of neuroscience brings hope to many people because its proponents and practitioners promise to resolve – somehow, at some time – many, if not most, of the ailments with which human experience is constantly consumed. From dealing with the horrors of cognitive debilitation to reveling in the wonder of understanding the most complex entity in the known universe (the human brain), neuroscientists, neurophilosophers, and neuro-enthusiasts believe a robust theory of the brain and nervous systems will bring not only practical utility but peace of mind and peace among people. Such hopes are inextricably moral. Yet our moral development has rarely if ever kept pace with our scientific and technical development. As we venture into the still largely unknown landscape of the human nervous systems, the need for a philosophical reconstruction of key human concepts becomes ever more urgent. Given the growing neuro-hype that surrounds the deluge of neuroscientific data, especially when it comes to the neuroscientific investigation of morality, a reconstruction of ethics – one that embraces both the historical development of ethical theories and the recent scientific investigations into whether and how various ethical theories operate in the brain – provides a platform from which to reach new moral vistas. Indeed, the rate at which our knowledge about how our brains and bodies work accelerates dramatically, leaving behind any hope of having a stable and fixed ethics that would afford us the means of dealing with the new problems such scientific advancement brings. While some may find this cause for nihilism, I contend that our best science points us toward a pluralistic ethics that is flexible enough to deal with new problems while rigid enough to anchor our resolution of such problems in our historical experiences of moral exigencies and their resolutions. Such a plasticity in ethics is available in a technology I have called *moral first aid*.[1]

My aim in this chapter is to extend my conception of moral first aid (as an ethical technology)[2] beyond its inception in the thought of Daniel Dennett,[3] beyond my modification through the pragmatism of John Dewey,[4] and toward its bearing on the developing relationship between ethics and neuroscience. To accomplish this task, my argument begins with two brief overviews. First, I consider the difference between two projects of relating science and commonsense, including morality. The first project is that of reconciliation and is the current orthodoxy. It, however, falls short in ways that the other project, reconstruction, does not. Second, I review moral first aid, starting with Dennett's idea of a manual to my addition of a kit. This relation serves to combat the Cartesian conceptions – often unacknowledged – in much contemporary neuroethics. Whereas the mind/body dualism is considered throughout this chapter, I take up at length the related dualism between fact and value in the final part of this chapter. There I consider the work of contemporary neuroethicists in order to reconstruct their positions into a pragmatically coherent ethical technology.

Tradition vs. science: how reconstruction evades the dilemma of reconciliation[5]

> Human life, tradition says, is infinitely valuable, and even sacred: not to be tampered with, not to be subjected to 'unnatural' procedures, and of course not to be terminated deliberately, except (perhaps) in special cases such as capital punishment or in the waging of a just war: 'Thou shalt not kill.' Human life, science says, is a complex phenomenon admitting of countless degrees and variations, not markedly different from animal life or plant life or bacterial life in most regards, and amenable to countless varieties of extensions, redirections, divisions, and terminations. The questions of when (human) life begins and ends, and of which possible variants 'count' as (sacred) human lives in the first place are, according to science, more like the question of the area of a mountain than of its altitude above sea level: it all depends on what can only be conventional definitions of the boundary conditions. Science promises – or threatens – to replace the traditional absolutes about the conditions of human life with a host of relativistic complications and the denial of any sharp boundaries on which to hang tradition.
>
> – Daniel Dennett[6]

A crucial yet underappreciated distinction made by John Dewey is between the projects of reconciliation and of reconstruction in conceiving the relationship between science and commonsense – what Wilfrid Sellars later called the scientific and manifest images of humanity. The commonsensical view holds that humans really are as we traditionally conceive ourselves to be: conscious, free, and morally responsible selves. The scientific image, according to Sellars, holds that what the most general and lowest-level science postulates is true reality. But we need not keep ourselves to the lowest-level of science: as the Dennett epigraph above suggests, science seems to eliminate the absolutes and the sacredness we traditionally conceive our moral lives to have. These two views – the scientific and the manifest – are not (obviously) compatible. The synoptic project of Sellars is one which grants the complete truth of each image while setting them in conflict with each other. As Sellars set the problem: 'How, then, are we to evaluate the conflicting claims of the manifest image and the scientific image thus provisionally interpreted to constitute *the* true and, in principle, *complete* account of man-in-the-world?'[7] The project of reconciliation comes to this: how can we reconcile our conception of ourselves with what science tells us about how the world really is independently of human activities or interests? This question becomes more pressing when we face the growing data coming from the neuroscience of morality.

Many claim that neuroscience is revealing or will soon reveal some sort of Rosetta Stone to morality and ethics. For neurophilosopher Patricia Churchland, 'morality originates in the neurobiology of attachment and bonding' as based in 'the oxytocin-vasopressin network in mammals' and as 'extended to others beyond one's litter of juveniles,' thus providing 'a backdrop [for] learning and problem-solving [for...] managing one's social life.'[8] Neuroscientists Michael Gazzaniga and Sam Harris go even further in declaring that science, especially neuroscience, can give us a universal ethics.[9] But in all of these claims, the brain and/or nervous systems play a much larger role than our commonsensical view that has our soul or a faculty of reason as the moral actor. Such scientific characterizations of morality seem to deflate the normative grip our self-conception claims we have – and ought to have – when it comes to morality. Conviction appears to fade away.

The traditional struggle between science and morality has been construed in terms of the fact/value dichotomy. As David Hume argued, one cannot *deduce* an 'ought' from an 'is.'[10] This inability to deduce carries with it an inability to bridge the world of facts with the world of values. From this, the argument goes, no matter how much empirical science tells us about the world (that is, no matter how many facts we accumulate), there is nothing it can tell us about how the world *ought* to be, how we *ought* to act, and so forth (that is, there is no new information about values contributed). Obviously, the likes of Gazzaniga, Churchland, and Harris disagree. Their reasons for doing so, if made explicit at all, are not uniform. An exhaustive treatment of their positions goes beyond the scope of this chapter. However, a brief consideration serves to further frame the conflict with which this chapter is presently concerned.

Of the three, Churchland provides the best and most pragmatic answer: we do not deduce value from fact, but we do infer it through the inferential process of *abduction*, or the inference to the best explanation, which Churchland believes should be thought of, when it comes to morals and ethics, as the inference to the best *decision*.[11] Harris's solution to the fact/value dichotomy is naïve and misses an important opportunity. Despite his admonishing pragmatism in earlier work for purportedly being relativistic,[12] his only significant philosophical citation is to the work of Hilary Putnam, a contemporary pragmatist – hence the missed opportunity.[13] Harris's naïveté comes down to his defining science as humans' 'best effort to understand what is going on in this universe, and the boundary between it and the rest of rational thought cannot always be drawn.'[14] Science may or may not be our best effort, but evaluating it as such does not tell us what it is about science that makes it fit to produce values. I return to Harris on science and this dichotomy later in the chapter.

The fact/value dichotomy is a reiteration of the dualisms we have inherited from ancient and modern philosophy. It parallels the conflict between nature and culture, body and mind, and naturalism and humanism. These conflicts, as Dewey saw, result from a specific conception of science that requires us to reconcile our otherwise traditional beliefs about the world with the new beliefs about the world produced by science.

At root of these problems of modern empiricism is the separation of mind or experience from the world by a veil of ideas. Pragmatists like William James and John Dewey rejected this separation. For Dewey especially, the influence of Darwin effected the need for reconstructing experience as the transaction of the organism and its environment.[15] Such a reconstruction gives a vital perspective on the relationship between fact and value.

On this view of experience, learning is dynamic and active. Instead of some spectator of a mind sitting back and enjoying the show in some sort of Cartesian Theater, experience is an active process of an organism, with adequate phenotypic plasticity, that can form patterns of behavior suitable to situations with which it becomes familiar through some sort of interaction.[16] This engagement of any organism with its environment is not only a product of evolution but also a participant in that process. Since evolution deals with replicability and survivability, there is, at the very least, retroactive valuation occurring. For the vast majority of the history of evolution, this process went about without any valuators, no designers, no minds to which a failed attempt at solving a survival problem mattered. Yet through an incomprehensible number of iterations, creatures evolved that were capable of not only becoming familiar with an environment such that they could better anticipate future events but were also capable of reevaluating these anticipations of events, of states of affairs, of factual matters, such that, through their imaginations these creatures could take greater control of their environments, of their interactions with their environments – that is, take greater control of their experiences.

Eventually this process of valuation and control evolved into the critical abilities of humans, best illustrated in philosophy, science, and art. With the development of inquiry, we see that this sort of experience – this organism–environment transaction – is not only value-laden but fact-producing. When we inquire we not only learn: we also create. We create new ways of being and doing, new tools and solutions for dealing with problematic situations. What we learn or discover are which of these creations work best for resolving problematic situations. Dewey often used the development of farming to illustrate the production of both norms and facts. For instance, it is a fact that a strategy of adequate water, sunlight, and shade is more effective for growing tomatoes than a strategy that involves gasoline, and excessive sunlight or shade: one strategy is *better* than another.

This entanglement of fact and value is also expressed neurally, as Harris points out.[17] The neural mechanisms responsible for fact-knowing and value-making are one and the same. That is, whenever we are critically assessing a situation, the same neural mechanisms are at work, regardless of whether we are considering facts or values. While Harris provides a wonderful token of continuity regarding this entanglement of fact and value, which Dewey would surely appreciate, another neuroscientist and neuroethicist, Eric Racine,[18] quotes Dewey directly from *Human Nature and Conduct*. Writes Dewey: 'Since morals is concerned with conduct, it grows out of specific empirical facts. Almost all influential moral theories, with the exception of the utilitarian, have refused to admit this idea.'[19]

This neglect of fact in value inquiry promotes the need for reconstruction. In a later section, I return to Harris's and Racine's treatments of the fact/value relation; for now, I elaborate further on the Deweyan view of reconstruction and experience as organism–environment transaction.

Dewey took issue with the conception of science at work in this project of reconciliation. For Dewey, the premise that science represents reality independently of human activity is what creates the conflict between science and commonsense in the first place.[20] Science, on Dewey's account, is the deliberate intervention of humans with other parts of nature with the aim of solving problems.[21] Indeed, since Dewey's conception of scientific activity is one that not only entails human activity and human interests but also technical artifacts, like beakers and theories, in scientific activity, we are better off rejecting, as Larry Hickman argues, the distinction between science, as pure, and technology, as merely applied.[22] Instead Hickman recommends we speak of *technoscience* to denote this deliberate and systematic intervention with nature. Following the lead of Dewey, Hickman states 'that what we now call science is in fact a type or branch of technology since it involves *the invention, development, and cognitive deployment of tools and other artifacts, brought to bear on raw materials and intermediate stock parts, to resolve perceived problems.*'[23]

Since technoscience is a deliberate and productive phase within human experience, it does not claim to divulge how the world or nature is independently of human experience. Rather technoscience is continuous with the rest of human experience, with commonsense. To be clear, commonsense, on this account, entails the traditional beliefs about the everyday objects and events of our everyday lives. Technoscience grows out of commonsense, for the initial problems in need of solving by inquiry are quotidian. For Dewey, the relationship between commonsense and technoscience is not necessarily oppositional. Ideally, these two are complementary because the results of technoscience *should* feedback into commonsense, improving both the lived experience of humans and helping to direct further technoscientific activity.

This is not to say, of course, that there is not a felt difficulty when it comes to this relationship between commonsense and technoscience. Those interested in reconciling science with commonsense may get their project off on the wrong foot, but their sense that there is a problem is not entirely unwarranted. The conflict that we see between science and commonsense is the result of the human neglect of intelligently incorporating the products of technoscience into our commonsensical self-conception.

In light of Darwin, Dewey rejected the standard dualisms of modern philosophy that have led to so many irreconcilable conflicts. One such divorce was between experience and nature, or the mental and physical. Since Darwin gave us a fresh perspective and set of tools with which to confront problems like the relation of mind and body, Dewey rejected the spectator theory of consciousness that was at the heart of the modern Cartesian conception of experience.[24] Dewey's evolutionary alternative was to conceive experience as the ongoing transaction between organism and environment. The veil of ideas that separated human mentation from the physical world was dismissed, and with it the belief that science was able to get behind the appearances. Nevertheless, with the evolution of human culture and technoscience, this ongoing organism–environment transaction produced beliefs, habits, and tools that have not always or necessarily afforded the specific productive organisms with the best means for achieving their ideals in all situations.

We see this conflict today when neuroscientists tell us that we lack freedom of the will because contemporary investigations show that there is no will in the brain, or that there is neural activity in the prefrontal cortex prior to the agent's knowing that he or she is about to do something.[25] These conclusions suffer from an impoverished philosophical context. Not only are such proclamations philosophically loaded, they illustrate the lack of respect for philosophy by the larger community.

Many have recently been lamenting over this failure of philosophy, such as Philip Kitcher, who draws from the pragmatism of William James and John Dewey.[26] My point in mentioning this is to draw attention to what Dewey saw as one of the main jobs for philosophers: being liaison officers between the special disciplines and to the general public so that better coordinated inquiries can proceed, especially as these inquiries are guided by a larger shared ideal by the community.[27] This coordinated effort is at the heart of reconstruction. Instead of seeing the products of science as eliminating or threatening our moral ideals, Dewey argued that the products of technoscience are the means for achieving those ideals. Instead of science as representing nature as it is independently of humans, humans can understand through technoscientific activity how nature *works*.[28] In understanding the operations of nature, we are able to guide our further behavior in ways more amenable to amelioration of our perceived problems and to the consummation of our ideals. Such a reconstructive attitude is all the more pressing when it comes to morality and ethics.

Moral first aid

Both the facts of the situation and the ideals and values we therein hold are tools for action. How we are able to employ such tools in the midst of a moral exigency is a central, if not the main, concern for understanding moral conflict. Given his evolutionary naturalism, Dennett presents his conception of morality as an outgrowth of our primate ancestry, from which we get the social need to get along with one another. Dennett calls his conception *Moral First Aid*.[29] Of the evolutionarily based ethics available, Dennett's is attractive for at least two reasons. First, there is his critique of utilitarianism and deontology as impractical, given the extensive time it takes to put into action whatever the maxim of utility or the categorical imperative dictate. Dennett offers as an alternative a pluralistic ethics that focuses on character development at times when a person is not in the midst of a moral exigency. The second reason is the name itself: moral first aid. It intimates medicine or health, which is useful for thinking about moral first aid – a point to which I return later. These two reasons emphasize the nature of moral experience. It is not detached from everyday life. Like the medical counterpart to which it alludes, moral first aid resonates with the idea that something needs to be done *now* and done *effectively*. The name also lends itself to the aforementioned crisis in philosophy: the failure of philosophers to be adequate liaison officers is a moral exigency in need of immediate first aid.

Dennett develops moral first aid by appealing to John Stuart Mill's metaphor of the nautical almanac as the means for navigating the seas without having to do much of the calculation on site. That is, most of the work is done ahead of time. In the same way, Dennett's *Moral First Aid Manual* is written and studied when the person is not in the midst of a moral difficulty. The lessons learned from the reading of a *Manual* are constitutive of the person's moral character, enabling that person to resolve the moral conflict without having to do the calculations of utility or the reasoning of the categorical imperative. Indeed, despite Dennett's disdain for absolute rules – after all, it is worth noting, as Owen Flanagan has argued,[30] that the human nervous system is ill-equipped to follow rules yet very well-equipped to recognize patterns – Dennett finds rules of thumb to be useful contents for the *Manual*. Moreover, given the plurality of contexts and cultures in which moral activity takes place, Dennett imagines there being several versions of moral first aid manuals available.

Some of the rules of thumb that Dennett has in mind include the following:[31] 'But that would do more harm than good.' 'But that would be murder.' 'But that would be to break a promise.' 'But that would be to use someone merely as a means.' 'But that would violate a person's *right*.' Elsewhere, Dennett offers another possible rule of thumb: 'It is better to think of the human capacity to rethink one's *summum bonum* as the possibility of extending the domain of the self.'[32] And in a rather reconstructive moment, Dennett elaborates on this rule of thumb when he writes that 'What these people ["spiritualists"] have realized is one of the best secrets of life: let your *self* go.'[33]

What is unclear about the technology of moral first aid, however, is how the manuals are produced. What rules of thumb belong in which manual, for instance? Given Dennett's concern with real-time constraints in moral experience and the inability of utilitarianism and deontology to meet them, the production of the manuals occurs when the authors are not in the midst of a moral exigency. Presumably, these authors are constructing rules of thumb in light of the history of ethics (so conceived as the technoscience that produces morality, that is, the system of norms to guide behavior). Indeed, versions of the Golden Rule, and other deontological guides are among what Dennett offers as rules of thumb. The same can be said about rules produced from the application of the maxim of utility. Short of simply granting some validity to these ethical theories without accounting for which is correct or superior, or, perhaps more importantly, without accounting for why there are these two factors in our moral experience, it is unclear what the production process is for the manuals, and why some manuals are better than others for reaching a moral end-in-view.

Dennett, years after he discusses *Moral First Aid Manuals* in *Darwin's Dangerous Idea*, takes another stab at the nature of ethical inquiry. In *Freedom Evolves*, he worries about whether our inquiries discover or construct something about the world.[34] He suggests that our conception of the Good is a product of ongoing inquiries into the Good akin to the progress we have made into inquiring into the Straight. That is, through iterations of generating and testing both articulations of the Straight and tools for measuring, utilizing, or producing straight things, we have become very accurate in our measurements such that the degree of error is very slight. While our progress in inquiring into the Good may not be as easily quantifiable as the Straight, Dennett believes that there is an affinity between these two inquiries.

The trouble, however, is that in order for this analogy to work, we would need at least a working definition of the Good akin to that of the Straight. From Plato to the present day, we have not been so fortunate to have a definition of the Good that is as straightforward as the

Straight's being the shortest distance between two points. Dennett's manuals, without some appeal to an a priori principle, some may argue, are without any grounding whatsoever. Since Dennett's approach is emblematic of most naturalists' approaches to the Good, it appears that any naturalistic account of moral life and ethics is doomed from the start. Any non-supernaturalist account of moral life that goes beyond the descriptive and into the normative cannot rely on non-natural sources of moral knowledge, such as a faculty of Reason or a mystical intuition. Ethics and its product, morality, must be reconstructed in an experimental fashion. Dennett's pluralism takes steps in this direction, but his ethics falls short without an account of inquiry.

To return to Dewey's conception of experience and inquiry, it is worth noting that Dennett's moral tool is in need of ethical technology. Indeed, we are beginning to get a detailed story about the origins of morality from the sciences of life and mind.[35] Importantly, this story is one that coincides with much of Dewey's own insights on the origins of morality and moral conflict, as William Casebeer has argued.[36] Dewey's experimental inquiry into the nature of morality is based in concrete experience – the organic-environmental transaction of the human being in a social context – and reconstructs the main competing ethical theories as tools produced through experience for resolving different sorts of human problems – indeed, as Hickman would remind us, these theories and their products become *stock parts for the resolution of perceived human problems*. Because these tools have different origins, Dewey thought, they cannot be reduced to one another. In fact, if there were one supreme or universal rule or law or principle to guide moral life, it is not clear why there would be any genuinely moral conflict within our experience in the first place.

Dewey's reconstruction of the history of moral development is not only rooted in concrete experience but is illustrative of the origin of moral conflict as one in which goods, rights, and virtues are in competition for guiding control over action. On this account, what is good is what effects some end, like happiness, pleasure, or self-realization – or simply the Good. This is simply a fact of life: all organisms have impulses and desires toward certain activities and goods. However, what distinguishes these natural goods from moral goods is the human ability to exercise foresight and compare. We recognize that certain goods are good *for* some specific end, such as flourishing. In recognizing that, we set out to pursue those goods for ourselves. Dewey saw the ancient Greeks as the strongest historical articulation of this teleological morality.

Of course, there is much conflict in sorting out competing goods. But there is another variety of experience that yields a different sort of competition. The concrete experience in which it is based is the fact that people demand something from those with whom they live. Whether it is help or nourishment (as children demand from parents) or obedience and conformity (as parents or rulers demand from children or subjects), demands on humans from other humans are often sources of conflict. These demands grow into the emphasis, found in Roman philosophy, on social order and harmony. Obviously, one may try to reduce these demands to teleological necessity. For instance, if I am to flourish, this reduction would go, I would need you to do this and that. But this sort of demand is an exercise of power of one individual over another. Dewey argued that for demands to be something other than an exercise of power, they must carry a different authority from the specific needs of a specific individual in a specific situation. In other words, to live in a society is not only to live among others, whose rights are equal, and where there are responsibilities for action on the behalf of individuals to maintain the rights of everyone; to live in a society is also to live in recognition and acceptance of this moral authority to strive toward the social order and harmony that goes beyond any rational self-interest of an individual.

To be clear, the difference between a morality of ends and a morality of laws comes to this, in Dewey's words: 'There is a ... difference ... between objects which present themselves as satisfactory to desire and hence good, and objects which come to one as making demands upon his conduct which should be recognized.'[37] As key as these two ethical theories are for the reconstruction offered here of moral experience, there is a third factor Dewey recognized. The first two factors have to do with determining what is good or right conduct to take; the third deals with evaluating the action once taken. This third factor concerns itself with questions such as, Is the individual's conduct praiseworthy? Does it deserve reward? Do we encourage others to do likewise? Affirmative answers to such questions produce the virtues; negative responses, vices.

Assessment of activities and objects is so natural in experience as to be reflexive – instinctive even.[38] There is no deliberate calculation with regard to commendation or condemnation, especially with regard to whether it is a matter of the Good or the Right. Only through experience does a person come to consider how others will respond to one's conduct, regardless of whether that conduct has to do with the Good or the Right.

Even though these three factors are independent, they are nevertheless entangled in our moral lives. It is this entanglement that brings about moral exigencies in the first place. Were only one of these three supreme, then ethics would be clear cut: do what is good, not evil, or do what is right, not wrong, or do what is virtuous, not vicious. But our moral experience is rarely if ever so clear. The uncertainty and confusion result from conflict across these factors. What is good is not always right or virtuous; what is right is sometimes undesired, if not vicious; what is virtuous, not always good and right.

To return to the reconstruction of experience and to Dewey's technological pattern of inquiry, we can now understand these three factors – goods, rights, and virtues – as products of experience that are now also sources or tools for further inquiry within the transaction of humans with each other. In short, these three factors are the stock parts of the technoscience of ethics, which produces moral first aid manuals and kits. This *origins* story of morality sheds light on the claims from Gazzaniga, Harris, and Churchland with which I opened, and sheds further light on Dennett's moral first aid.

The claim, made by both Harris and Gazzaniga, that neuroscience is shedding light on a universal ethics is too ambitious. Surely, our understanding of the nervous system is enlightening our understanding of morality as much as it raises new ethical questions. However, an all-encompassing ethical theory does a disservice to our moral experience, if it tries to pin down the one and true morality to one factor over the other two. For much of our moral conflict results from an unresolved tension between goods, rights, and virtues. Churchland's claim that morality originates in social experience as the oxytocin-vasopressin network in mammals modulates it, specifically as it becomes a means of problem-solving activity is an improvement on Gazzaniga's temptation to universalize. But her position also runs the similar risk of Gazzaniga and of Harris: that morality has a single origin in primate's sociality or in the mammalian brain. Dewey argued – which Kitcher has elaborated recently[39] – that morality has several origins within the history of human social transactions.

This returns me to Dennett's moral first aid. The clues we have for the production process of the *Moral First Aid Manuals* attend to some sort of utility in deontology and something right about utilitarianism. What it does not account for is how well-executed both the production of the manuals and the characters effected by the manuals are in navigating moral exigencies. In other words, Dennett only accounts for two of the three of Dewey's independent factors in morality. Since they are independent yet entangled in moral inquiry – through conflict and resolution – a better analogy is necessary for moral first aid than Mill's nautical almanac. I have already suggested that medical first aid is what comes to mind when I hear 'moral first aid,' to which I now turn.

For every medical first aid manual there is a medical first aid kit. The manual is especially helpful for those unskilled in using the first aid kit. When more experience is gained with the kit in medical exigencies, the less the need there is for the manual. The parallel with moral first aid should be clear. The manuals provide instruction on how to use the various tools of goods, rights, and virtues in a variety of problematic situations. Of course, before productive studying can go forth, the student of moral first aid, just as the student of medical first aid, must be made interested through the proper control of the conditions of the learning environment. In short, teachers of morality and of medicine must be able to organize the tools, resources, and environments in such a way that enforces growth in the student. In medicine, the working standard that has developed to measure growth is a conception of *health*. This concept is always being revised in light of new experiences and has a clear historical development (which medical historians help clarify further). When a problem is resolved in medicine, its resolution is attained through a measure of health, especially in this sense: is the patient healthier than he or she was before medical intervention?

This self-referring measurement is an instance of what Dewey called *growth*.[40] It is a matter of whether an organism is adjusting itself and/or its environment to the situation at hand. Ongoing and successful adjustments yield less stress and greater viability. The application of this conception of growth to *health* is relatively straightforward. What remains in my drawing this parallel between medical and moral first aid is to offer a guiding ideal for Dennett's moral first aid. My suggestion is Dewey's conception of *democracy* as a way of life.

Democracy, for Dewey, is much more than the periodic visit to the polls.[41] Briefly, democracy is a coordinative organization of social life in which individuals are cultivated, and the lived experience is enriching. Here is Dewey:

> ...democracy is belief in the ability of human experience to generate the aims and methods by which further experience will grow in ordered richness. Every other form of moral and social faith rests upon the idea that experience must be subjected at some point or other to some form of external control; to some 'authority' alleged to exist outside the processes of experience. Democracy is the faith that the process of experience is more important than any special result attained, so that special results achieved are of ultimate value only

as they are used to enrich and order the ongoing process. Since the process of experience is capable of being educative, faith in democracy is all one with faith in experience and education. All ends and values that are cut off from the ongoing process become arrests, fixations. They strive to fixate what has been gained instead of using it to open the road and point the way to new and better experiences.[42]

Instead of looking toward the Good without any clear working definition – though emphatically not a conception of Good that is an 'authority' alleged to exist outside the process of organism–environment transaction – Dennett's moral first aid may now evade ambiguity by taking up a deliberately experimental stance. Where medicine is concerned with whether the patient is improving, ethics is concerned with whether society is improving in its ability to edify through technoscientific inquiry into our cherished ideals. Dewey's three independent factors in morality are useful tools in the moral first aid *kit*, and the manuals provide rules of thumb on the possibilities of standard operation. Nevertheless, as any experienced moral person can attest, often times resolving a moral exigency requires more creativity and imagination than strict adherence to any set of rules. Such creativity and imagination are integral to both the experimentalism of Dewey's pragmatism and the ideal of democracy. Yet these traits are not the sorts of things found in a manual. Rather they are the sorts of things one illustrates through acting with what one has on hand.

Dennett uses products of deontology and utilitarianism as rules of thumb within the *Moral First Aid Manual*. I agree with Dennett that these rules of thumbs are not for consultation during a moral exigency but for consultation as preparation for moral exigencies. The preparation is for acting with these theoretical tools as if they were second nature; a person, in the midst of a moral exigency, acts morally, without hesitation, because that person sculpted a moral character with these theoretical tools presented in the manuals. In addition to Dennett's use of the tools of goods and rights, Dewey has provided a third tool, the virtues, and a moral ideal, democracy, that orients the general use of moral first aid in the myriad and particular situations in which moral problems arise.

Democracy as a way of life that goes beyond the structure of government is the moral ideal by which moral first aid operates. Not only does it serve to account for the different types of moral situations in a nonreductive fashion, it also promises a way of reaching rapprochement between scientific naturalism and commonsense humanism through its pragmatic experimentalism.

Within moral first aid, a kit is unique to a moral individual. The idiosyncrasies of a person's life, history, and situation afford opportunities for using the tools of the moral first aid kit to resolve moral exigencies. Given the complexity of a moral exigency and the plurality of diverse persons who could be in such a situation, we should expect different solutions to an exigency – just as we would expect there to be more than one way to treat a health ailment or there to be more than one dish to be prepared from the same set of ingredients. The freedom and diversity that moral first aid affords and permits are characteristic of the democratic orientation Dewey advocated. A democratic culture and a democratic person advocate experimentation in living well. There may not be one true way to live, but that does not imply that there cannot be many ways for humans to live morally. In encouraging both diversity and experimentation, the practice of moral first aid allows for both creativity in moral life and the subsequent refinement of moral lives, just as innovations in medicine take time to become refined and ubiquitous due to their ability to address human needs in a superior fashion to the alternatives.

The brain and the moral landscape

> To see the organism *in* nature, the nervous system in the organism, the brain in the nervous system, the cortex in the brain is the answer to the problems which haunt philosophy. And when thus seen they will be seen to be *in*, not as marbles are in a box but as events are in history, in a moving, growing never finished process.
>
> – John Dewey[43]

> In short the problem of reconstruction in philosophy, from whatever angle it is approached, turns out to have its inception in the endeavor to discover how the new movements in science and in the industrial and political human conditions which have issued from it, that are as yet only inchoate and confused, shall be carried to completion. For a fulfillment which is consonant with their own, their proper direction and momentum of movement can be achieved only in terms of ends and standards so distinctively human as to constitute a new moral order.
>
> – John Dewey[44]

The promise of neuroscience is not simply the resolution of the historical problems of philosophy. It is much more: Dewey is effectively calling the defeat of Cartesianism by experimentalist means the greatest moral concern of our age.

The meeting of neuroscience and morality has generated a vast amount of literature in recent years. I have already briefly discussed Churchland and Gazzaniga. Here I focus on two others also briefly mentioned earlier. The first is Eric Racine, whose book, *Pragmatic Neuroethics: Improving Treatment and Understanding of the Mind-Brain*, explicitly recognizes Dewey as an important predecessor to what would become bioethics and neuroethics. While Racine's discussion on Dewey's ethical theory is limited, Racine does note several themes central to pragmatism, like the rejection of the fact/value dichotomy and the emphasis on solving particular problems. The second neuroscientist is Sam Harris. His book, *The Moral Landscape: How Science Can Determine Human Values* likewise has similar pragmatist themes as Racine's book but without explicitly recognizing classical pragmatists like Dewey.

These two recent books are concerned with the ethical consequences of science, especially the sciences of mind and brain, by either focusing on specific ethical problems or by imagining a way of life on a very general level. For both of these writers, there are pragmatist themes, like the rejection of the dichotomy between fact and value. This denial is set in some sort of naturalistic framework that for one of these authors is seen as pragmatic, following Dewey, while the other is clearly, if implicitly and partially, influenced by pragmatism.

Regardless of whether the influence of pragmatism is explicitly acknowledged or not, both positions could benefit immensely from an explicitly pragmatist conception of technoscientific inquiry, which is at work in the technology of moral first aid. My aim here is to bridge the particular view of Eric Racine's pragmatic neuroethics with the general view of Sam Harris as a further elaboration of moral first aid. To do this, I first briefly discuss the entanglement of fact and value (to borrow Putnam's felicitous phrase) before critically turning to Racine's sketch of strong vs. moderate naturalism in which I set Harris's extremely broad conception of science as a producer of human values. Through this relating of Racine's and Harris's positions, I highlight where their conception of science is problematically narrow or problematically broad. Having bridged these positions, I relate Harris's notion of the moral landscape with the practice of moral first aid.

Racine distinguishes between two types of naturalism by contrasting their respective epistemological commitments. I take these commitments in turn, critically reviewing both the strong naturalist and moderate naturalist positions as Racine sees them. As he aligns himself with moderate naturalism, which he also refers to, unfortunately, as pragmatic naturalism, his conception of the relation between fact and value demands consideration, especially in contrast to strong naturalism. I aim to collapse Racine's distinction as it is based on a faulty conception of experience at odds with a pragmatic reconstruction of experience.

Regarding the first epistemological commitment, the strong naturalist sees 'no distinction between "is" and "ought."'[45] The moderate naturalist grants the distinction with qualifications.[46] This presentation of the relationship between fact and value or 'is' and 'ought' is disingenuously simplistic and threatens to reintroduce the dualisms that pragmatists like Dewey worked to eliminate. The two distinct epistemological commitments Racine articulates for both strong and moderate naturalism collapses once we recognize that talk of distinctions between fact and value or 'is' and 'ought' is an extremely complex affair that is determined less by syntax and semantics than by the pragmatic considerations of the particular situation.[47] If facts are values, and values facts, then whether we consider a proposition as an imperative depends on a host of other factors germane to the particular situation. Such factors include other background beliefs and the specifics around the problem in need of resolution.

The second epistemological commitment of each naturalism sets ethical predicates in some relation with natural properties. The strong naturalist reduces the ethical to the natural, which is never clearly defined by Racine but is taken to mean nonhuman or noncultural, or, as many recent naturalists use the term, to convey lowest-level physical properties. Racine's moderate naturalism holds that '[e]thical predicates are properties that cannot be reduced to natural properties but are understood within a fact-value continuum.'[48] Given this statement of continuity, one surmises that at one end of the spectrum is fact but no value, and at the other end value but no fact. When pragmatists talk about continuity and the entanglement of fact and value, it is not in this way.[49] Rather, there is valuation going on all the way down. Valence electrons illustrate this entanglement to a degree far simpler than, for instance, the emotional valence of humans. The continuity is between the cultural and the natural, not value and fact, which are entangled along the whole of that continuity.[50]

The third epistemological commitment regards the nature of ethical knowledge. The strong naturalist sees '[e]thical knowledge...[as] an outgrowth of empirical knowledge.'[51,52] The moderate naturalist sees empirical knowledge as constraining the ethical capacities of humans. That is, empirical knowledge does not justify ethical norms. Racine is not clear how these two positions conflict. Surely each emphasizes different aspects of ethical knowledge. Moreover, as Racine's quotation from Dewey states, morals does indeed grow out of experience. What Racine seems to fail to understand is that growth has its own constraints. These constraints account for human capacities: actual growth in one direction (as opposed to once-potential growth in another) both opens and closes possibilities for action.

When it comes to the nature of ethical principles, the strong naturalist sees ethical norms as natural laws, whereas the moderate naturalist denies there being any ethical laws. Rather ethical norms apply strictly to human social life. Again, the distinction here does not withstand analysis. What counts as a natural law is left unclear, but given Racine's desire to resist both dualism and eliminative reductionism, it is safe to assume that there are principles that are to be discovered and mathematically expressed. Such principles hold universally and ground all subsequent knowledge claims. But this view of what science produces does not account for much of what scientists actually do. The developments in the life and mind sciences over the last century stand on their own and resist any clear reduction to physics. Besides, the terminology of *laws* is an atavistic hangover. At best, talk of laws is talk of regularities that permit humans to make reliable predictions in problem solving. The real business of science is the production of theories that connect, make use of, and open new possibilities within the regularities of nature. Racine does not make clear why moral behavior should be exempt from such treatment. If all he means to say is that the empirical knowledge of physics is insufficient for making ethical judgments, then he is setting up a straw man. Only when decisions are required about the role of physical research in a larger context will that specific empirical knowledge be useful, although only when it is combined with other bits of empirical knowledge, from experiences in politics, law, and so forth. But to consider such 'non-natural' cultural activities as empirical seems to be too much a stretch for Racine's rather limited conception of experience.

The last issue I consider here is the epistemological commitments over the source of ethical principles: are they a priori or a posteriori? For Racine's strong naturalist, 'ethical norms stem a posteriori and from experience and observation.'[53] The moderate naturalist holds that 'ethical norms do not simply follow from reason or experience but from their interaction.'[54] Here the dualism that pragmatists like Dewey wish to eliminate is clearest. By no means does the strong naturalist claim that the stemming or growth of norms from experience happens in a simple, straightforward fashion. But this is the least egregious of Racine's errors here. In the contrasting claim, the moderate naturalist – the position Racine himself advocates – makes a distinction between reason and experience: just the distinction Hume makes that leads to the problems of skepticism, including the dichotomy of fact and value.

As I discussed, the evolutionary reconstruction of experience as the transaction of organism and environment evades many of the difficulties that Racine's sketch faces. Culture is continuous with nature. The sort of distinctions that drive Racine to see ethics as specific to cultural activity and divorced from natural or biological or physical activity is at odds with Dewey's principle of continuity as well as Racine's larger concern for carving out a niche for neuroethics. Furthermore, the call for an interaction between reason and experience not only fails to appreciate both the role of continuity and the reconstruction of experience, it fails to see the crucial insight of this reconstruction: that intelligence or cognitive activity is a phase of experience that grows out of non-cognitive, 'non-rational' experience. Indeed the sort of experience that Racine values in quoting Dewey is just the sort of interaction Dewey called culture. The critical or rational inquiry that Racine sees as 'reason,' which must interact with 'experience,' is already engaged with experience as reasoning is a specific way of experiencing.

Sam Harris no doubt sees science as just this sort of reasonable engagement of reason and experience, especially since he argues that it can produce human values as well as facts. Like Racine, he gives a standard rejection of the fact/value dichotomy. He emphasizes especially the role of epistemic values in scientific inquiry, as well as the value-ladenness of experience. Unlike Racine, however, Harris gives a conception of science that is not muddled by various naturalisms. Rather he gives a very broad definition of science in his defense of the claim that there are indeed moral truths. Here are his own words:

> Some people maintain this view [that there is no moral truth] by defining 'science' in exceedingly narrow terms, as though it were synonymous with mathematical modeling or immediate access to experimental data. However, this is to mistake science for a few of its tools. Science simply represents our best effort to understand what is going on in this universe, and the boundary between it and the rest of rational thought cannot always be drawn. There are many tools one must get in hand to think scientifically – ideas about cause and effect, respect for evidence and logical coherence, a dash of curiosity and intellectual honesty, the inclination to make falsifiable predictions,

etc. – and these must be put to use long before one starts worrying about mathematical models or specific data.⁵⁵

It seems that for Harris, science is simply the label we give to our best thinking or understanding, which we evaluate by the tools or values he lists. This rather broad view of science may strike many as too broad. For classical pragmatists, it is mistaking the whole of inquiry for a very specific and powerful sort. Harris seems sensitive to this issue of boundary drawing when he again states the definition of science in the context of its justification. He writes 'Science is defined with reference to the goal of understanding the processes at work in the universe. Can we justify this goal scientifically? Of course not.'⁵⁶

At first glance, it may seem that Harris is advocating what Racine calls strong naturalism. Science gives us both facts and values. Well-being and health are empirical for Harris. Moreover, if we understand Racine's implicit conception of science as one in which the conditional 'If there is scientific truth, then there is scientific or natural law' holds, then Harris's assertion that there is moral truth implies moral natural laws. On such a view, Harris is offering a very strong form of naturalism. Another reading, however, of both Harris's conception of science and Racine's is that Harris's broad view is problematic for the otherwise constrained view of Racine. In other words, Harris's broadening of science passes over the distinctions Racine needs to maintain his landscape of naturalisms. But in so broadening what science is, Harris passes over what makes science the evolutionary and cultural achievement that it is, while opening himself to accusations of scientism that hinder progress more than anything else.

Instead of conceiving of science as our best efforts at thinking about or understanding the world, I have argued we take up the pragmatist conception of science *qua* technology. Science is a highly sophisticated and highly organized social activity, but it shares general traits with other forms of inquiry. Here is a slightly modified version of Hickman's conception of technoscience: it is 'the invention, development, and cognitive deployment of tools and other artifacts (such as rules of inference), brought to bear on raw materials (such as data) and intermediate stock parts (such as the results of previous inquiries), to resolve and reconstruct situations which are perceived as problematic.'⁵⁷ This conception of inquiry leaves room for a wide spectrum of endeavors, from artistic inquiry like writing a novel to medical inquiry like investigating neurodegenerative disease. I believe this accounts for both Harris's general conception of science as well as Racine's efforts to restrain scientific and ethical inquiry to particular human problems, and that this does so through the pragmatic reconstruction of experience I introduced earlier.

Furthermore, this transactional conception of experience fits well with what Harris has called *the moral landscape*. Harris explains:

> ...I make reference to a hypothetical space that I call 'the moral landscape' – a space of real and potential outcomes whose peaks correspond to the heights of potential well-being and whose valleys represent the deepest possible suffering. Different ways of thinking and behaving – different cultural practices, ethical codes, modes of government, etc. – will translate into movements across this landscape, and, therefore, into different degrees of human flourishing. I am not suggesting that we will necessarily discover one right answer to every moral question or a single best way for human beings to live. Some questions may admit of many answers, each more or less equivalent. However, the existence of multiple peaks on the moral landscape does not make them any less real or worthy of discovery. Nor would it make the difference between being on a peak and being stuck deep in a valley any less clear or consequential.[58]

In reconstructing experience as the transaction of organism and environment, and in conceiving of science in Hickman's technological sense, ethical inquiry – *moral first aid* – is a matter of engineering new tools of climbing or navigating the moral landscape. It is also a matter of *engineering new ways of altering the landscape itself*[59] through our technical interactions with it. Depending on the platforms or cranes on which we stand,[60] we may find new insights into particular ethical problems, such as those surrounding the disorders of consciousness with which Racine is concerned.[61] Yet these vistas only become available through the actions we take in critically reflecting on the sorts of experiences we have had, are capable of having, and desire to have. Such efforts of the imagination are central to the larger reconstruction of experience, of the relation between mind/brain, that Dewey saw as indispensable, not only to solve the problems that have long haunted philosophy but to address the everyday problems of everyday people, which primarily have to do with the question of how to live well.

Conclusion: moral first aid in the age of neuroscience

The ethical technology of moral first aid is well suited for navigating the ever-changing moral landscape. As our technoscience grows, so must our moral capacity. While we have solved the moral exigencies of the past, such solutions may be worth revisiting with the new tools provided by the sciences of life and mind. Advances in our understanding of how people learn skills, for instance, come from increasing study of mirror neuron systems.[62] Such advancement promises not only new means of improving moral education but also new means of dealing with those who have committed some wrongdoing. The more we learn about how our moral cognition works in the brain, through the body, and into the world – and back again in a ceaseless looping – the greater the need for reconstruction in philosophy. For the problems we will face are not always clearly anticipated, or easily or promptly resolved. As we delve further into the cranium, the greater the consequences are for life beyond it. Our best hope for dealing democratically with this new landscape includes our cultivating a facility with the ethical technology of moral first aid.

Notes

1. Tibor Solymosi, 'Three Tools for Moral First Aid,' *Essays in the Philosophy of Humanism*, 20(2) (2012): 61–77.
2. By 'ethical technology,' I do not primarily mean a technology that is ethical (though I hope this technology of moral first aid is ethical itself). Rather, I mean that moral first aid is a technology for doing ethics. Ethics, as I use the term here, is the inquiry into morality. Morality, subsequently, is the established but modifiable set of norms, goods, rights, duties, and virtues for living well among others.
3. Daniel C. Dennett, 'The Moral First Aid Manual,' in *Tanner Lectures on Human Values*, Vol. VIII, ed. Sterling M. McMurrin (Salt Lake City: University of Utah Press, 1988), pp. 120–147. And Dennett, *Darwin's Dangerous Idea: Evolution and the Meanings of Life* (New York: Simon & Schuster, 1995/1996).
4. Solymosi, 'Three Tools for Moral First Aid.'
5. This section and the next are significantly modified (if not extended) versions of sections from Solymosi, 'Three Tools for Moral First Aid.'
6. Dennett, 'How to Protect Human Dignity from Science,' in *Human Dignity and Bioethics: Essays Commissioned by the President's Council on Bioethics*, 2008. Available at http://bioethicsprint.bioethics.gov/reports/human_dignity/chapter3.html. Last accessed May 21, 2008. See also Dennett's 'Manifest Image and Scientific Image' in *Intuition Pumps and Other Tools for Thinking* (New York: W.W. Norton, 2013), pp. 69–72; and his 'Aching Voids and Making Voids: A Review of *Incomplete Nature: How Mind Emerged from Matter* by Terrence W. Deacon,' *The Quarterly Review of Biology* 88(4) (2013): 321–324.
7. Wilfrid Sellars, *Science, Perception and Reality* (Atascadero, CA: Ridgeview Publishing Company, 1963), p. 25.
8. Patricia S. Churchland, *Braintrust: What Neuroscience Tells Us About Morality* (Princeton, NJ: Princeton University Press, 2011), p. 71.

9. Michael S. Gazzaniga, *The Ethical Brain* (New York: The Dana Press, 2005), pp. 163–178; and Sam Harris, *The Moral Landscape: How Science Can Determine Human Values* (New York: Free Press).
10. The British philosopher G. E. Moore also argued against a naturalistic fallacy. He thought that any attempt to define *good* in natural terms or properties was a failed enterprise. The good, instead, was some sort of non-natural property intuited by the mind. Much ink has been spilled on this issue, which to the eyes of an evolutionary pragmatic naturalist is much ado about nothing (cf. Philip Kitcher, *The Ethical Project* (Cambridge, MA: Harvard University Press, 2011)). For the notion of any intrinsicality, such as the intrinsic value of the good that Moore was after, seems dubious after Darwin. In other words, words and language are tools that have evolved and have been constructed for ameliorative action in problematic situations.
11. See Patricia S. Churchland, 'Inference to the Best Decision,' in *The Oxford Handbook of Philosophy and Neuroscience*, ed. John Bickle (New York: Oxford University Press, 2009), pp. 419–430. I am not concerned with the exact nature of abduction. But it is worth noting that Charles Sanders Peirce's conception of abduction (Peirce introduced the notion) is distinct from the derivative inference to the best explanation that is prominent in philosophy of science today. For more on Peirce and abduction, see Mark Tschaepe, 'Gradations of Guessing: Preliminary Sketches and Suggestions,' *Contemporary Pragmatism*, 10(2) (2013): 135–154; 'The Creative Moment of Scientific Apprehension: Understanding the Consummation of Scientific Explanation through Dewey and Peirce,' *European Journal of Pragmatism and American Philosophy* 5(1) (2013): 32–41; and 'Guessing and Abduction,' *Transactions of the Charles S. Peirce Society* 50(1) (2014), pp. 115–138.
12. See Harris, *The End of Faith: Religion, Terror, and the Future of Reason* (New York: W.W. Norton & Company, 2005), pp. 279–283, n. 23. To be clear, Harris's attack is on the neopragmatism of Donald Davidson and Richard Rorty. Davidson never liked the label of pragmatist, and Rorty's readings of James and Dewey are notorious for his neglect of experimentalism.
13. See Harris, *The Moral Landscape*, p. 202, n. 16.
14. Harris, *The Moral Landscape*, p. 29.
15. See Solymosi, and John Shook, 'Neuropragmatism and the Culture of Inquiry: Moving Beyond Creeping Cartesianism,' *Intellectica* 60(2) (2013): 137–159; and Solymosi, 'Neuropragmatism on the Origins of Conscious Minding,' in *Origins of Mind in Nature*, ed. L. S. Swan (Dordrecht: Springer Verlag, 2013), pp. 273–287.
16. Dewey described this view of the mind as the spectator theory. Dennett more recently called it the Cartesian Theater. On the affinity between Dewey and Dennett, see Jerome A. Popp, *Evolution's First Philosopher: John Dewey and the Continuity of Nature* (Albany, NY: State University of New York Press, 2007), and Solymosi, 'Neuropragmatism, Old and New,' in *Phenomenology and the Cognitive Sciences*, 10(3) (2011): 347–368. For these specific theories of mind in the original, see Dewey, *Experience and Nature*, in *The Later Works of John Dewey*, Vol. 1, ed. Jo Ann Boydston (Carbondale, IL: Southern Illinois University Press, 1925/1981–1991), and *The Quest for Certainty*, in *The Later Works of John Dewey*, Vol. 4, ed. Jo Ann Boydston (Carbondale, IL: Southern

Illinois University Press, 1929/1981–1991); and Dennett, *Consciousness Explained* (Boston: Little, Brown, 1991).
17. Harris, *The Moral Landscape*, p. 11. More specifically, this entanglement appears to involve the medial prefrontal cortex (MPFC). But the neurological picture is far from clear: the interconnectivity of neural processes makes it difficult to generate a very general picture of moral reasoning in the brain (see pp. 93ff). Other attempts to localize ethical cognition in the brain run the same risks, as does Harris's approach (*Cf.* Joshua Green, *Moral Tribes: Emotion, Reason, and the Gap Between Us and Them* (New York: Penguin Press, 2013), which has far too many shortcomings to list here; though his kidnapping 'pragmatism' as the new name for his 'brainy' utilitarianism is clearly absurd). The first problem such approaches face is the presumption that morality is modular and not a global affair in both the brain and its interaction with the body and world. The second issue is the heavy reliance on various brain imaging technologies as though they were adequate mirrors of what's 'really' going on in the brain. An excellent critical perspective, which makes use of philosophical pragmatism, is Robert G. Shulman, *Brain Imaging: What It Can (and Cannot) Tell Us About Consciousness* (New York and London: Oxford University Press, 2013).
18. See Racine's contribution to this volume, Chapter 11.
19. Eric Racine, *Pragmatic Neuroethics: Improving Treatment and Understanding of the Mind-Brain* (Cambridge, MA: MIT Press, 2010), p. 63; Dewey, *Human Nature and Conduct* in *The Middle Works of John Dewey*, Vol. 14, ed. Jo Ann Boydston (Carbondale, IL: Southern Illinois University Press, 2010/1981–1991), p. 204.
20. Dewey, *The Quest for Certainty*, p. 35.
21. See Dewey, *Logic: The Theory of Inquiry* in *The Later Works of John Dewey, Volume 12*, ed. Jo Ann Boydston (Carbondale, IL: Southern Illinois University Press, 1938/1981–1991); Larry A. Hickman, *John Dewey's Pragmatic Technology* (Bloomington and Indianapolis: Indiana University Press, 1990); 'Philosophical Tools for Technological Culture,' *Philosophy of Education 2001* (Urbana, IL: Philosophy of Education Society, 2001), pp. 25–35; *Philosophical Tools for Technological Culture: Putting Pragmatism to Work* (Bloomington and Indianapolis: Indiana University Press, 2001); and *Pragmatism as Post-Postmodernism: Lessons from John Dewey* (New York: Fordham University Press, 2007); and Tschaepe, 'John Dewey's Conception Scientific Explanation: Moving Philosophers of Science Past the Realism-Antirealism Debate,' *Contemporary Pragmatism* 8(2) (2011): 187–203.
22. Hickman describes the situation as this: '[that the] popular notion [of technology is that...] technology is chronologically later than, and even ontologically inferior to science, and this because science is theoretical and technology is "merely" practical. The intuition behind this common usage is that theory takes both temporal and ontological precedence over practice. It is apparently what people have in mind when they contrast technology to the "pure (could they possibly mean non-technological?) research" that they think takes place in scientific laboratories. I shall contend that nothing of the sort exists' (*Philosophical Tools for Technological Culture*, p. 10). A page later he goes on, 'it is now apparent that the scientific revolution of the seventeenth century would have been impossible without the major technological

advances that produced glass beakers, the telescope, the microscope, the air pump, and many other types of instruments. So much for chronological priority' (Ibid., 11).
23. Hickman, *Philosophical Tools for Technological Culture*, p. 43.
24. See note 16 above.
25. Cf. Benjamin Libet, 'Neural Destiny: Does the Brain Have a Free Will of Its Own?' in *The World and I* (1990), pp. 32–35, and 'Do We Have Free Will?' in *The Oxford Handbook of Free Will*, ed. Robert Kane (New York: Oxford University Press, 2002), pp. 551–564; Gazzaniga, *Who's In Charge? Free Will and the Science of the Brain* (New York: HarperCollins Publishers, 2011); Harris, *Free Will* (New York: Free Press, 2012); and Solymosi, 'A Reconstruction of Freedom in the Age of Neuroscience: A View from Neuropragmatism,' *Contemporary Pragmatism* 8(1) (2011): 153–171.
26. See Kitcher, 'Philosophy, Inside Out,' *Metaphilosophy* 42 (2011): 248–260.
27. Dewey, *Experience and Nature*, p. 306.
28. Hickman, *Philosophical Tools for Technological Culture*.
29. Dennett 'The Moral First Aid Manual,' and *Darwin's Dangerous Idea*.
30. Owen Flanagan, 'The Moral Network,' in *Churchlands and Their Critics*, ed. R. N. McCauley (Cambridge, MA: Blackwell, 1996), pp. 192–215.
31. Dennett, *Darwin's Dangerous Idea*, p. 507.
32. Dennett, *Freedom Evolves* (New York: Viking, 2003), p. 180. Dennett goes on: 'I can still take my task to be looking out for Number One while including under Number One not just my own living body, but my family, the Chicago Bulls, Oxfam…you name it.'
33. Dennett, *Breaking the Spell: Religion as a Natural Phenomenon* (New York: Viking, 2006), p. 303. Dennett continues: 'If you can approach the world's complexities, both its glories and its horrors, with an attitude of humble curiosity, acknowledging that however deeply you have seen, you have only just scratched the surface, you will find worlds within worlds, beauties you could not heretofore imagine, and your own mundane preoccupations will shrink to *proper* size, not all that important in the greater scheme of things. Keeping that awestruck vision of the world ready to hand while dealing with the demands of daily living is no easy exercise, but it is definitely worth the effort, for if you can stay *centered*, and *engaged*, you will find the hard choices easier, the right words will come to you when you need them, and you will indeed be a better person. That, I propose, is the secret to spirituality, and it has nothing at all to do with believing in an immortal soul, or in anything supernatural.'
34. Dennett, *Freedom Evolves*, p. 303.
35. Two of the most recent and worthwhile are Kitcher, *The Ethical Project*, and Kim Sterelny, *The Evolved Apprentice: How Evolution Made Humans Unique* (Cambridge, MA: MIT Press, 2012).
36. William D. Casebeer, *Natural Ethical Facts: Evolution, Connectionism, and Moral Cognition* (Cambridge, MA: MIT Press, 2003).
37. Dewey, 'Three Independent Factors in Morals,' in *The Later Works of John Dewey, Volume 5*, ed. Jo Ann Boydston (Carbondale: Southern Illinois University Press, 1930/1981–1988), p. 285.
38. Ibid., p. 286.
39. See Kitcher, *The Ethical Project*.

40. There is much debate over what Dewey exactly meant in his use of the term *growth*. In this chapter, I am not interested in becoming involved in this scholarly debate. My position is, however, close to the interpretation offered by Jerome Popp in *Evolution's First Philosopher*, pp. 97–101, and in 'John Dewey's Ethical Naturalism,' *Contemporary Pragmatism* 5(2) (2008): 149–163, pp. 152–155.
41. Dewey, 'Creative Democracy – The Task Before Us,' in *The Later Works of John Dewey, Volume 14*, ed. Jo Ann Boydston (Carbondale: Southern Illinois Press, 1939).
42. Ibid., p. 229.
43. Dewey, *Experience and Nature*, p. 224.
44. Dewey, 'Introduction: Reconstruction as Seen Twenty-Five Years Later,' in *Reconstruction in Philosophy*, in *The Middle Works of John Dewey, Volume 12*, ed. Jo Ann Boydston (Carbondale: Illinois University Press, 1945/1982–1991), p. 275.
45. Racine, *Pragmatic Neuroethics*, p. 64.
46. Ibid., p. 65.
47. See Tschaepe, 'Pragmatics and Pragmatic Considerations in Explanation,' *Contemporary Pragmatism* 6(2) (2009): 25–44; and Solymosi, 'Pragmatism, Inquiry, and Design: A Dynamic Approach,' in *Origin(s) of Design in Nature: A Fresh, Interdisciplinary Look at How Design Emerges in Complex Systems, Especially Life*, ed. R. Gordon, L. S. Swan, and J. Seckbach (Dordrecht: Springer Verlag, 2012), pp. 143–160.
48. Racine, *Pragmatic Neuroethics*, p. 65.
49. Make no mistake: I am not equivocating the technical term of physics and chemistry with the (folk) psychological homonym. My point is simply that as humans navigate nature at all its levels, from particles to persons, we not only do so with the use of values and established facts but also through finding a continuity of patterns. But continuity *is not* identity. For more on continuity, see Solymosi, 'Neuropragmatism, Old and New,' pp. 352–354; Mark Johnson, *The Meaning of the Body: Aesthetics of Human Understanding* (Chicago: University of Chicago Press, 2007), pp. 122–123; Popp, *Evolution's First Philosopher*; Dewey, *Logic*, pp. 29–65, especially 29, pp. 30–31.
50. Indeed, the ability to identify retrospectively facts and values is unique to the phase of nature we call human culture (see Dewey, *Experience and Nature*, and Hickman, *Pragmatism as Post-Postmodernism*, p. 139).
51. Racine, *Pragmatic Neuroethics*, p. 64.
52. Ibid, p. 65.
53. Ibid., p. 64.
54. Ibid., p. 65.
55. Harris, *The Moral Landscape*, p. 29.
56. Ibid., p. 37.
57. Hickman, *Pragmatism as Post-Postmodernism*, p. 159.
58. Harris, *The Moral Landscape*, p. 7.
59. Harris's use of the word 'discovery' in the quoted passage could be read in a strongly realist manner, suggesting that the moral landscape is fixed and waiting to be discovered. I think such a reading is too strong; after all, if

humans do anything to a landscape, it is modify it upon first encounter, for better or worse.
60. On platforms, see Hickman, *Philosophical Tools for Technological Culture*, p. 16; on cranes, see Dennett, *Darwin's Dangerous Idea*, pp. 75ff. The parallels between the two are significant but go well beyond the scope of this chapter.
61. See Racine, *Pragmatic Neuroethics*, Chapters 7, 8, and 9.
62. See Sterelny, *The Evolved Apprentice;* and Marco Iacoboni, *Mirroring People: The New Science of How We Connect with Others* (New York: Farrar, Straus and Giroux, 2008). Also, see Bywater and Piso's chapter in this volume that draws on Sterelny and Iacoboni.

Index

abduction
  cognition for, 72, 79–80, 95, 147
  pragmatism and, 313
action
  cognition and, 3, 11, 15, 21, 43, 61, 71–5, 80–4, 90–3, 96–7, 188, 190, 219–27, 233–5, 257, 281–3
  coordination of, 8, 59, 80–1, 88, 96, 176, 187, 195, 215, 251, 265
  perception and, 84–6, 89–90, 94–5, 107–8, 110–17, 133–5, 142–3
Adams, F., 65, 70n. 19, 119, 120, 121nn. 3, 6, 8, 124nn. 59–60
aesthetics, 11, 145–6
  somaesthetics, 16
affordances, 23, 74, 107–17, 135, 268–9
Aizawa, K., 65, 70n. 19, 119, 120, 121nn. 3, 6, 8, 124nn. 59–60
Alexander, Thomas, 187, 211n. 14
amygdala, 41, 85, 130, 151, 202
analytic philosophy, 127–8
anthropology, 4, 6, 24–5, 174, 185, 191, 196–8, 204, 259, 270, 272
Arendt, Hannah, 127, 138n. 8
Aristotle, 246, 265, 275, 278
artificial intelligence, 59, 60, 74, 83, 127, 130
Austin, J. L., 37, 55n. 1
autonomic nervous system, 47, 144, 151
autonomy, 25, 219, 277

Bain, Alexander, 83, 100n. 54, 166
Beauchamp, Tom L., 254, 263n. 45
Bechtel, William, 33n. 26, 47, 48, 54, 55n. 22, 56n. 32, 122n. 15
Beer, R., 108, 109, 121n. 1, 122n. 17
behavior
  adaptive, 8, 78, 141, 169–70, 181, 251
  cognition and, 3, 11, 15, 21, 43, 61, 71–5, 80–4, 90–3, 96–7, 188, 190, 219–27, 233–5, 257, 281–3

behaviorism, 5, 58, 125–6, 131–5
Bennett, M. R., 40, 41, 55n. 9
bioethics, 243
biology, 9, 24, 31, 74–5, 97, 166, 181, 245
  allostasis, 40
  homeostasis, 40
brain, *see also* cortex
  as a computer, 52
  connectedness, 7–8, 11, 27, 40–2, 46–7, 88–94, 129, 151, 201–3, 222–3
  development, 128–30, 192–4, 222–4
  environment and, 8–10, 19–25, 29, 39–41, 47–8, 72, 131, 134–5, 143–7, 228, 293–6, 309
  evolution of, 4, 44, 72, 75–80, 93–7, 128–9, 170–1, 174–6, 178–9
  mind and, 10, 13–16, 38, 44, 65–8, 83–5, 126–33, 248–9, 305–6
  -stem, 129, 202
brain imaging, 46–7, 52, 90, 137, 191, 247
Brandom, Robert, 33n. 36
Brothers, Leslie, 129, 130, 139n. 13
Bywater, Bill, 24, 35n. 52, 317n. 62
Cannon, Walter B., 40, 55n. 7
Cartesianism, 11, 16, 27, 31, 62, 74, 297
Casebeer, William, 54, 246, 247, 261n. 23, 300, 315n. 36
causation, 14, 48, 283
  efficient, 73
  levels of, 48, 54, 285
  mechanistic, 14
Changeux, Jean-Pierre, 244, 248, 261n. 2, 262n. 41
Chemero, Anthony, 6, 33n. 25, 70n. 26, 107, 121, 121nn. 1, 9, 122nn. 15, 18, 20, 288nn. 6, 9
Childress, James F., 254, 263n. 45

Churchland, Patricia, 26, 27, 36nn. 56–7, 38, 53, 54, 55n. 3, 234, 238n. 62, 239n. 68, 246, 261n. 21, 270, 288n. 12, 293, 294, 302, 306, 312n. 8, 313n. 11
Churchland, Paul, 26, 27, 36nn. 57–8, 54, 57, 58, 61–3, 65–9, 69n. 1, 70nn. 11, 16–18, 22–3, 30–1, 239n. 69, 290n. 35
Clark, Andy, 32n. 23, 65, 98n. 11, 99nn. 40–1, 47, 102nn. 108, 110, 112, 121nn. 1–2, 5
cognition
 Cartesian theater, 11, 74, 297
 embodied, 4, 7–8, 19–21, 27–9, 42, 52–3, 72–3, 81–3, 114–18
 emergent features of, 61
 evolution of, 24
 executive, 11, 203, 216, 222–5, 228, 233, 236
 extended, 7, 14–16, 27, 29, 64–8, 96, 105–9, 115–17, 119–20
 functional nature of, 7–10, 13, 52–4, 58, 72–3, 300, 308–9
 infant, 12, 17, 64, 129–30, 145, 152, 199–200, 205
 social, 8–11, 27, 76, 79, 84–6, 96, 125–34, 235
 unconscious, 7, 12, 73, 83, 128, 130, 194–5, 198, 202–3, 226–7, 268
cognitive science, 52, 82–3
 embodied, 4, 10, 16, 107, 120
 neuroscience and, 10, 25, 38–9, 73, 166, 235
common sense, 5, 71, 83
 experience and, 29
 science and, 29, 54
communication, 11, 15–16, 20, 49, 97, 129, 134
 animal, 135
 emotion and, 144–5, 152
 intelligence and, 27, 52, 63, 77, 89, 143
community, 25, 78, 187–8, 197, 209, 264, 267, 270
consciousness
 brain activity and, 7, 13–15, 23–4, 87–8, 171–3
 disordered, 311

 as natural, 27–8, 145–6
 self, 11, 13, 18, 22, 279, 297
 as stream, 42, 279
 subjective, 11, 13–14, 64, 271, 279
 unconsciousness and, 7, 12, 73, 83, 128, 130, 194–5, 198, 202–3, 226–7, 268
cortex
 auditory, 91
 frontal, 75, 85, 92, 171, 223
 motor, 80, 89, 92–3, 135–7
 neocortex, 80, 87–90
 parietal, 85, 92, 268
 prefrontal, 128, 130, 203, 216, 222–30, 234, 297
 temporal, 85, 92, 171, 223
 visual, 41, 85–6, 268
Crick, Francis, 45, 171, 182n. 9
culture
 cognition and, 10–12, 17, 23–5, 28, 129, 137
 experience and, 10–12, 21, 208–9, 309
 nature and, 12, 294, 309
 science and, 4, 26, 74, 245
cybernetics, 16

Damasio, Antonio, 32n. 10, 40, 41, 44, 46, 54, 55n. 6, 11, 18, 21, 56n. 30, 127, 128, 130, 133, 136, 137, 138n. 10, 139nn. 16, 19, 24, 156n. 2, 160n. 75, 227, 237nn. 39–40, 261n. 14
Darwin, Charles, 31, 75, 85, 86, 95, 100nn. 64, 66, 140, 141, 142, 143, 156n. 3, 195, 294, 297, 313n. 10
Davidson, Donald, 313n. 12
De Jaegher, H., 109, 122n. 19
democracy, 25, 63, 97, 189, 208–10, 251, 304–5
Dennett, Daniel, 19, 25, 26, 27, 31n. 6, 34nn. 38, 41, 36n. 60, 179, 292, 293, 298–300, 302–4, 312nn. 3, 6, 313–14n. 16, 315nn. 29, 31–4, 317n. 60
Descartes, René, 27
deSousa, Ronald, 127

Dewey, John, 3, 5, 6, 8, 9, 14, 19, 20, 24, 25, 27, 29, 30, 31n. 2, 32nn. 11, 15, 34nn. 37, 41, 35n. 50, 36nn. 64–6, 70, 39, 41, 42, 43, 45, 47, 51, 53, 55nn. 5, 12–13, 15, 56nn. 27, 31, 58–63, 65, 66, 68, 69nn. 4, 7, 9, 70nn. 13–15, 27, 72–5, 78, 81, 83, 86–9, 93–5, 97, 97nn. 1, 4, 98nn. 7, 9–10, 12–13, 15, 21, 25, 99nn. 31, 33–4, 37, 100nn. 57, 65, 68, 101n. 84, 102n. 104, 109, 111, 113, 115, 125, 126, 131, 134, 139nn. 17, 21, 140–61, 165–8, 172, 173, 181, 182n. 2, 185–98, 203, 210–14, 218, 220, 221, 230, 232, 236nn. 2, 4, 10, 13–15, 238n. 53, 243, 249–51, 255–60, 263, 275, 282, 289, 292–7, 300–9, 311, 313nn. 11–12, 16, 314nn. 19–21, 315nn. 27, 37
DiPaolo, E. A., 109, 122n. 19
dualism
  Cartesian, 31, 62
  epistemic, 12, 45, 125–6
  fact–value, 28, 71, 295, 307
  mind–world, 4–5, 10, 13–16, 27, 38, 44, 65–8, 83–5, 126–34, 248–9, 292, 305–6
  semantic, 249
dynamic system
  1/f scaling of, 118–19
  brain as, 65–8, 228
  cognition as, 4, 7, 9, 13–16, 22–3, 29, 40–1, 106–20, 153–4, 181, 246, 264–7, 282
  moral situation as, 284–7

ecological psychology, 4, 9, 16, 20, 89, 106–8, 268
Eddington, Arthur, 127
Edelman, Gerald, 54, 158n. 51
education, 12, 17, 19, 24, 63, 196–7, 203–10, 217–36
  moral, 304, 312
  neuroscience and, 185–93
embodied cognition, 4, 7–8, 19–21, 27–9, 42, 52–3, 72–3, 81–3, 114–18
emotion, 25, 207, 226, 269, 307
  cognition and, 7, 41, 45–7, 62, 86–7, 129–30, 140–6, 220–2, 226–30

expression, 140–2, 144–55, 215
  morality and, 275–80
  reason and, 127–8, 275
empathy, 246
empiricism
  history of, 5, 19, 27, 59, 125–6, 131–2, 251
  radical, 63, 73, 97, 107, 143
  sensationalistic, 63, 71
  skepticism and, 294
Engelhardt, Tristram, 246, 261n. 20
epistemology, 11–12, 19, 46, 51, 58, 126, 306–8
  pragmatist, 61–3, 80, 84, 243, 260
  social, 16–17, 131–4
ethics, 5, 97
  deontological, 298–9, 301–4
  naturalized, 12, 243, 252–6
  neuroscience and, 244–9
  pragmatic, 250–6
  social, 251–2
  utilitarian, 251, 295, 298–9, 302, 304
evolution
  culture and, 12, 17, 21, 297
  hominid, 170–6
  philosophy and, 5, 166
experience
  biological basis of, 8, 27–8, 74, 82, 168
  culture and, 10, 18, 21, 24, 200
  dynamic, 9, 13–15, 20, 29, 45, 71–4, 113, 205–6, 230
  experimental, 19, 20, 22, 50–3, 72–4, 191–2, 300–2
  first-person, 11, 13, 20, 50, 146, 257, 259
  in life-world, 9, 15–16, 18–19, 22–5, 34, 39, 42, 52–3, 72–3, 166–8
  knowledge and, 5, 42, 49–53, 62–3, 73, 251
  qualitative, 19, 42–3, 62–3, 143
  social, 12, 27–8, 76, 96–7, 152–5, 197–8, 201, 222, 302–3, 310–12
  transactional, 8–10, 19–25, 29, 39, 47–8, 72, 131, 134–5, 143–7, 293–6, 309

fact–value difference, 28, 45, 71,
    295, 307
fallibilism, 21
Farah, Martha, 245, 258, 261n. 10,
    263n. 55
Fesmire, Steven, 187, 211n. 17
Flanagan, Owen, 26, 27, 36n. 59,
    54, 298, 315n. 30
Fleschig, Paul, 245
fMRI, 90, 137, 247, 249
Fodor, Jerry, 6, 31n. 7, 59, 69n. 6
folk psychology, 26, 246
Forel, August, 245
Fowler brothers, 245
Franks, David, 25, 32n. 22, 33n. 27,
    35n. 53, 194, 212nn. 55, 57, 226,
    227, 237nn. 38, 41, 239nn. 70–1
free will, 249, 280, 297, 299, 305
freedom
    moral, 25, 26, 258–9, 277
    political, 218
Freeman, Walter J., 33n. 26, 67,
    124n. 54
functionalism, 52, 68, 74

Galileo, 127
Gall, Franz Joseph, 245
Gallagher, Shaun, 33n. 24, 54, 109,
    122n. 19, 157n. 29, 158n. 50,
    159n. 64, 198–201, 213nn. 78,
    80–1, 93, 97
Gallese, Vittorio, 54, 102n. 100,
    159n. 62, 160n. 66
Gazzaniga, Michael, 32n. 10, 98n.
    13, 100n. 71, 138n. 7, 139n. 12,
    194, 212n. 56, 235, 239n. 65, 244,
    261n. 4, 293, 294, 302, 306, 313n.
    9, 315n. 25
Gibson, J. J., 32n. 19, 66, 98n. 17,
    99n. 38, 107, 121n. 12, 122n. 20,
    124n. 47, 267, 268, 288n. 3
Goethe, Johann Wolfgang von, 205,
    208, 209, 214n. 122
Goldie, Peter, 144, 156n. 24,
    157nn. 25–6
Goodin, Robert, 273,
    289n. 20
Greene, Joshua, 246,
    262nn. 24–5

habit
    belief as, 80, 203, 284
    formation of, 7, 11–12, 20–2, 83,
        146, 207, 217–22, 230, 254
Hacker, P. M. S., 40, 41, 55n. 9
Harris, Sam, 293–6, 302, 306,
    309–11, 313nn. 9, 12–14,
    314n. 17, 315n. 25, 316nn. 55,
    58–9
Heidegger, Martin, 10, 160n. 74
Heisenberg, Werner, 127
Hickman, Larry, 33n. 36, 34n. 41,
    36n. 70, 54, 102n. 105, 211n. 19,
    296, 300, 310, 311, 314nn. 21–2,
    315nn. 23, 28, 316nn. 50, 57, 317n.
    60
hippocampus, 130, 203
Holt, E. B., 107
homeostasis, 40
hominids, 75–6, 78, 165, 168, 170–7,
    197
humanism, 26–30
humanities, 29–30, 73, 245, 247
Hume, David, 18, 19, 27, 63, 293,
    309
Hutto, Daniel, 65

Iacoboni, Marco, 126, 138n. 4, 192,
    199, 200, 211n. 39, 213nn. 84, 96,
    317n. 62
idealism, 5, 126, 132–3
imagination, 11, 45, 90, 153, 168,
    186–9, 201, 205–8, 304
Industrial Revolution, 30
infant development, 12, 17, 64,
    129–30, 145, 152, 199–200, 205
intellectualism, 51
intelligence
    for action, 7, 19–20, 59, 75, 174,
        220–2, 232–3, 281
    in inquiry, 20, 23, 29–30, 47–9, 54,
        72, 83, 170, 186, 197, 300, 302,
        309–10
    social, 7, 12, 78–9, 170, 251, 277
intentionality, 18, 24, 82, 154, 200,
    283, 287

Jackall, Robert, 278, 289n. 29
Jackson, Frank, 62

James, William, 5, 7, 8, 25, 31n. 9, 32nn. 15, 19, 34nn. 41–3, 45–7, 50, 51, 55nn. 14, 19, 56n. 26, 58, 61, 63, 69nn. 3, 10, 71–4, 81, 87, 88, 90, 94, 95, 97, 97nn. 3–4, 98nn. 6, 8, 10, 99n. 36, 107, 120, 121nn. 1, 10, 124n. 61, 139n. 17, 140–3, 147, 148, 152, 156nn. 4, 17, 166, 219, 220, 221, 235, 235nn. 6–7, 9, 12, 16, 238n. 64, 267, 288n. 2, 294, 297, 313n. 12
Jesus, 278
Johnson, Mark, 6, 75, 89, 98n. 23, 101n. 83, 128, 133, 134, 138n. 11, 139n. 26, 198–200, 205, 213nn. 76–7, 80, 94, 214n. 121, 316n. 49
Juarrerro, Alicia, 285

Kahneman, Daniel, 18, 22–5, 33n. 33, 34n. 45, 35n. 52, 185, 194–7, 202, 203, 210n. 2, 212nn. 52, 58, 265, 267, 280, 283, 289n. 18, 290n. 34
Kandel, Eric, 222, 236n. 18, 239n. 66, 261n. 10
Kant, Immanuel, 34n. 39, 51, 63, 83, 100n. 51, 272
kinesthesia, 193
Kitcher, Philip, 297, 302, 313n. 10, 315n. 26, 316n. 39
knowledge
　education and, 12, 17, 19, 24, 63, 196–7, 203–10, 217–36
　experience and, 5, 42, 49–53, 62–3, 73, 251, 257–8
　learning and, 16, 20, 23, 29–30, 47–9, 54, 72, 83, 170, 186, 197, 300, 302, 309–10
　practice and, 59–61, 89, 201, 207–8, 258–9
　truth and, 11, 19, 81
Koestler, Arthur, 168, 172, 173

Lakoff, George, 52, 53, 55nn. 4, 20, 56n. 28, 89, 101n. 83, 128, 133, 134, 138n. 11, 139n. 26
language
　cognition and, 10, 11, 14, 60–3, 131, 136, 169–71
　emotional, 151

culture and, 21, 28, 78, 129
　ordinary, 28
　philosophy of, 5, 6, 28, 52
Laughlin, Robert, 127, 138n. 6
learning
　habit and, 7, 11–12, 20–2, 83, 146, 207, 217–22, 230, 254
　by inquiry, 7, 20, 23, 29–30, 47–9, 54, 72, 83, 167–70, 186, 197, 300, 302, 309–10
LeDoux, Joseph, 156n. 2, 160n. 75, 214nn. 114, 117, 234, 238n. 61
Lee, David, 108, 121n. 13
Lewontin, Richard, 10, 32n. 13
Libet, Benjamin, 315n. 25
life world, 9–10, 18–22, see also Umwelt
Locke, John, 25
logic, 11, 12, 43, 48, 60, 74, 80, 83
logical positivism, 27–8

MacLean, Paul, 246, 261n. 19
materialism
　Cartesian, 11, 16, 27
　reductive, 5
Mead, George Herbert, 5, 25, 31n. 4, 88, 94, 97n. 4, 98n. 18, 102n. 111, 125, 126, 131–4, 139n. 17, 166, 221, 222, 231, 236n. 17, 238n. 55, 264, 267, 279, 280, 282, 290n. 33
memory
　cognition of, 11, 78–80, 107, 178, 201–3
　emotional, 130
Merleau–Ponty, Maurice, 101n. 85, 115, 123n. 38, 145, 146, 157nn. 30–3, 35, 160n. 77
metaphysics, 13, 17, 19, 43–5, 50, 69
　of ethics, 267–9
Meynert, Theodor, 245
Mill, John Stuart, 298, 302
mind
　body and, 3, 8–10, 13, 20, 27, 68
　brain and, 10, 13–16, 38, 44, 65–8, 83–5, 126–33, 248–9, 305–6
　Cartesian, 11, 16, 27, 31, 62, 74, 297
　environment and, 8–10, 19–21

mind – *continued*
  world and, 4–5, 10, 13–16, 27, 38, 44, 65–8, 83–5, 126–34, 248–9, 292, 305–6
mirror neurons, 24, 125–8, 131, 134–8, 188, 192, 198–200, 205, 312
Mithen, Steven, 168, 170–3, 181, 182nn. 4, 8, 183nn. 17, 18, 184n. 23
Moore, G. E., 313n. 10
moral psychology, 246–7, 253–9, 283–7
morality, *see also* ethics
  agency and, 279–84
  cognition for, 244–7, 269–73, 291–4, 304–6
  freedom and, 25, 293
  as natural, 244–8, 251–2, 310–11
  as social, 12, 25, 97, 251, 264–8, 269–79, 283–7, 297–304
  truth of, 308–11

naturalism
  ethics and, 12, 243, 250–6, 300, 306–10
  evolutionary, 141–3, 248, 298
  nonreductive, 4–5, 9, 44–5
  pragmatic, 5–9, 12, 97, 257–60, 306
  scientific, 304
naturalistic fallacy, 249, 307
neural network, 62, 72, 88, 92, 95, 246
neuroessentialism, 247
neuroethics, 243–60
neurophenomenology, 16
neurophilosophy, 4–5, 12–13, 26–8, 38, 63, 234, 246, 291
neuropragmatism, 6–12, 38–40, 49–51, 54–5, 181–2
  as ethics, 248–57
  neuroscience and, 49, 54–5, 68–9, 94, 168
  pedagogy and, 188–92, 203–7, 217–19, 235–6
  as philosophy, 3–6, 12–13, 26–31
neuroscience, 3–4, 14, 31, 37
  dualistic, 45–6
  eugenics and, 246
  history of, 245–6
  reductive, 41, 47, 67–8

neurosociology, 16, 24, 125–34, 235
Nietzsche, Friedrich, 167
Noë, Alva, 32n. 16, 33n. 23, 54, 65, 98n. 11, 99nn. 41, 47, 101nn. 82, 85
noise
  1/f scaling, 118–19
  pink noise, 118

ontology, 5, 16, 20, 44–6
oxytocin, 191, 273, 302

Peirce, Charles Sanders, 4, 33n. 29, 35n. 48, 57, 58, 69n. 2, 72, 74, 79, 80–4, 88, 93–6, 97n. 4, 98n. 20, 99nn. 44–5, 48–50, 100nn. 52–3, 56, 58, 102n. 108, 139n. 17, 166, 182n. 1, 313n. 11
perception
  action and, 84–6, 89–90, 94–5, 107–17, 133–5, 142–3
  knowledge and, 5, 42, 49–53, 62–3, 73, 251
  sensory, 19, 42–3, 50–1, 62–3, 143
personality, 17, 218–19
personhood, 249, 258
phenomenology
  of cognition, 58–9
  of experience, 14, 27, 42, 47, 63, 143–5, 148
  history of, 5, 152
phrenology, 245
Plato, 299
Popp, Jerome, 31n. 4, 313n. 16, 316nn. 40, 49
positivism, 5, 19, 58, 69, 71, 243, 247–8, 257
postmodernism, 132
pragmatism, *see also* neuropragmatism
  naturalism and, 5–9, 12, 97, 257–60, 306
  neurophilosophy and, 3–12, 38–40, 49–51, 54–5, 181–2
  origins of, 4–5, 57–8, 74, 80–3, 94–5, 166
primates, 75–6, 78, 84–5, 92, 168, 170–7, 197
problem-solving, *see* learning
psychologist's fallacy, 50, 51, 146

psychology
    behavioral, 5, 58, 125–6, 131–5
    cognitive, 18, 37, 38, 74, 87, 223–4
    cultural, 5, 74, 223
    ecological, 4, 9, 16, 20, 89, 106–8, 268
    evolutionary, 5, 165–71, 181
    faculty, 52
    folk, 14, 26, 246, 249
    functional, 58
    pragmatic, 5, 46, 63, 72–4, 95, 107
    social, 8–11, 16, 24, 27, 76, 84–6, 96, 125–34, 225, 235
Putnam, Hilary, 6, 31n. 5, 36n. 63, 52, 56n. 24, 97n. 4, 101n. 85, 294, 306

Quine, W. V., 6, 26

Racine, Eric, 295, 296, 306–11, 314nn. 18–19, 316nn. 45, 48, 51, 317n. 61
radical empiricism, 63, 73, 97, 107, 143
rationalism, 6, 15, 27, 58–9, 63
reasoning, 11
    experience and, 5, 74, 309
    concepts and, 11, 44, 120, 174
    emotion and, 45–7, 127–8, 227
    limits of, 127
reciprocity, 245, 270–2, 281, 287
reductionism, 4, 5, 13, 14, 43, 74
    scientific, 9, 47–8, 53–4, 69, 246–9, 308
reflexes, 12, 88, 202, 250, 301
    reflex arc, 83, 87, 89, 95, 195
representationalism, 5, 7, 11, 13–14, 50–1
representations
    experience and, 23, 50–2, 107, 268
    knowledge and, 11, 21, 57, 66–7, 229
    neuroscience and, 8, 14–15
    purposive, 81–2, 91–2
Rizzolatti, Giacomo, 101n. 92, 102nn. 96, 100, 126, 135, 138n. 5, 139n. 23, 160n. 66
Rockwell, W. Teed, 6, 27, 33n. 26, 36nn. 58, 62

Rorty, Richard, 6, 25, 34n. 41, 51, 52, 58, 69n. 4, 70n. 14, 99n. 36, 313n. 12
Roskies, Adina, 246, 261n. 22
Rupert, Robert D., 65, 119, 121nn. 4, 6–7, 124n. 58

Samuels, Boba, 189, 190, 191, 210n. 3, 211nn. 23, 35
Scheler, Max, 144–6, 157n. 28
Schulkin, Jay, 40, 54, 55, 188, 201, 202, 211n. 20, 213n. 98, 214nn. 107, 110, 116
science
    humanities and, 29–30, 73, 245, 247
    methods of, 12, 48–53, 186
    unity of, 14
    worldview of, 5, 14, 26–31, 190–2, 279, 292–3, 296–7
Searle, John, 171, 182n. 10
Sellars, Roy Wood, 97n. 2
Sellars, Wilfrid, 26, 27, 29, 36n. 61, 62, 69n. 5, 70n. 14, 293, 312n. 7
semantics, 33, 62–3, 87, 92, 136–7, 174, 249, 307
semiotics, 17, 21, 81, 83
Sherrington, Charles, 247
Shook, John R., 31n. 1, 33n. 27, 35n. 47, 54, 196, 212n. 67, 313n. 15
Shulman, Robert G., 314n. 17
Sinigaglia, Corrado, 126, 135, 138n. 5, 139n. 23
Snow, C. P., 26, 35n. 55
sociology
    cognitive, 8–11, 27, 76, 79, 84–6, 96, 125–34, 235
    neuroscience and, 16, 24, 125–34, 235, 245
    pragmatic, 5
Solymosi, Tibor, 6, 31nn. 1, 8, 35n. 47, 38, 54, 55n. 2, 68, 70n. 29, 196, 212n. 67, 312nn. 1, 4–5, 313nn. 15–16, 315n. 25, 316nn. 47, 49
somaesthetics, 16
Sperry, Roger, 133, 246
Sterelny, Kim, 24, 33n. 34, 35n. 50, 197, 198, 200, 201, 204, 207, 213n. 69, 214n. 119, 315n. 35, 317n. 62

subjectivity, 11, 13, 18, 20, 22, 50, 146, 257, 259, 279, 297
symbolic interactionism, 17
systems 1 and 2, 18, 22–5, 33, 194–7, 202, 281–2
and system 3, 22, 25

Taoism, 64
Taylor, K. R., 110, 111, 117, 122n. 24, 123n. 41, 124n. 50
technology
culture and, 12–13, 17, 30, 292–3
scientific, 12, 30
stages of, 175–9
teleology
moral, 300–1
organic, 5
psychological, 14, 133, 141
science and, 14
Thagard, Paul, 54
thalamus, 202
Thompson, Evan, 32n. 13, 33n. 24, 54, 160n. 67
truth
knowledge and, 12, 17, 28, 57, 59
necessity and, 12, 44, 50
reality and, 61–2
Tschaepe, Mark, 190, 211n. 27, 313n. 11, 314n. 21, 316n. 47
Tucker, Don, 43, 54, 55n. 17, 56n. 30

Turing machine, 83
Turing test, 36

Uexküll, Jacob von, 9, 32n. 13
Umwelt, 9–10

value
cognition and, 16, 43, 249
fact and, 28, 45, 71, 295, 307, 310
science and, 293–4, 306, 310
Varela, Francisco, 6
Vogt, Oskar and Cécile, 245

Wagman, Jeffrey B., 122nn. 24, 28, 123nn. 30–2, 34–6, 41, 43, 124nn. 44–5, 48, 50
Walker, David, 209, 214n. 123
Warren, W. H., 110, 111, 122nn. 22, 25, 123n. 35
Watson, John, 131
Whang, S., 110, 111, 122nn. 22, 25, 123n. 35
Willingham, Daniel, 189–91, 211n. 24
Wilson, Robert A., 10, 32n. 21, 65, 121n. 5
Wright, Chauncey, 75, 98n. 24

Zull, James, 201–2, 213n. 102

GPSR Compliance
The European Union's (EU) General Product Safety Regulation (GPSR) is a set of rules that requires consumer products to be safe and our obligations to ensure this.

If you have any concerns about our products, you can contact us on

ProductSafety@springernature.com

In case Publisher is established outside the EU, the EU authorized representative is:

Springer Nature Customer Service Center GmbH
Europaplatz 3
69115 Heidelberg, Germany

www.ingramcontent.com/pod-product-compliance
Lightning Source LLC
Chambersburg PA
CBHW071616100426
42873CB00004B/56